Astronomers' Observing Guides

For other titles published in this series, go to
www.springer.com/series/5338

Other Titles in This Series

Star Clusters and How to Observe Them
Mark Allison

Saturn and How to Observe it
Julius Benton

Nebulae and How to Observe Them
Steven Coe

The Moon and How to Observe It
Peter Grego

Supernovae and How to Observe Them
Martin Mobberley

Double & Multiple Stars and How to Observe Them
James Mullaney

Galaxies and How to Observe Them
Wolfgang Steinicke and Richard Jakiel

Total Solar Eclipses and How to Observe Them
Martin Mobberley

Cataclysmic Cosmic Events and How to Observe Them
Martin Mobberley

Richard Schmude

Comets and How to Observe Them

with 175 Illustrations

Richard Schmude
109 Tyus Street
Barnesville, GA 30204
USA
Schmude@gdn.edu

Series Editor
Dr. Mike Inglis, BSc, MSc, Ph.D.
Fellow of the Royal Astronomical Society
Suffolk County Community College
New York, USA
inglism@sunysuffolk.edu

ISBN 978-1-4419-5789-4 e-ISBN 978-1-4419-5790-0
DOI 10.1007/978-1-4419-5790-0
Springer New York Dordrecht Heidelberg London

Library of Congress Control Number: 2010935726

Printed on acid-free paper

Springer is part of Springer Science+Business Media (www.springer.com)

This book is dedicated to the many people who have helped me along the way. First, to my father and mother, Richard and Winifred Schmude, who first showed me the stars and answered my many science questions; next to the many fine teachers, professors and school administrators who guided me during my childhood and early adulthood; next, to the many librarians at the Hightower Library (Gordon College), the Sterling Evans Library (Texas A&M University) and the Georgia Tech Library who helped me obtain critical information related to this book; and finally, to the American taxpayers who made it possible to send spacecraft to several of the comets described in this book.

Author's Note

I became interested in astronomy initially when I saw what appeared to be a countless number of stars from my parent's home in Cabin John, Maryland. I was no older than six when I had this life-changing view of the night sky. I purchased my first telescope at age 15 and shared it with siblings and a neighbor girl - Kathy. This was my first experience with public outreach.

My first view of a comet was on March 21, 1986. My brothers James, Fred, and I got up on a cold morning to view Halley's Comet. Since then, I have viewed several comets. What I have come to realize in my studies of comets is that each one is unique.

This book is broken down into two major parts. The first one (Chapters 1 and 2) summarizes our current understanding of comets. In Chapter 2, I have chosen to describe in detail our current knowledge of four comets - 9P/Tempel 1, 1P/Halley, 19P/Borrelly, and 81P/Wild 2. The second part describes observational projects that one may carry out with the unaided eye and binoculars (Chapter 3), small telescopes (Chapter 4) and large telescopes (Chapter 5). Finally, an appendix, a bibliography and an index are included.

The three organizations which are engaged in serious comet studies and with which I am most familiar are the Association of Lunar and Planetary Observers (ALPO), the British Astronomical Association (BAA), and the organization that publishes the *International Comet Quarterly.*

Various websites are quoted in this book. All of them existed in July 2009; some however, may change or be discontinued by the time that this book is read. Changes in websites or addresses are bound to happen. What I have often seen, though, is that a discontinued website is replaced by something even better.

About the Author

Dr. Richard Willis Schmude, Jr. was born in Washington, DC, and attended public schools in Cabin John, Maryland; Los Angeles, California; and Houston, Texas. He started his college career at North Harris County College and graduated from Texas A&M University with a Bachelor of Arts degree in Chemistry. Later, he obtained a Master of Science degree in Chemistry, a Bachelor of Arts degree in Physics, and a Ph.D. in Physical Chemistry, all from Texas A&M University. He worked at NALCO Chemical Company as a graduate co-op student and at Los Alamos National Laboratory as a graduate research assistant.

Since 1994, Dr. Schmude has taught astronomy, chemistry, and other science classes at Gordon College in Barnesville, Georgia. He is a tenured Professor at this college and continues to teach his students (and others) in these areas. He has published over 100 scientific papers in many different journals, and has given over 500 talks, telescope viewing sessions, and workshops to over 20,000 people.

Contents

Chapter 1

Comets: An Overview

Introduction

When I think of a comet, I imagine a bright object with a long tail. Many comets fit this description; however, most are very faint and lack visible tails. In the dictionary, a comet is defined as a celestial body moving about the Sun and having a solid portion which is surrounded by a misty envelope which may or may not have a tail. In many cases, the misty envelope may be faint, and, hence, a comet could resemble an asteroid. There are a large number of known comets, perhaps the most famous one being "Halley's Comet" which orbits our Sun approximately every 76 years.

Most comets have four visible parts. See Fig. 1.1. The central condensation is the bright central part of the coma. The coma is a gaseous envelope that surrounds the central condensation. It can have either a circular, elliptical or parabolic shape. The dust tail lies beyond the coma and is usually the brightest portion of the tail. It is made up of dust. The gas tail is made up of ions, gas atoms and gas molecules. It often has a bluish color.

This chapter describes the characteristics of a comet. Its sections are "Naming Comets", "Comet Orbits", "Comet Orbits and Kepler's Second and Third Laws of Planetary Motion", "Classification of Comets", "Sources and Movement of Comets", "Comet Brightness and Some Statistics", "Parts of a Comet", "Brightness Changes of Comets over Time" and "Comet Impacts in the Near Past".

Naming Comets

In early 1995, a new system of naming comets was established. If the comet is believed to be newly discovered, its designation would be constructed from the steps in Fig. 1.2. Table 1.1 lists the time intervals and the corresponding letter sequences mentioned in the second box in Fig. 1.2. The last name of the person(s) who discovered the comet may be added to this designation, limited, however, to three last names. For example, Comet Hale-Bopp (C/1995 O1) is named after Alan

R. Schmude, *Comets and How to Observe Them*, Astronomers' Observing Guides,
DOI 10.1007/978-1-4419-5790-0_1, © Springer Science+Business Media, LLC 2010

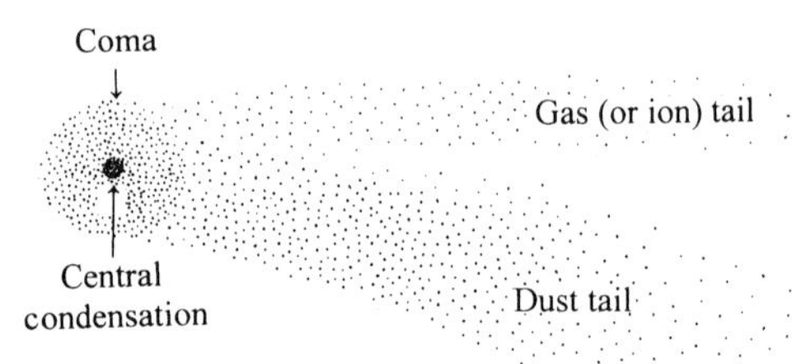

Fig. 1.1. Visible parts of a comet (credit: Richard W. Schmude Jr.).

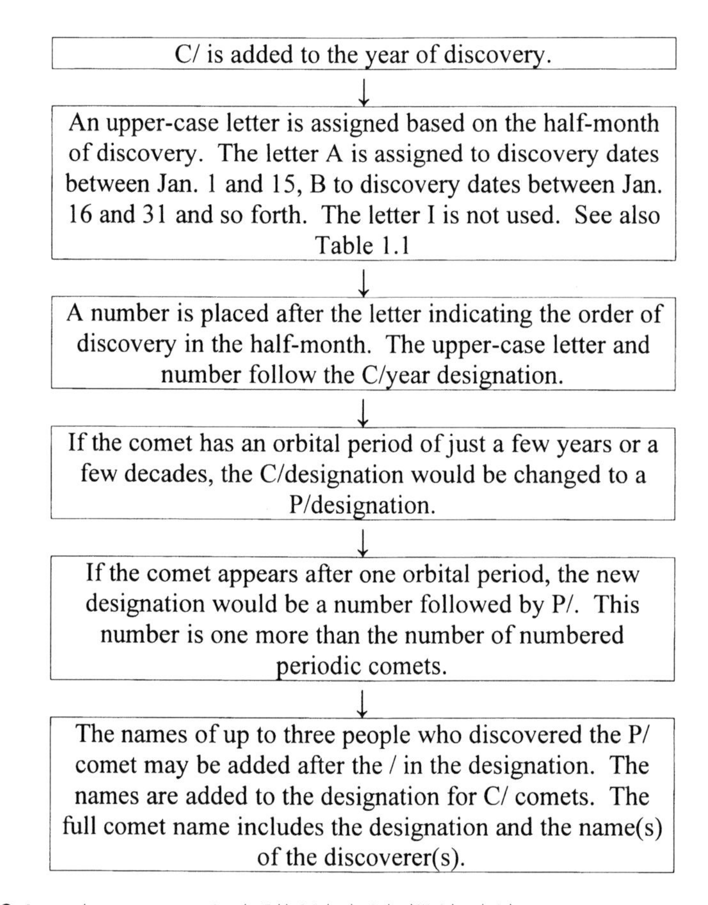

Fig. 1.2. Steps used in naming a comet. See also Table 1.1 (credit: Richard W. Schmude Jr.).

Hale and Thomas Bopp. The full name is Hale-Bopp C/1995 O1, and its designation is C/1995 O1. If two people who have the same last name discover a comet, such as Gene and Carolyn Shoemaker, it would be given the last name of both individuals (Shoemaker). If several people are involved with a discovery at an observatory, the comet may be named after the observatory instead of the indi-

Table 1.1. Time intervals and letter designations which are used in naming a comet

Time interval	Letter	Time interval	Letter	Time interval	Letter
Jan. 1–15	A	May 1–15	J	Sept. 1–15	R
Jan. 16–31	B	May 16–31	K	Sept. 16–30	S
Feb. 1–15	C	June 1–15	L	Oct. 1–15	T
Feb. 16–28 (or 29)	D	June 16–30	M	Oct. 16–31	U
Mar. 1–15	E	July 1–15	N	Nov. 1–15	V
Mar. 16–31	F	July 16–31	O	Nov. 16–30	W
Apr. 1–15	G	Aug. 1–15	P	Dec. 1–15	X
Apr. 16–30	H	Aug. 16–31	Q	Dec. 16–31	Y

viduals. An example of this is Comet Siding Spring (C/2007 K3). Many comets are named after the all-sky survey which resulted in their discoveries. An example would be Comet LINEAR (C/2008 H1). A few of the all sky surveys include the Lincoln Near Earth Asteroid Research (LINEAR), the Near-Earth Asteroid Tracking Program (NEAT), the Catalina Sky Survey (CSS), the Lulin Sky Survey (Lulin) and the Lowell Observatory Near-Earth Object-Search (LONEOS).

A comet is named after its discoverer (or discoverers) provided that it is not a returning comet or it was not discovered previously as another object, such as an asteroid. In rare cases, a comet is named also after the person who determines its orbit. The English astronomer, Edmund Halley, for example, determined that the comets which appeared in 1456, 1531 and 1607 had orbits which were similar to the comet which appeared in 1682, and he predicted that one 1682 comet would return about 1758. (See *Comet Science* ©2000 by Jacques Crovisier and Thérèse Encrenaz.) The comet returned in 1759 and, hence, it bears his name – Comet (1P/Halley).

If the same person discovers more than one comet a number would follow the name. For example, E. W. L. Tempel discovered two different Short-period comets, one in 1867 and another in 1873. These comets are named Tempel 1 and Tempel 2, respectively. This rule applies also to comets with the names of two or three co-discoverers. For example, Comet Shoemaker-Levy 9 is the ninth comet discovered by the Shoemaker-Levy team.

Most of the Sungrazing comets are discovered by individuals who analyze images made by the Solar and Heliospheric Observatory (or SOHO) probe. These comets are designated in Fig. 1.2 and are named after the SOHO probe instead of the individual who made the particular discovery. A few Sungrazing comets are also named after the Solar Maximum Mission (or SMM) spacecraft, the SOLWIND spacecraft and the Solar Terrestrial Relations Observatory (or STEREO) spacecraft.

The letter C in Fig. 1.2 may be replaced by a P, D or X depending on the nature of the comet. An explanation follows.

If a comet is found to have a short orbital period, it would be given the P/ designation until it is recovered in a second apparition. At this point, the P/Year designation would be replaced with a number followed immediately by an upper case P; and a slash followed by the name of the person who discovered it. The number here is one more than the number of known periodic comets that have reappeared. For example, Comet Hug-Bell (P/1999 X1) was given the full name 178P/Hug-Bell after it reappeared in 2007. Since 177 periodic comets had been assigned numbers before Hug-Bell reappeared, this comet was given number 178.

If a comet is destroyed, or if it fails to appear after several apparitions, it would be given the designation of D/ followed by the year of its discovery. For example, Comet Shoemaker-Levy 9 has been assigned D/1993 F2 since it was discovered in the second half of March in 1993 and was destroyed when it crashed into Jupiter in 1994.

Comets that are believed to be real but lack sufficient position measurements for an orbital determination are given the designation of X/ followed by the year of discovery and the appropriate letter and number code. For example, Comet X/1896 S1 was first seen on Sept. 21, 1896, by L. Swift at Lowe Observatory in California. Apparently he was the only one to see the comet and was unable to determine enough positions for an orbital determination. Kronk lists 18 comets with the X/ designation that people had seen in the nineteenth century.

An unusual situation occurs when a comet's nucleus splits. In this case, each fragment is given the comet designation followed by A, B, C, etc. For example, the nucleus of comet LINEAR C/2003 S4 split into two parts. Astronomers named the two parts LINEAR C/2003 S4-A and LINEAR C/2003 S4-B. Each part is treated as a separate comet.

Astronomers have discovered many objects which were believed to be minor planets but were later determined to be comets. These objects retain their minor planet designation but are given either a C/ or P/ prefix. For example, astronomers imaged a magnitude 19.3 object on Feb. 18, 2004, and reported it as a minor planet. This object was given the minor planet designation 2004 DZ_{61}. A few weeks later, a different group of astronomers at Mauna Kea in Hawaii imaged this object but found that it had a small coma with a faint tail. This object was reclassified as a comet and was given the designation C/2004 DZ_{61}.

The name of the discoverer of the comet will be used whenever possible for the remainder of this book. For periodic comets the P/number designation is used. For C/ comets, the designation is placed in parentheses. Older designations which include a year and a roman numeral or a year and a lower-case letter will not be used.

Comet Orbits

Like the Earth, comets in our Solar System move under the influence of the Sun's gravity. Comets follow one of four paths, namely, circular, elliptical, parabolic or hyperbolic; and a comet's path is determined by six quantities which are called orbital elements. In this section, comet paths, orbital elements and how astronomers refine a comet's orbit are discussed.

Comets under the gravitational influence of a single large object like our Sun can move in one of four paths. Each of these is illustrated in Fig. 1.3. Comets which move in a circular orbit maintain a constant Sun distance, the Sun being located in the center of such orbit. Comets which move in an elliptical orbit have a varying Sun distance. Ellipses can have different shapes, ranging from nearly circular to very stretched out. See Fig. 1.4. The eccentricity defines how the ellipse is stretched out. If the eccentricity is low the ellipse would have an almost circular shape,

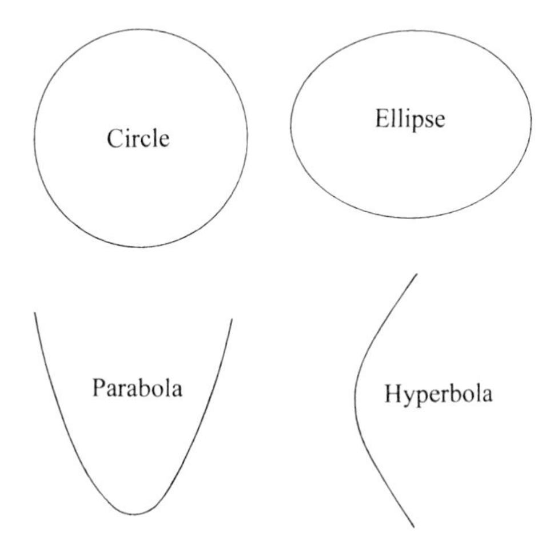

Fig. 1.3. A comet's path can be any one of the four modes shown here (credit: Richard W. Schmude Jr.).

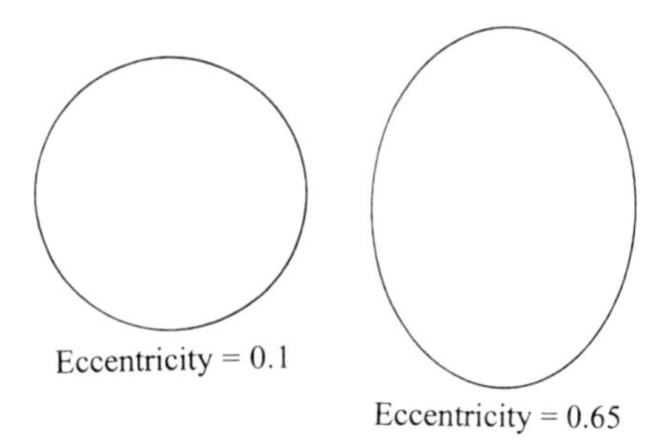

Fig. 1.4. The ellipse on the *left* has an eccentricity of 0.1 and is nearly circular. The one on the *right* has an eccentricity of 0.65 and is much more squashed (credit: Richard W. Schmude Jr.).

whereas if the eccentricity is high, it would be stretched out. The eccentricity is defined as the distance between the two foci points (F1 and F2) divided by the length of the major axis (line segment A B) in the ellipse. See Fig. 1.5. Ellipses can have an eccentricity of just above 0.000 to just below 1.000. Comets which move in circular or elliptical paths will orbit our Sun. Comets which move in a parabolic or hyperbolic path may never return to the Sun. A parabolic orbit has an eccentricity of exactly 1.0. The eccentricity of a hyperbolic orbit is greater than 1.0.

As for orbital elements, which are discussed in the subsection below, one will need to note that all orbital elements have some uncertainty. One source of uncertainty is in the position measurements. Each measurement of a comet's position has an uncertainty in time, right ascension and declination. A second source of orbital uncertainty is due to gravity. Gravitational forces exerted by other Solar System bodies will change a comet's orbit and, hence, its orbital elements. A third source of uncertainty results from several types of non-gravitational forces. These non-gravitational forces are described later in this chapter. As a result of these

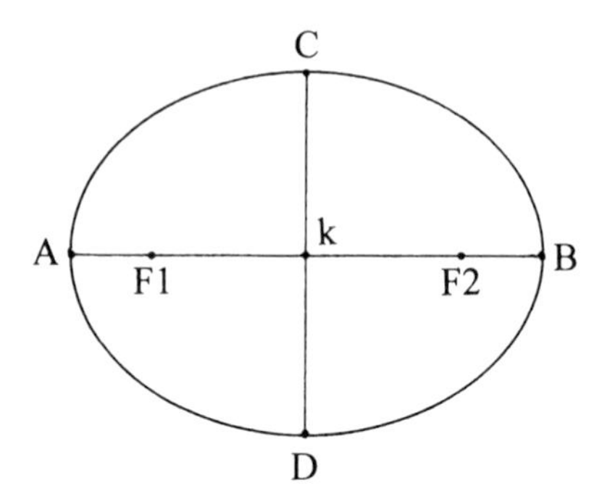

Fig. 1.5. The major axis runs from points A to B, and the semi major axis has half the length (A – k or B – k) of the major axis. The two foci points are at F1 and F2. The eccentricity is the distance between the two foci points divided by the length of the major axis. In this example, the eccentricity equals 0.66 (credit: Richard W. Schmude Jr.).

uncertainties, a comet may have a computed orbital eccentricity just over 1.0 but, in reality, it may be just below 1.0.

The elliptical and hyperbolic paths have a range of eccentricities, whereas circular and parabolic paths have eccentricities of exactly zero and one, respectively. In nature, comets will follow paths having eccentricities which will be between either zero and one or above one; but never *exactly* zero or one. Because of this, it is believed that comets should be broken down into three groups which are Short-period, Long-period and No future predictions. Comets in the Short-period group have orbital periods of less than 200 years and will have orbital eccentricities below 1.0. Comets in the Long-period group include both periodic comets with orbital periods over 200 years and comets which are not returning to the inner Solar System. Comets in this group have orbital eccentricities near 1.0 or just over this amount. They will follow either elliptical orbits which are very stretched out or hyperbolic orbits. Comets falling into the third category (No future predictions) are either lost or have been destroyed and are given the designation D/. Table 1.2 gives a breakdown of the percentage of each type of comet.

Can a comet following one path be pushed into another one? Yes! The most common situation occurs when a comet enters our Solar System following an elliptical orbit and makes a close approach to one of the planets. As a result, its path would change to a hyperbolic one, and the comet would leave our Solar System.

Of the almost 1,000 comets listed with a C/ designation in the *Catalogue of Cometary Orbits 2008* (17th edition) by Brian G. Marsden and Gareth V. Williams, 23% of them have an eccentricity above 1.000000 and 77% of them have an eccentricity below 1.000000.

Orbital Elements

A comet's path is determined mathematically by six quantities which are called orbital elements. These quantities are used in determining the different groups, families and sub-groups of comets. They also yield information on the source of comets. A description of these quantities follows.

Table 1.2. Percentages of comets in different groups, families and sub-groups as of mid-2008, with all of the percentages expressed in terms of the total number of C/, P/ and D/ comets. The writer computed the percentages from data in *Catalogue of Cometary Orbits 2008, 17th edition* ©2008 by Brian G. Marsden and Gareth V. Williams

Group	Percentage
Short-period group (C/ and P/ comets)	13.8%
Encke Family	0.6%
Jupiter Family	11.1%
Chiron Family	0.3%
Short-period nearly isotropic NI Family	1.6%
Kracht 2 Family	0.1%
Long-period group (C/ and P/ comets)	84.7%
Sungrazer Family	51.9%
Kreutz sub-group	44.6%
Meyer sub-group	3.1%
Marsden sub-group	1.1%
Kracht 1 sub-group	1.1%
Kracht 3 sub-group	0.1%
Anon 1 sub-group	0.1%
Anon 2 sub-group	0.1%
Unclassified (Sungrazer)	1.7%
Long-period nearly isotropic NI Family	32.8%
No future predictions (D/ comets)	1.5%

The first quantity is the date and time of perihelion passage. Perihelion is the point on the path where the comet is closest to the Sun. If a comet follows an elliptical orbit, this date and time would be specified for each perihelion passage. The symbol for the date and time of perihelion passage is T.

The comet-Sun distance at perihelion is the second quantity needed for determination of a comet's path. This distance is given in astronomical units with the symbol of q. This quantity is the distance between the Sun and point q in Fig. 1.6.

A third quantity is the angle between the plane containing the comet's path and the plane containing the Earth's orbit. This angle is called the orbital inclination, and it has the symbol of i. It can range from 0° up to 180°. An inclination of 180° is the same as 0° except that the comet moves in the opposite direction in which the Earth moves around the Sun. If a comet moves in the same direction as the Earth and remains in Earth's orbital plane, its orbital inclination would equal 0°. The orbital inclination is illustrated in Fig. 1.6.

A fourth quantity is the comet's orbital eccentricity. As mentioned earlier, this determines the shape of the path which the comet follows. The symbol for its orbital eccentricity is e.

The fifth and sixth quantities needed for determination of a comet's path are the longitude of the ascending node (symbol = Ω) and the argument of perihelion (symbol = ω).

The two points where the comet's path intersects the Earth's orbital plane are called nodes. The ascending node is the point where the comet moves from south to north of the Earth's orbital plane. The descending node is the point where the comet moves from north to south of Earth's orbital plane. The longitude of the ascending node is the angle between a line connecting the Sun and equinox point

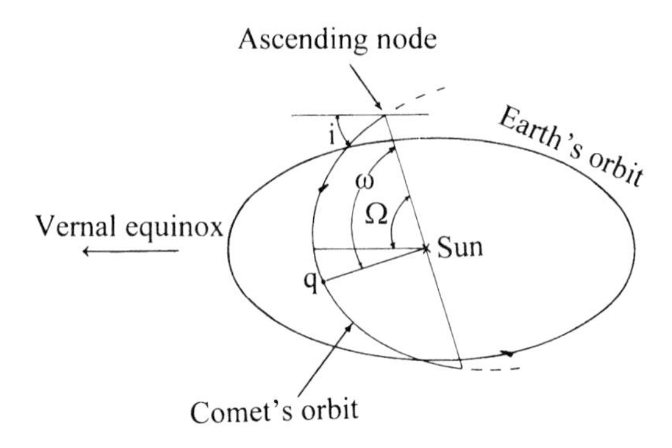

Fig. 1.6. Orbital elements of a comet. The orbital inclination (i), perihelion point (q), longitude of the ascending node (Ω), and argument of perihelion (ω) are illustrated in the drawing. Note that the perihelion distance is the distance between the Sun and point q (credit: Richard W. Schmude Jr.).

Table 1.3. Orbital elements and other characteristics of Comet Shoemaker-Levy 9 (D/1993 F2) listed in various International Astronomical Union Circulars

IAU circular number	5744	5800	5892 and 5893	5906	6017
Date of IAU circular	April 3, 1993	May 22, 1993	November 22, 1993	December 14, 1993	July 9, 1994
Orbital eccentricity (e)	0.07169	0.065832	0.206613	0.207491	–
Orbital inclination (i) (degrees)	2.206	1.3498	5.7864	5.8254	–
Orbital period (years)	11.45	11.728	17.670	17.685	–
Predicted date of closest approach to Jupiter in 1992	A distinct possibility	July 8.8	July 8.0	July 7.8	–
Predicted date of first impact (1994)	No mention of an impact	July 25.4	July 18.7	July 17.6	July 16.826
Predicted date of last impact	No mention of an impact	July 25.4	July 23.2	July 22.3	July 22.330

and a second line connecting the Sun and the ascending node. The equinox point is the location of the Sun in the sky at the first moment of spring in the Earth's northern hemisphere. The argument of perihelion- ω is the angle between the line connecting the Sun and the perihelion point and a second line connecting the ascending and descending nodes. See Fig. 1.6.

Determination and Refinements of a Comet's Orbital Elements

Once a comet is discovered, one of the first things that astronomers do is to measure its position. Repeated position measurements are used in computing its path and future positions. Future positions will serve as an aid for making more position measurements and a refinement of the comet's path. The accuracy of the path improves as the number of position measurements increases and as the time interval of the position measurements increases. For example, Table 1.3 shows the computed orbital eccentricity, the orbital period and other predictions of Comet Shoemaker-Levy 9 (D/1993 F2). This table illustrates that as astronomers made more position measurements of this comet, they learned that it made a close passage to Jupiter in early July of 1992. With more refinement, they determined that it would crash into Jupiter and were able eventually to predict the time of the crash to an accuracy of a few minutes.

Comet Orbits and Kepler's Second and Third Laws of Planetary Motion

If a comet moves in an elliptical orbit around the Sun without pull from other bodies, its orbit would not change. It would not grow larger or smaller. This type of motion is called regular motion and its orbit is referred to as a Kepler-type orbit; and, in this context, it would prove the correctness of Kepler's Second and Third Laws of Planetary Motion. However, all comets which enter or are in our Solar System experience gravitational tugs by the planets, which are called perturbations. As a result of perturbations, Kepler's Second and Third laws are compromised, but, in many cases, they may be used as a *good approximation* of a comet's movement.

Figure 1.7 shows the orbit of Comet 1P/Halley. Kepler's Second Law states that a line connecting the Sun to an object that orbits the Sun sweeps out equal areas in equal intervals of time. If for example, the area swept out by this line in going from point A to point B equals the area swept out in going from point D to point C then the time it takes for the comet to go from point A to point B will equal the time it takes in going from point D to point C. Obviously the distance between point A and point B is much greater than the distance between point D and point C. Therefore this shows that the comet moves much faster when it is close to the Sun. Figure 1.8 shows approximate velocities of this comet at different points along its elliptical orbit. It moves fastest when it is at perihelion and moves slowest when it is at aphelion – its farthest distance from the Sun. One consequence of Kepler's Second Law is that a comet moves fastest when it is closest to the Sun (due to stronger gravitational pull).

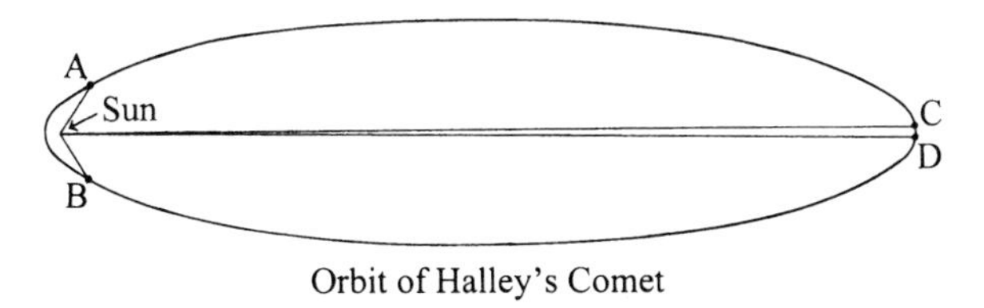

Fig. 1.7. A drawing illustrating Kepler's Second Law of Planetary Motion. It takes Halley's Comet 1 year to travel from points A to B and from points C to D. Since this comet is close to the Sun at points A and B, it moves much faster than when it is at points C and D. This is due to Kepler's Second Law which states that a line connecting the Sun to an orbiting planet (or comet) sweeps out equal areas in equal time (credit: Richard W. Schmude Jr.).

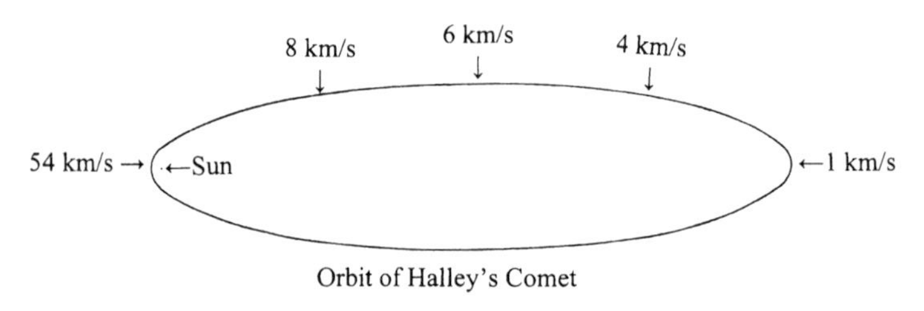

Fig. 1.8. A drawing showing the approximate orbital velocities of Comet 1P/Halley at different locations in its orbit (credit: Richard W. Schmude Jr.).

Comets in an elliptical orbit obey Kepler's Third Law of Planetary Motion. This law states that for objects orbiting our Sun, there is a relationship between the orbital period and the average distance from the Sun. Essentially, the farther an object is from the Sun, the longer that it will take for that object to orbit the Sun. Kepler's Third Law in equation form is:

$$p^2/r^3 = 1\ \text{year}^2/\text{au}^3 \quad \text{(for objects orbiting our Sun)} \tag{1.1}$$

or

$$r = \left[p^2 \times \text{au}^3/\text{year}^2\right]^{1/3} \tag{1.2}$$

In these equations, p is the orbital period, r is the average comet-Sun distance (or the semimajor axis of the comet's orbit) and au is an abbreviation for astronomical unit. By knowing the orbital period, one can compute the comet's average distance from the Sun.

Classification of Comets

Table 1.2 shows the percentage breakdown for different families and sub-groups of comets. Comets with an X/ in their designation are not included in Table 1.2. (In all cases, the orbital statistics quoted in this Section are based on comets known as of mid-2008. Orbital parameters are from *Catalogue of Cometary Orbits 2008, 17th edition* ©2008 by Brian G. Marsden and Gareth V. Williams.) Long-period comets make up 84.7% of all known comets. This group contains two large families of comets – the sungrazer family, with 51.9% of all known comets, and the long-period nearly isotropic family with 32.8% of all known comets. The short-period group contains 13.8% of all known comets, with the Jupiter family making up most of this group.

Comets in a family have similar orbits and are controlled by similar gravitational forces. In many cases, comets in a family also have a similar source. The mathematics used in dividing comets into different families is complex.

Many astronomers rely on the Tisserand Parameter (T) to divide comets into families. According to *Encyclopedia of the Solar System, 2nd edition* ©2007 by Lucy-Ann McFadden et al., the Tisserand Parameter, T, is defined as:

$$T = (j/r) + 2 \times \left[(1 - e^2) \times r / j\right]^{1/2} \times \cos(i) \qquad (1.3)$$

In this equation j = 5.20280 au and is Jupiter's average distance from the Sun; r is the comet's average distance from the Sun; e is the orbital eccentricity of the comet's path; i is the orbital inclination of the plane containing the comet's path in degrees; and cos is the cosine function.

The Tisserand Parameter (T) is a useful tool in establishing comet families in the short-period group. Figure 1.9 contains a flowchart illustrating different comet families in this group. If T is less than 2.0, the comet would belong to the nearly isotropic family. If T is between 2.0 and 3.0, the comet would be in the Jupiter family. If T exceeds 3.0, it would be in either the Encke or Chiron family depending on the size of its orbit. Table 1.4 lists several comets, certain orbital elements, Tisserand Parameter values and the families to whom they belong. Figures 1.10–1.12 illustrate the orbits of Jupiter along with the orbits of Comets 2P/Encke, 17P/Holmes and 95P/Chiron, respectively. Comet 2P/Encke lies always inside of Jupiter's orbit with the Sun, and it never gets close to Jupiter. Comet 95P/Chiron lies always outside of Jupiter's orbit, and, hence, it does not get close to Jupiter. Comet 17P/Holmes, on the other hand, may get close to Jupiter, and, hence, Jupiter's gravity would affect its orbit. In the next section, short-period comets and the families in this group are described.

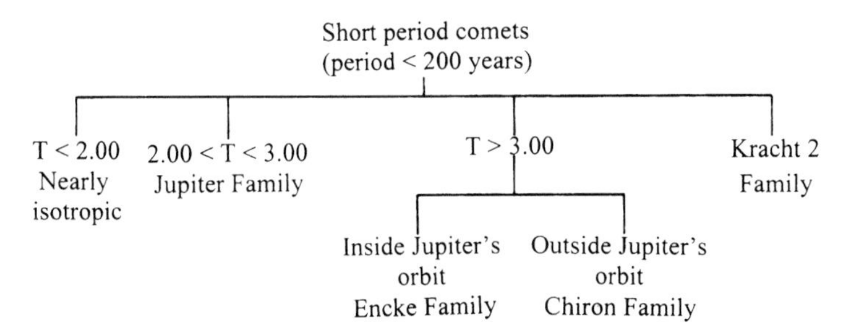

Fig. 1.9. A flowchart illustrating the different families of comets within the Short-period group of comets (credit: Richard W. Schmude Jr.).

Table 1.4. Orbital elements, the Tisserand parameter (T) and assigned family for a few comets. The general orbital elements of a comet are illustrated in Fig. 1.6; all data in this table are from *Catalogue of Cometary Orbits 2008, 17th edition* ©2008 by Brian G. Marsden and Gareth V. Williams

Comet	Perihelion distance – q (au)	Orbital eccentricity – e	Orbital inclination – i (degrees)	Tisserand parameter – T	Family
1P/Halley	0.5871	0.9673	162.2	–0.607	Short-period nearly isotropic
2P/Encke	0.3393	0.8470	11.75	3.03	Encke
9P/Tempel 1	1.506	0.5175	10.53	2.97	Jupiter
17P/Holmes	2.053	0.4324	19.11	2.86	Jupiter
95P/Chiron	8.454	0.3831	6.93	3.36	Chiron

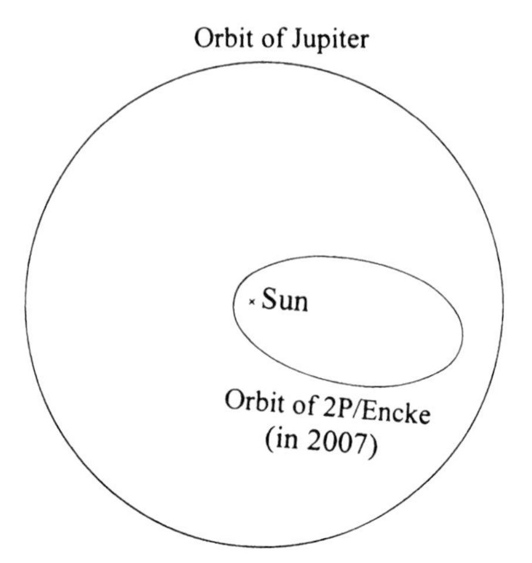

Fig. 1.10. A comparison of the orbits of Comet 2P/Encke in 2007 and Jupiter. Note that this comet never gets as far from the Sun as does Jupiter. As a result, it never gets close to Jupiter (credit: Richard W. Schmude Jr.).

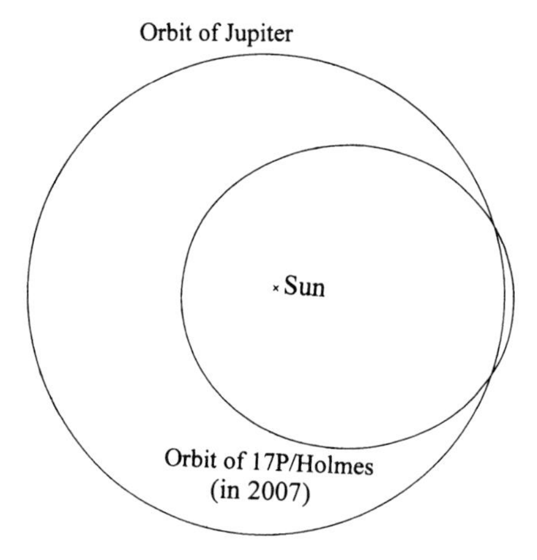

Fig. 1.11. A two-dimensional comparison of the orbits of Comet 17P/Holmes in 2007 and Jupiter. Note that the orbits of the two bodies overlap slightly. This means that they may come closer to one another in time (credit: Richard W. Schmude Jr.).

Short-Period Group

The short-period group contains the Encke, Chiron, Jupiter, Short-period nearly isotropic and Kracht 2 families. The orbital characteristics of these families are discussed below.

Comets in the Encke family have orbits inside that of Jupiter. Their movements, however, are not controlled by Jupiter's gravity. Another of their characteristics is that they have fairly low inclinations. All of them have orbital inclinations below 12°. They, however, may have a wide range of orbital eccentricities, as illustrated

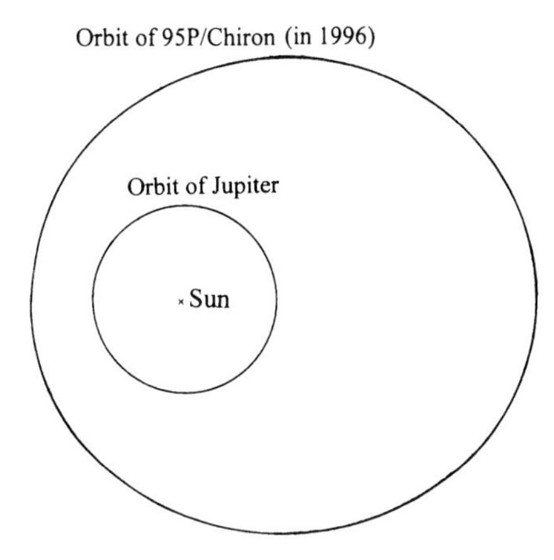

Fig. 1.12. A comparison of the orbits of Comet 95P/Chiron in 1996 and Jupiter. Note that the comet does not get close to Jupiter (credit: Richard W. Schmude Jr.).

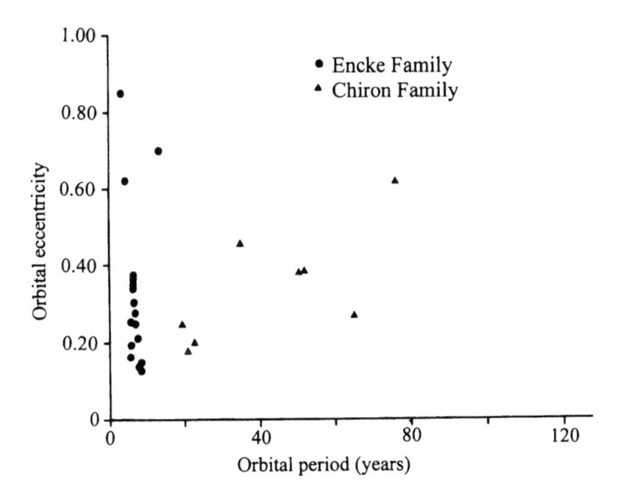

Fig. 1.13. A graph of the orbital eccentricity versus the orbital period (in years) for comets in the Encke and Chiron Families. Data are from *Catalogue of Cometary Orbits 2008, 17th edition* ©2008 by Brian G. Marsden and Gareth V. Williams (credit: Richard W. Schmude Jr.).

in Fig. 1.13. This figure shows a graph of the orbital eccentricity versus the orbital period of comets in the Encke and Chiron families. Therefore, comets in the Encke family follow orbits that range from being nearly circular to those that are stretched out.

Comets in the Chiron family follow paths which lie beyond Jupiter's orbit. Like the Encke family, these comets are not controlled by Jupiter's gravity. At least one comet in this family, 95P/Chiron, crosses Saturn's orbit and, hence, it may make a close pass to that planet. Comets in the Chiron family have low orbital inclinations. All of them have inclinations below 24°, and 63% of them have inclinations below 12°. Comets in this family have a wide range of eccentricities like those in the Encke Family. The orbital periods range from 20 to almost 80 years. See Fig. 1.13.

Most of the Short-period comets fall in the Jupiter family. This means that Jupiter's gravitational field affects strongly their movement. Jupiter family comets can pass close to Jupiter. Jupiter's orbital inclination is 1.3°, and, hence, most comets in this family have paths that are not in Jupiter's orbital plane. In spite of this, some of these comets cross Jupiter's path. Comet Shoemaker-Levy 9 (D/1993 F2) had an orbital inclination of 6.0° but in July 1994, it not only crossed Jupiter's path but collided with it. Figure 1.14 shows the percentages of Jupiter family comets with different orbital inclinations. Most of these comets have an orbital inclination below 24° which is similar to the Encke and Chiron families. Figure 1.15 shows the orbital period versus orbital eccentricity for Jupiter family comets. Jupiter family comets have a wide range of orbital eccentricities. Those with periods longer than about 25 years tend to have high eccentricities. High eccentricities are the only way that these comets can get close to Jupiter and, hence, come under its gravitational influence.

Comets in the short-period, nearly isotropic (NI) family have a different distribution of orbital inclinations than those in the Encke, Chiron and Jupiter families. Figure 1.16 shows the percentage of Short-period NI comets with different orbital inclinations. The figure shows that there is a fairly even distribution of orbital inclinations which is in contrast to comets in the Encke, Chiron and Jupiter families. The difference in inclination suggests that there is a different source for the Short-period NI comets compared to the other three families of comets discussed above. Astronomers believe that the Oort Cloud is the main source of short-period NI comets because of the almost even distribution of orbital inclinations. The Oort Cloud, named after Dutch astronomer Jan Hendrix Oort, is a spherical shell extending from 3,000 au (or 0.05 light years) to about 100,000 au (or about 1.6 light years) from the Sun. There may be as many as 10^{12} icy objects in this cloud that are big enough to become comets. Because of the abundance of low orbital inclinations for the Encke, Chiron and Jupiter families, astronomers believe that their main source of origin is the region just beyond Neptune.

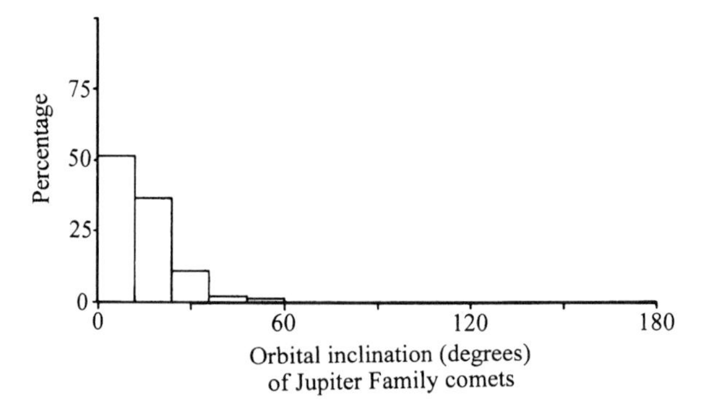

Fig. 1.14. A graph of the percentage of comets in the Jupiter Family having different orbital inclinations. Data are from *Catalogue of Cometary Orbits 2008, 17th edition* ©2008 by Brian G. Marsden and Gareth V. Williams (credit: Richard W. Schmude Jr.).

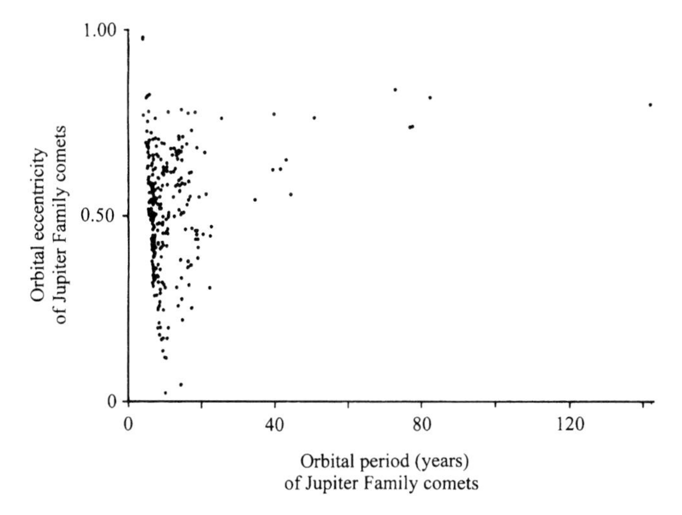

Fig. 1.15. A graph of the orbital eccentricity versus the orbital period for comets in the Jupiter family. Note that comets having an orbital period of less than about 15 years have a wide range of orbital eccentricities whereas those with an orbital period above 30 years have orbital eccentricities above 0.5. Data are from *Catalogue of Cometary Orbits 2008, 17th edition* ©2008 by Brian G. Marsden and Gareth V. Williams (credit: Richard W. Schmude Jr.).

Fig. 1.16. This graph shows the percentage of comets in the Short-period nearly isotropic (NI) family with different orbital inclinations. Note the fairly even distribution of orbital inclinations. Data are from *Catalogue of Cometary Orbits 2008, 17th edition* ©2008 by Brian G. Marsden and Gareth V. Williams (credit: Richard W. Schmude Jr.).

Figure 1.17 shows a graph of the orbital eccentricity plotted against the orbital period of Short-period NI comets. (Comets 153P [period = 364 years, eccentricity = 0.990062] and C/1937 D1 [period = 187 years, eccentricity = 0.981] are not included in Fig. 1.17.) In all cases, these comets have orbital eccentricity values greater than 0.60. This means that comets in this family follow stretched out orbits. They will get close to the Sun for a short period of time and will remain far from the Sun for the remainder of the time.

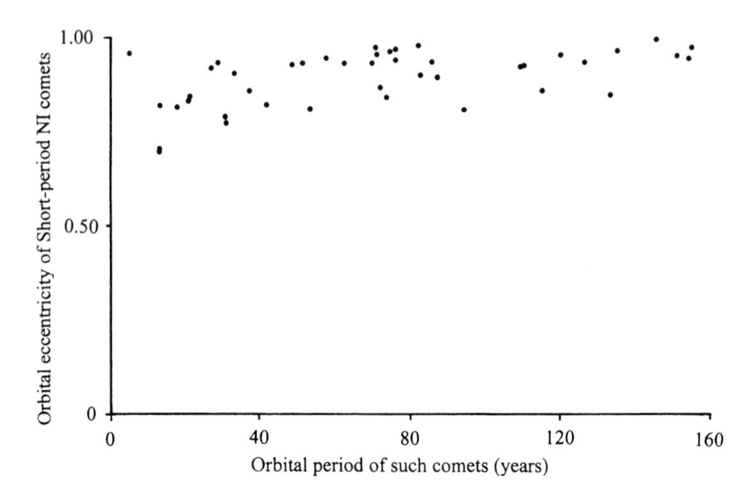

Fig. 1.17. A graph showing the orbital eccentricity for different orbital periods of comets in the Short-period nearly isotropic (NI) family with different orbital periods. Note that all of the comets have orbital eccentricities above 0.6. Data are from *Catalogue of Cometary Orbits 2008, 17th edition* ©2008 by Brian G. Marsden and Gareth V. Williams (credit: Richard W. Schmude Jr.).

Table 1.5. Average orbital characteristics of seven different sub-groups of Sungrazer Family of Long-period comets and one sub-group of Kracht 2 Short-period comets. In some cases, the standard deviation is given in parentheses. All values are based on data in *Catalogue of Cometary Orbits 2008, 17th edition* ©2008 by Brian G. Marsden and Gareth V. Williams

Sub-group	Argument of perihelion – ω (degrees)	Longitude of ascending node – Ω (degrees)	Orbital inclination – i (degrees)
Kracht 1	58.6 (7.7)	44.2 (7.4)	13.4 (0.8)
Marsden	24.2 (6.5)	79.0 (8.6)	26.5 (2.5)
Meyer	57.4 (2.4)	73.3 (4.1)	72.8 (2.3)
Kreutz	79.9	360.4 or 0.4	143.3 (~2)
Anon 1	86.9 (2.3)	238 (12.7)	87.8 (0.9)
Anon 2	81.6	243.1	27.48
Kracht 3	179.8	326.3	55.1
Kracht 2	45.2	6.0	13.5

Finally the Kracht 2 family of comets consists of four members as of mid-2008. These comets may get very close to the Sun. They have an average perihelion distance of 0.054 au. Three of the four Kracht 2 comets have eccentricities of between 0.977 and 0.979, which is consistent with a semimajor axis of about 2.5 au. These comets may get close to Jupiter at aphelion and if so, their orbits may get changed by it. Average values of other orbital characteristics for the Kracht 2 family are listed in the bottom row of Table 1.5.

Long-Period Comets

There are two families of comets in the long-period group, namely the sungrazer comets and the long-period nearly isotropic (NI) comets. These families along with the sub-groups in the sungrazer family are described below. See Table 1.2.

Figure 1.18 shows a flowchart illustrating different comet families and sub-groups in the long-period group.

The sungrazers are comets that move in stretched-out orbits and approach the Sun at distances of usually less than 0.05 au. (An astronomical unit is the average Earth-Sun distance and it equals 149.6 million kilometers.) Figure 1.19 shows the orbit of a typical Kreutz family comet – Comet C/2007 V13. The orbit is very stretched out because of the high orbital eccentricity (~0.9999). Essentially this comet is just 0.0057 au from the Sun's center, or just 157,000 km above the Sun's photosphere at perihelion. At its farthest point, it is over 172 au from the Sun (assuming an orbital period of 800 years). Figure 1.20 shows a close-up view of the path of Comet C/2007 V13 as it approached the Sun in 2007. The comet's path is the dashed curve.

Do sungrazing comets burn up when they pass close to the Sun? Oftentimes they do. Many burn up because of their small sizes. Astronomers believe that their nuclei are between a few meters to a few tens of meters across. Others collide with the Sun. At least 26 Kreutz group comets had perihelion distances of less than the radius of the Sun (as of mid-2008), and, hence, they collided with the Sun. Many more sungrazers make such a close approach that they either spiral into the Sun or perhaps sublime or explode during their close pass to the Sun. Some Sungrazers, however, survive their close passage of the Sun. How can this be? This may be answered with a question: Have you ever moved your finger quickly through a candle flame? If you have, it probably did not hurt. This is because you did not allow enough time for your finger to heat up. Many sungrazers also pass

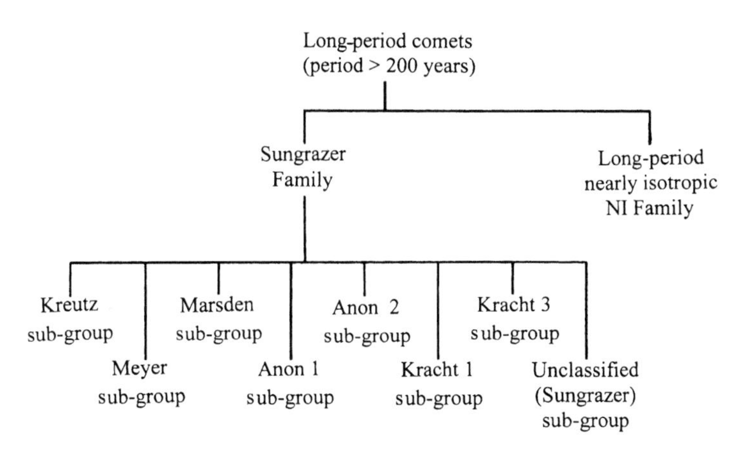

Fig. 1.18. A flowchart illustrating the different families of comets and their sub-groups within the Long-period group of comets (credit: Richard W. Schmude Jr.).

Comet C/2007 V13

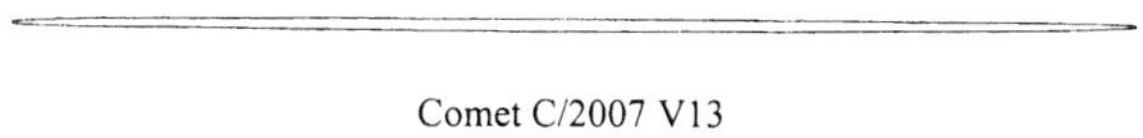

Fig. 1.19. The approximate path that Comet C/2007 V13 follows. Note that this path is very stretched out. This comet spends almost all of its time far from the Sun (credit: Richard W. Schmude Jr.).

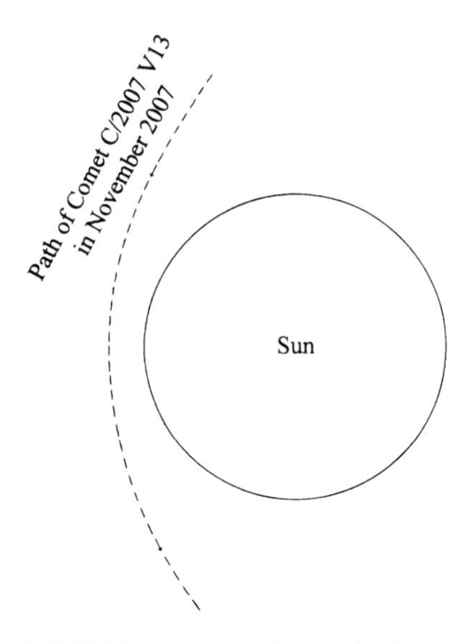

Fig. 1.20. The path that Comet C/2007 V13 followed as it made its close approach to the Sun in November of 2007 (credit: Richard W. Schmude Jr.).

by the Sun fast enough to remain intact. These comets move quickly when near the Sun because of Kepler's Second Law of Planetary Motion. In Fig. 1.20, it took Comet C/2007 V13 about 1.5 h to move from one end of the dashed curve to the other end. Therefore, even though sungrazers get very close to the Sun, they do not get as hot as one might expect.

Figures 1.21 and 1.22 show the distribution of perihelion distances for the Meyer and Kreutz sub-groups, respectively. Comets in the Kreutz sub-group have a narrow range of perihelion distances. Most of these comets pass within 0.01 au of the Sun's photosphere. Comets in the Meyer sub-group do not get as close to the Sun. Most of these comets pass within 0.04 au of the Sun's photosphere.

Figure 1.23 shows the distribution of orbital inclinations of a few larger sub-groups of sungrazing comets. All of the sub-groups have a narrow distribution of orbital inclinations. This is evidence that the different sub-groups have different sources. Table 1.5 summarizes average values of three orbital quantities in degrees (Argument of perihelion – ω, Longitude of the ascending node – Ω and the Orbital inclination – i) for comets in several sub-groups of the sungrazer family. Comets in the Kreutz sub-group are believed to be fragments of a single comet. Comets in each of the other sub-groups may be fragments of different comets which broke up some time in the past.

I have analyzed the brightness of the sungrazing comet SOHO (C/2001 C5), based on brightness values reported in the International Astronomical Union Circular Number 7585. The brightness values of it are plotted in Fig. 1.24 on different decimal dates in February 2001. The decimal date is the date followed by the fraction of the day that has elapsed since the start of the day (0:00

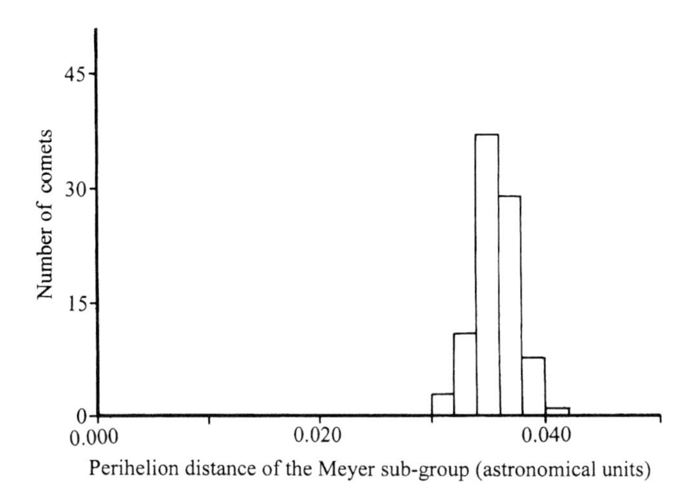

Fig. 1.21. This graph shows the distribution of perihelion distances for the Meyer sub-group of comets. Note that there is a fairly narrow range of perihelion distances for these comets. Data are from *Catalogue of Cometary Orbits 2008, 17th edition* ©2008 by Brian G. Marsden and Gareth V. Williams (credit: Richard W. Schmude Jr.).

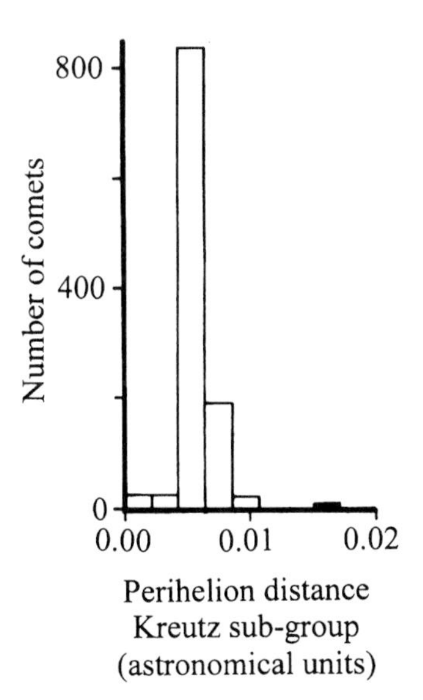

Fig. 1.22. This graph shows the distribution of perihelion distances for the Kreutz sub-group of comets. Note that there is a narrow range of perihelion distances for these comets. Several of these comets crashed into the Sun. Data are from *Catalogue of Cometary Orbits 2008, 17th edition* ©2008 by Brian G. Marsden and Gareth V. Williams (credit: Richard W. Schmude Jr.).

Universal Time). This comet has a normalized magnitude of $H_{10} = 24.58$ (normalized magnitudes are described later in this chapter). This value is almost 18 stellar magnitudes fainter than the average H_{10} value of comets with a C/ designation. This means that comet SOHO (C/2001 C5) would be just over 10

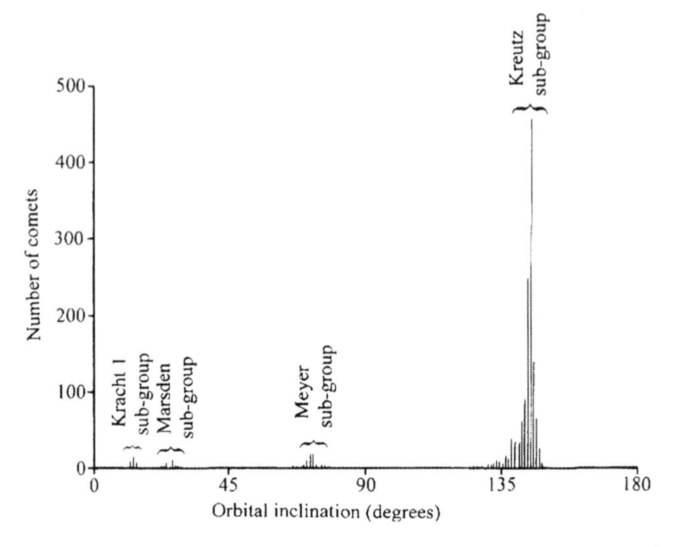

Fig. 1.23. This graph shows the distribution of orbital inclinations for different sub-groups of Sungrazing comets. Note that each sub-group has a narrow range of orbital inclinations. Data are from *Catalogue of Cometary Orbits 2008, 17th edition* ©2008 by Brian G. Marsden and Gareth V. Williams (credit: Richard W. Schmude Jr.).

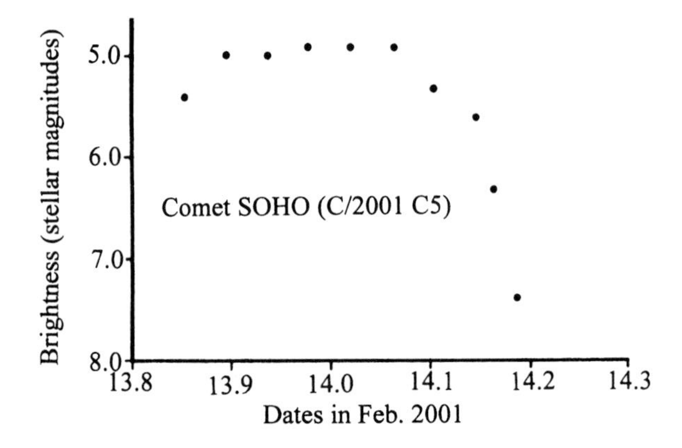

Fig. 1.24. This graph shows the brightness of the Sungrazing Comet (C/2001 C5) in stellar magnitudes on different decimal dates in February, 2001. Data are from IAU Circular Number 7585 (credit: Richard W. Schmude Jr.).

million times fainter than a typical comet with a C/ designation at the same distance from the Earth and Sun. The only reason why comet SOHO (C/2001 C5) reached a stellar magnitude of 4.9 is because of its close distance to the Sun at the time the measurements were made.

The long-period nearly isotropic (NI) family of comets has a wide range of orbital inclinations. This is similar to the short-period NI comets. The percentages of orbital inclinations for all NI comets are plotted in Fig. 1.25. The orbital inclinations are fairly evenly distributed between 40° and 150°. The numbers drop off at orbital inclinations of less than 40° and above 160°.

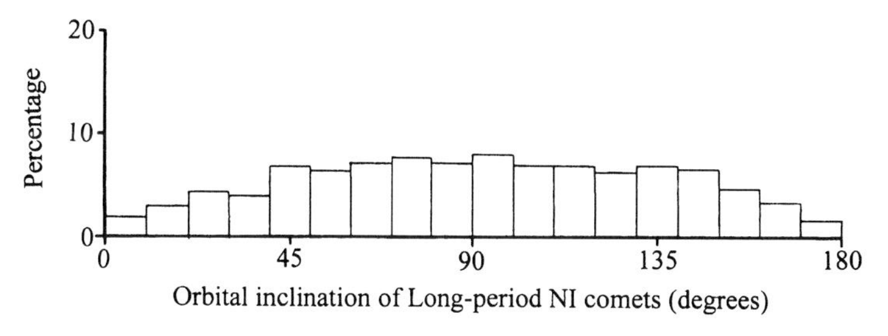

Fig. 1.25. This graph shows the percentage of Long-period and Short-period nearly isotropic (NI) comets having different orbital inclinations. Data are from *Catalogue of Cometary Orbits 2008, 17th edition* ©2008 by Brian G. Marsden and Gareth V. Williams (credit: Richard W. Schmude Jr.).

As of mid-2008, 50 sungrazer comets do not belong to any of the seven subgroups of the sungrazer family listed in Table 1.5. These comets fall into the unclassified (sungrazer) category in Table 1.2. I plotted the orbital inclination versus the argument of perihelion for these comets. After a close study of this plot, along with the longitude of the ascending node values, I concluded, based on the data collected up to mid-2008, that there are no new families for the unclassified comets. Two of these unclassified comets – SOHO (C/1999 O4) and SOHO (C/2004 Y10) – have respective orbital values of ($\omega = 104.04°$, $\Omega = 107.87°$, $i = 133.32°$) and ($\omega = 99.02°$, $\Omega = 129.68°$, $i = 131.61°$). Of all of the unclassified Sungrazers, these two have the closest orbital constants. Because of their large difference in the Ω value, it is doubtful that they represent a new family of comets.

Comets with No Future Predictions

As of May 2008, almost four dozen comets were declared lost or destroyed. Astronomers gave these comets a D/ as the first two symbols in their designation. Most of these comets had orbital inclinations below 20°. All but three of them had orbital periods of less than 10 years. The majority of these comets would have been in the Jupiter family of the short-period group had they survived.

Sources and Movement of Comets

Over the last 2,000 years, hundreds of comets have streaked across our sky. Where did they come from? What forces control their movement? These questions and others related to comet sources and comet movement are answered below. This section discusses first the source of comets; next, gravity and what pushes comets into the inner Solar System; and last, some of the many non-gravitational forces which affect comet movement.

The first object discovered beyond the orbit of the planet Neptune was the newly-classified dwarf planet Pluto. Since the early 1990s, astronomers have

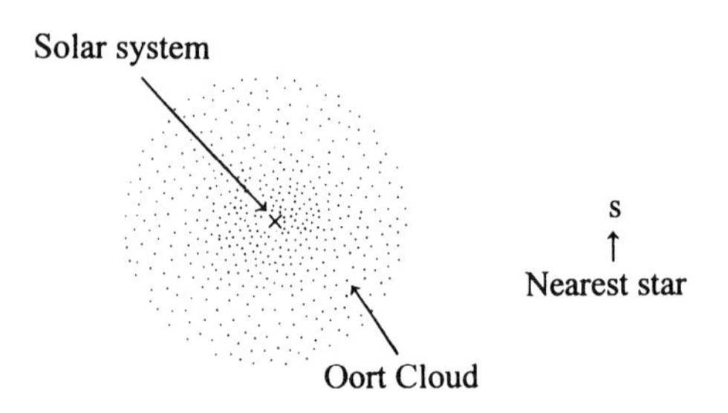

Fig. 1.26. This drawing shows the Oort cloud, and the distance to the nearest star (point s) outside of our Solar System (point ×) (credit: Richard W. Schmude Jr.).

discovered over 1,200 objects beyond Neptune's orbit; and there may be numerous other objects with diameters exceeding a few dozen km between 30 and 50 au from the Sun. These bodies usually have low orbital inclinations and are believed to be a source of short-period comets. A second source of comets is believed to be the Oort Cloud described above. Unlike the objects lying just beyond Neptune, those in the Oort Cloud have a nearly random distribution of orbital inclinations. See Fig. 1.26. These objects cannot become comets as defined, until they pass close to the Sun because, without the Sun's warmth, they will not develop comas.

An icy body in the trans-Neptune area (30–50 au from the Sun) may be pulled inwards by Neptune's gravity. Objects in the more distant Oort Cloud may be pulled inward by gravitational forces exerted by stars passing close to the Sun or by galactic perturbations. When an icy body is pushed inward (and into warmth) it may develop a coma and, hence, become a comet. Forces which may push or pull an icy body towards the Sun are gravity from the Sun, gravity from planets, gravity from nearby stars, and gravity from the galactic disk. These forces will be discussed after a brief discussion of Newton's Law of Universal Gravitation.

Newton's Law of Universal Gravitation

The force of gravity (F_g) between two objects according to Newton's Law of Universal Gravitation is: $F_g = (G \times m_1 \times m_2)/D^2$ where G is the gravitational constant, 6.67×10^{-11} $m^3/(kg\,s^2)$, m_1 is the mass of the first object (let's say the comet), m_2 is the mass of the second object which can be the Sun or a planet, and D is the distance between the centers of m_1 and m_2. The important trend to note is that the force of gravity (F_g) changes with the square of the distance. Therefore, at small distances, F_g gets large quickly whereas for large distances, F_g gets small quickly. In short, an object's gravitational force depends on both its mass and distance to a second object.

Although the Sun is the most massive object in our solar system, its gravity may not always exceed that of a nearby planet. Another consequence of Newton's Law of Universal Gravitation is that the Sun's gravity grows weaker outside of our solar system. This is because gravity drops off with the square of the distance. For example, the Sun's gravity is 900 times stronger at a distance of 1.0 au compared to a distance of 30.0 au.

Gravity and Its Influence on Comets

Gravity is the reason why some bodies in our outer solar system end up passing close to the Sun and becoming comets. Any large object, like a planet, can exert a gravitational force. In this section, I will discuss how the Sun, large planets, nearby stars and our galaxy pull, push or nudge objects into or out of our inner solar system.

The Sun is the most massive object in our solar system. As a result, comets are drawn towards it. Once an icy object passes into our solar system, it will most likely be drawn towards the Sun. If the conditions are right, it will orbit close to the Sun, develop a coma and become a comet.

Planets can exert a strong gravitational force on a nearby comet and, in many cases, can change the orbital period or direction of a comet. Figure 1.27 illustrates a comet making a close pass to Jupiter. As the comet approaches Jupiter, it speeds up (represented by a long arrow) and, as it moves away, it slows down (represented by a short arrow). The comet speeds up as it approaches Jupiter because Jupiter's gravity is pulling the comet towards it. As the comet moves away, Jupiter's gravity pulls it back causing it to slow down. Since Jupiter is also moving, operation of the gravitational forces becomes complicated.

It is, however, possible to compute the trajectory of the comet as a result of its close pass to a planet. In most cases, a nearby planet will change both the direction in which a comet moves and its speed. In Fig. 1.27, Jupiter's gravitational force causes a change in comet speed. Since Jupiter has such a large mass, the gravitational force of a comet has little effect on Jupiter's movement. The net result of these gravitational forces causes orbital change of the comet.

Each of the planets can change a comet's orbit. Figure 1.28 shows how the orbital period of Comet 1P/Halley has changed since 239 BC. The period was as high as 79.3 years in 451 and 1066 AD and was as low as 76.0 years in 1986. Much of the change in the orbital period is due to small planetary perturbations. For several periodic comets, Table 1.6 lists the range of their orbital periods, their average orbital periods and the related time intervals. In many cases, a planetary perturbation can cause a comet to change course and head to the inner solar system. Neptune plays an important role for trans-Neptune objects. In a similar manner, a planetary perturbation can eject a comet from our solar system.

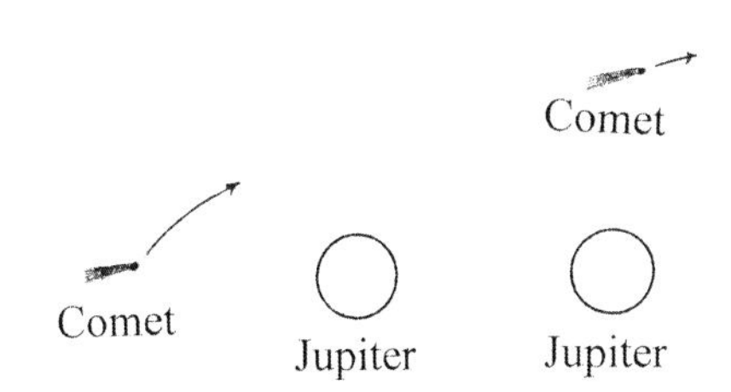

Fig. 1.27. As a comet approaches Jupiter, its orbital speed increases as a result of the pull effect of Jupiter's gravity. This is illustrated with a *long, curved arrow*. As a comet moves away from Jupiter its orbital speed decreases because of the pull of Jupiter's gravity against the outward movement from Jupiter. This is illustrated with a *short, curved arrow* (credit: Richard W. Schmude Jr.).

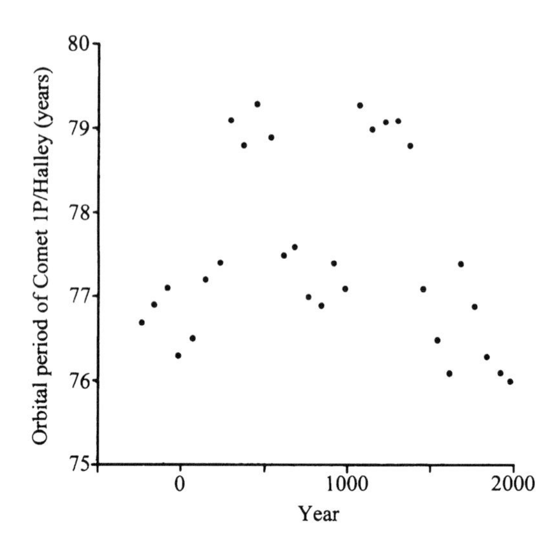

Fig. 1.28. This graph shows the orbital period of Comet 1P/Halley since 239 BC. Note that the orbital period has changed over the last 2000-plus years. Data are from *Catalogue of Cometary Orbits 2008, 17th edition* ©2008 by Brian G. Marsden and Gareth V. Williams (credit: Richard W. Schmude Jr.).

Table 1.6. The range of orbital periods and average orbital periods for several Short-period comets. Data are based on values in *Catalogue of Cometary Orbits 2008, 17th edition* ©2008 by Brian G. Marsden and Gareth V. Williams

Comet	Range of orbital periods (years)	Average orbital period (years)	Time interval
1P/Halley	76.0–79.3	77.5	239BC–1986
2P/Encke	3.28–3.32	3.30	1786–2007
4P/Faye	7.32–7.59	7.44	1843–2006
7P/Pons-Winnecke	5.55–6.38	6.05	1819–2008
8P/Tuttle	13.5–13.9	13.7	1790–2008
9P/Tempel 1	5.49–5.98	5.60	1867–2005
10P/Tempel 2	5.16–5.48	5.28	1873–2005
14P/Wolf	6.77–8.74	7.88	1884–2009
15P/Finlay	6.50–6.97	6.77	1886–2008
19P/Borrelly	6.76–7.02	6.89	1905–2008
81P/Wild 2	6.17–6.40	6.30	1978–2003

Our galaxy can also exert a gravitational force on potential comets. Our solar system moves once around the galaxy every 220 million years or so. In addition, our solar system moves through the galactic plane where there is a higher density of matter. The extra mass can exert small gravitational tugs on icy objects in the Oort Cloud. In many cases, objects may be directed to the inner solar system and become comets. One astronomer reported that a group of 152 long-period comets do not have random orbits. This person suggested that gravitational tugs from the galactic disc were responsible for the non-random distribution. (See *Astronomy and Astrophysics*, Vol. 187, pp. 913–918 by A. H. Delsemme for more information.)

Nearby stars can also nudge objects in the Oort Cloud towards the inner solar system. One group of astronomers reported that about 12 stars pass within 1.0 parsec (or 3.26 light years) of the Sun every million years. Stars passing this close should exert gravitational tugs on some of them in the Oort Cloud causing a few of them to enter and pass through the inner solar system and become comets.

To sum up, a comet may be pushed or drawn into the inner solar system by the Sun, planets, nearby stars, or our galaxy; and the Sun and planets may continue to modify a comet's path inside of our solar system. Once a comet is in the inner solar system, non-gravitational forces can also play a role in the path that it takes. These forces include those caused by volatile substances escaping from the comet's nucleus, meteor impacts, the Yarkovsky Effect, sunlight, the solar wind, occasional solar explosions and occasional solar storms. These non-gravitational forces are discussed below.

Movement of Comets: Non-gravitational Forces and Their Influence

Since comets move in the vacuum of space, the application of slight forces, over time, can operate to change a comet's movement or orbit. Such forces are described here as non-gravitational.

When an icy body approaches the Sun, its surface warms up and volatiles escape from its nucleus. When this occurs, the nucleus recoils slightly in the opposite direction. I call this the "rocket force", because a rocket works on the same principle. This recoil is due to Newton's Third Law of Gravitation which states that for every action, there is an equal and opposite reaction. See Fig. 1.29. The magnitude of the rocket force depends on several factors, including the amount of material leaving the nucleus, the comet's distance from Sun, orientation of spin axis of the nucleus and the rotation rate of the nucleus.

Impacts are a second source of non-gravitational force. During an impact, the impacting body exchanges momentum with the cometary nucleus, causing it to move a little in the direction of the momentum. An object's momentum is the product of its mass and velocity. The larger the mass or velocity of an object, the

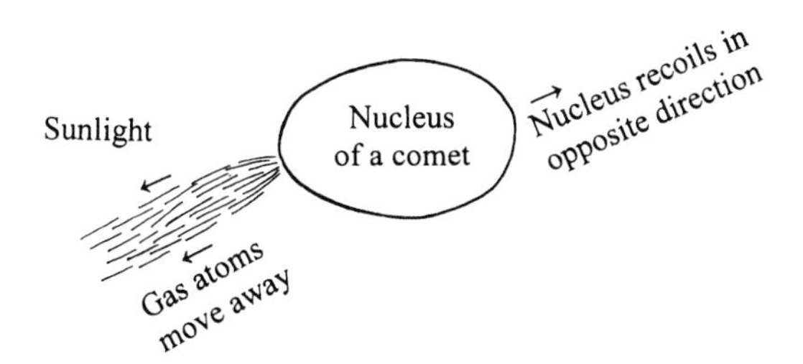

Fig. 1.29. As gas escapes from the nucleus, it moves in one direction. Due to Newton's Third Law of motion, the nucleus recoils slightly in the opposite direction. This movement is one example of a non-gravitational force. Note that sunlight is coming from the left side (credit: Richard W. Schmude Jr.)

greater will be its momentum. Impacting objects may also expose buried ices. These ices will then sublime when the comet approaches the Sun. This, in turn, may lead to an additional "rocket force." Impacting objects may affect also the rotation rate of the nucleus.

A third type of non-gravitational force results from the non-random reflection of sunlight for a comet. The Yarkovsky Effect is the result of a spinning comet absorbing sunlight and then re-radiating it in a different direction. This non-random reflection of sunlight may lead to small pushes in the direction away from where the radiation is released, provided that the nucleus rotates around a single axis. One group reports that this effect shifted the position of the minor planet Golevka 15 km between 1991 and 2003. Comets with small orbits which rotate around a single axis will be most susceptible to the Yarkovsky Effect because a comet in a small orbit receives more sunlight. Furthermore, the Yarkovsky Effect builds up quicker for a nucleus which rotates around a fixed axis compared to one that does not do so.

Sunlight also exerts a small push. This is called solar radiation pressure. Solar radiation pressure pushes in the opposite direction as the Sun's gravity. Essentially, light possesses momentum which may alter a comet's momentum. Solar radiation pressure is small compared to the gravitational force of the Sun. Nevertheless, this pressure over many centuries may affect a comet's movement.

A fifth source of non-gravitational force is exerted by the solar wind. The solar wind consists of protons and other particles leaving the Sun, including gamma rays and X-rays. It generally moves rapidly from the Sun and, hence, exerts a force in a direction opposite to the Sun's gravity. While solar wind exerts a small force on a comet, over long periods of time, it may result in changes in a comet's movement or orbit.

In the *Catalogue of Cometary Orbits 2008*, these non-gravitational forces are lumped together into two parameters, A1 and A2. The radial component is described by A1 and the transverse component is described by A2. The values of A1 and A2 can change over time. For example, the values of A1 and A2 for Comet 2P/Encke were −0.03 and −0.0030 in 1977 but were −0.01 and −0.0007 in 2007.

Finally, but not least, the random occurrences of solar explosions and solar storms cannot be overlooked. These phenomena exert slight forces which could affect over time the orbit of a comet within their ranges.

It is very difficult to analyze precisely a comet's movement and orbit. This is because cometary orbits are not only chaotic, but because they may be affected by so many varying forces – gravitational and non-gravitational – which affect their movement and orbit. Even small forces caused by the "rocket force", impacts, the Yarkovsky Effect, solar radiation pressure and solar wind pressure may be of importance.

Comet Brightness and Some Statistics

What is a typical comet? In this section, some statistical trends which will help answer this question are described. More specifically, trends in the distribution of perihelion distances (q), orbital inclinations (i) and values of the parameters H_{10}, H_0 and n for comets are summarized. The last three parameters describe comet brightness in visible light.

Figure 1.30 shows the distribution of orbital inclinations of short-period comets (excluding those with a D/ designation). Most of these comets have orbital inclinations of less than 30° due to the large number of Jupiter family comets in this group. The low inclination of these comets suggests that most of them came from the zone of icy objects just beyond Neptune where orbital inclinations are usually low.

Figure 1.31, plots the distribution of orbital inclinations for all comets known as of mid-2008 except those named SMM, SOHO, SOLWIND, STEREO and those that have been given a D/ designation. There is a fairly even distribution of orbital inclinations except those near 0° and 180°. The peak near 0° is due to the large number of Jupiter family comets.

Does Fig. 1.31 give an accurate picture of comet inclinations? Probably. Over the past few centuries, technology has played an important role in changing our view of the Universe. Major technological breakthroughs or advances in photography, digital imaging, the Internet and space probes have occurred. During the latter part of the nineteenth century, photography was developed to the point where it was used successfully in astronomy. Comet Barnard 3 (D/1892 T1) was the first comet discovered with photography. Since then, astronomers have discovered dozens of comets with this technique. During the late twentieth century, digital cameras, the SOHO space probe and the internet have allowed astronomers to discover over 1,000 new comets. In fact, more comets were discovered between 2000 and 2008 than in the previous 2,000 years. In essence, our under-

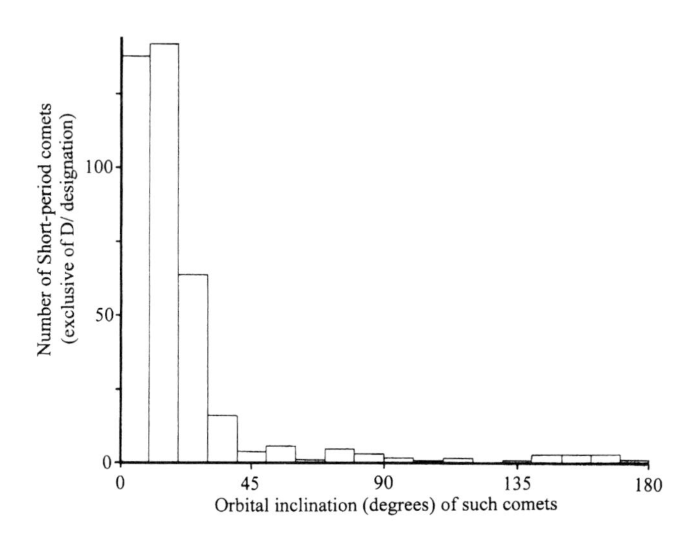

Fig. 1.30. Except for short-period comets with a D/ designation, this graph shows the number of all other Short-period comets with different orbital inclinations. Data are from *Catalogue of Cometary Orbits 2008, 17th edition* ©2008 by Brian G. Marsden and Gareth V. Williams (credit: Richard W. Schmude Jr.).

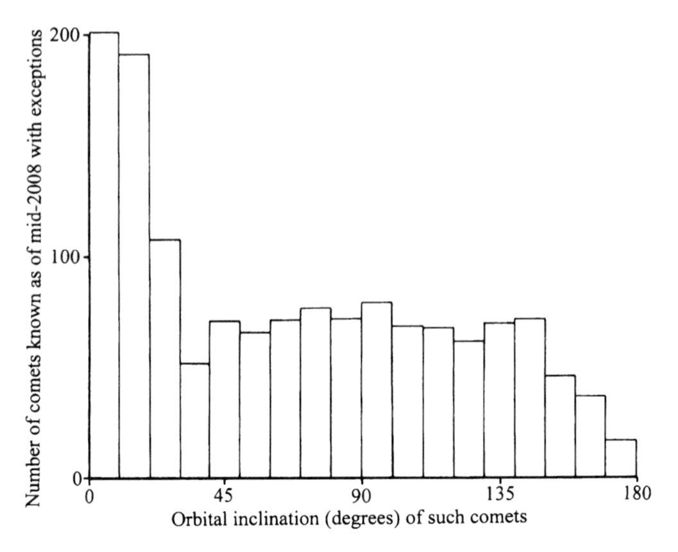

Fig. 1.31. This graph shows the different orbital inclinations of all comets known as of mid-2008 except those named SMM, SOHO, SOLWIND, STEREO and those with a D/ designation. Data are from *Catalogue of Cometary Orbits 2008, 17th edition* ©2008 by Brian G. Marsden and Gareth V. Williams (credit: Richard W. Schmude Jr.).

standing of comets and the distribution of their orbital inclinations have changed as a result of technological advances. Therefore, in an attempt to answer the question posed, our knowledge of comet inclinations as of 1892 and 1989 is presented below.

Orbital Inclination

Figure 1.32 shows, as of early 1892, the distribution of orbital inclinations of all known comets excluding those with a D/ designation. All of the comets in this Figure are visual discoveries. On average, about 25 comets lie in each 10° bin of orbital inclinations. Two noteworthy trends in Fig. 1.32 are (1) the number of comets with orbital inclinations below 30° which is about average compared to those with inclinations between 30° and 160° and (2) the number of comets with an orbital inclination above 160° which is below average. With the advent of high-quality photographs, astronomers discovered many comets during the first 90 years of the twentieth century.

Figure 1.33 shows the distribution of orbital inclinations of all known comets as of December 31, 1989 except for comets named SMM, SOHO, SOLWIND and those with a D/ designation. There is a higher than normal abundance of comets with orbital inclinations below 30° in Fig. 1.33 than with inclinations above 30°. There is also a definite drop of comets with orbital inclinations over 160°. Therefore, the first trend in Fig. 1.32 did not hold the test of time, but the second one did. By 1989, it was apparent that orbital inclinations were not evenly distributed. Figure 1.31 shows the distribution of orbital inclinations up to mid-2008. The trend of low orbital inclinations held up between 1989 and 2008. The scarcity of comets with

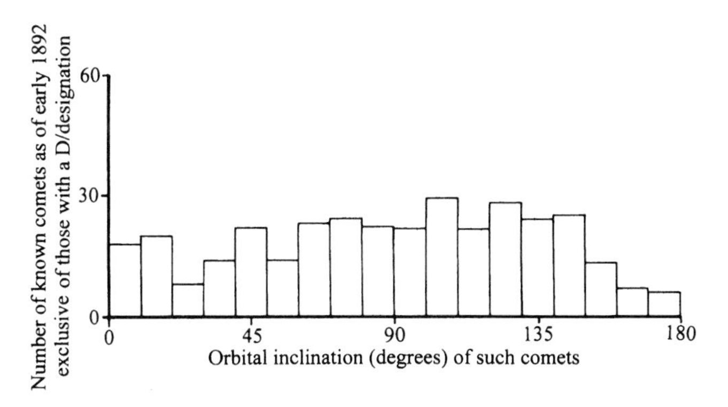

Fig. 1.32. This graph shows the number of comets known as of 1892 (excluding those with a D/ designation) with different orbital inclinations. Data are from *Catalogue of Cometary Orbits 2008, 17th edition* ©2008 by Brian G. Marsden and Gareth V. Williams (credit: Richard W. Schmude Jr.).

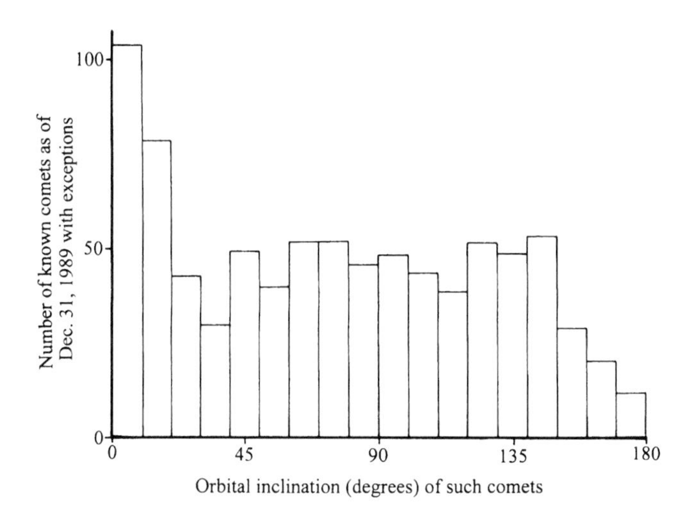

Fig. 1.33. This graph shows the number of comets with different orbital inclinations. All comets known as of December 31, 1989 are included except those named SMM, SOHO, SOLWIND, and those with a D/ designation. Data are from *Catalogue of Cometary Orbits 2008, 17th edition* ©2008 by Brian G. Marsden and Gareth V. Williams (credit: Richard W. Schmude Jr.).

orbital inclinations above 160° is still apparent in 2008. This trend was evident in 1892.

Will the two trends of a higher-than-average number of comets with orbital inclinations below 30° and a lower-than-average number of comets with orbital inclinations above 160° hold up in the future? Probably they will. It is believed that the first trend will hold up because it did between 1989 and 2008. Furthermore, there are two sources of comets with low orbital inclinations, and just one source of comets with high ($i > 30°$) orbital inclinations. This will lead to a higher number of comets with low orbital inclinations. It is believed that the trend of a lower-than-average number of comets at orbital inclinations above 160° will hold in the future because it held between 1892 and 2008.

Perihelion Distance

Another critical comet characteristic is the perihelion distance (q). Figure 1.34 shows the distribution of perihelion distances for all known comets except for a few of them having $q > 7$ au and those with a D/ designation. It is apparent that there are two maxima in this figure. The largest maxima is at $q < 0.1$ au. This is due to the large number of sungrazing comets. A second maxima is at about $q = 0.9$ au. Figure 1.35 is similar to Fig. 1.34 except that it excludes comets with $q < 0.1$ au and it also excludes comets named SMM, SOHO, SOLWIND and STEREO. This figure shows the second maxima of $q = 0.9$ au better. There is a definite drop of comets with q between 0.2 and 0.6 au.

Brightness

In addition to orbital inclination and perihelion distance, a third comet characteristic is brightness. There are three common equations which are used in describing a comet's brightness. One deals with the bare nucleus, while the other two deal with the coma.

When a comet is far from the Sun, it loses its coma. All that is left of the comet is its bare nucleus. It is always very faint. Its brightness, in stellar magnitudes, follows (1.4).

$$\text{V filter brightness} = V(1,0) + 5\log[r \times \Delta] + c_V \times \alpha \tag{1.4}$$

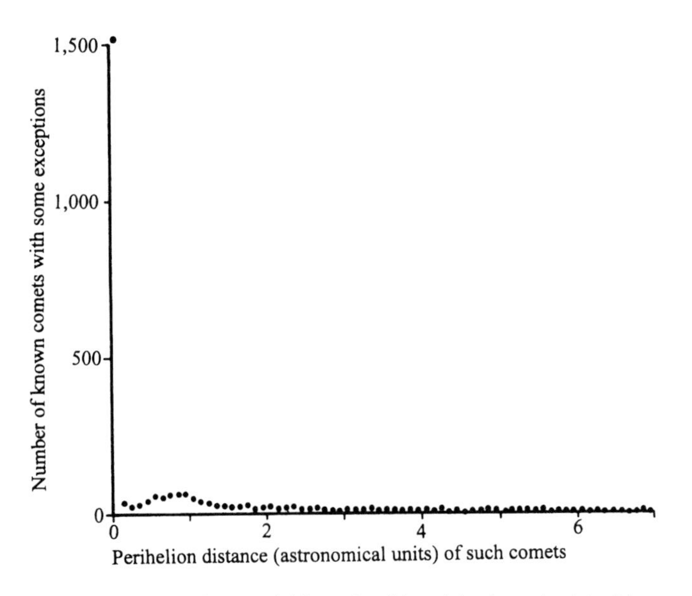

Fig. 1.34. This graph shows the number of comets with different values of the perihelion distance. It includes all known comets as of mid-2008 except those with perihelion distances exceeding 7.0 au and those with a D/ designation. Data are from *Catalogue of Cometary Orbits 2008, 17th edition* ©2008 by Brian G. Marsden and Gareth V. Williams (credit: Richard W. Schmude Jr.).

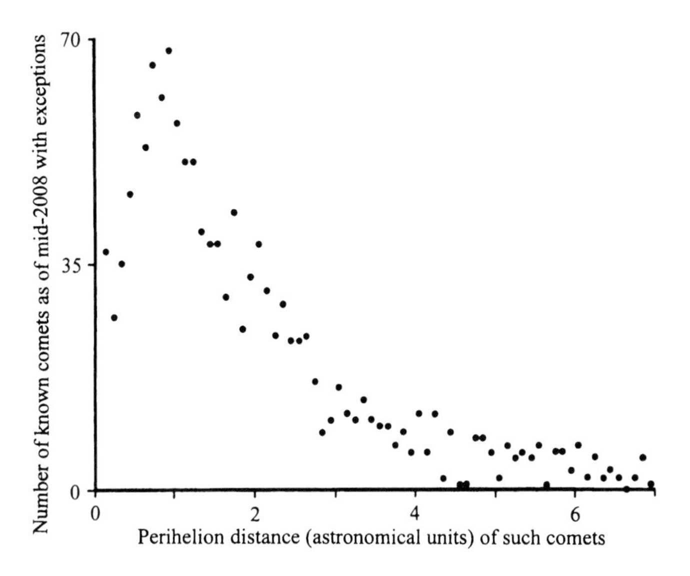

Fig. 1.35. This graph shows the number of comets with different perihelion distances. It includes all comets known as of mid-2008 except for those with perihelion distances exceeding 7.0 au, those named SMM, SOHO, SOLWIND and STEREO and those with a D/ designation. Data are from *Catalogue of Cometary Orbits 2008, 17th edition* ©2008 by Brian G. Marsden and Gareth V. Williams (credit: Richard W. Schmude Jr.).

In this equation, V(1,0) is the absolute nuclear magnitude in a filter transformed to the Johnson V system, r and Δ are the comet-Sun and comet-Earth distances in astronomical units, c_V is the solar phase angle coefficient, in units of magnitudes/degree, and α is the solar phase angle in degrees. The solar phase angle is the angle between the Sun and the observer measured from the target which, in this case, is the nucleus. The solar phase angle coefficient varies with different nuclei. Ferrín (2005, 2007) reports values of the solar phase angle coefficients for ten nuclei ranging from 0.025 magnitudes/degree up to 0.063 magnitudes/degree.

The absolute nuclear magnitude, V(1,0), is the brightness of the nucleus, in stellar magnitudes, when that object is 1.0 au from both the Earth and the Sun and is at a full phase. The value of V(1,0) depends on the size of the nucleus and on its albedo. The albedo is the fraction of light that an object reflects compared to the total amount of light falling on it. If we assume that comet nuclei have nearly the same albedo, the value of V(1,0) would be a measure of the size of the nucleus. One group of astronomers has compiled values of absolute nuclear magnitudes for 118 different Jupiter family comets. Most of the absolute nuclear magnitude values fall between 15 and 18. The average absolute nuclear magnitude is 16.7, with a standard deviation of 1.3 stellar magnitudes.

Once a nucleus begins developing a coma, (1.4) is no longer applicable. The brightness depends on both the nucleus and the coma. At this point, the comet's brightness depends on its distance from both the Earth and Sun, along with the size and density of the coma. Any sudden event, such as an explosive release of gas and dust, will affect also the brightness. When a comet begins developing a coma, its brightness will depend on both the nucleus and the coma. Once a large coma develops, the nucleus has almost no effect on the comet's brightness. At this point, (1.5) and (1.6), below, are used to describe comet brightness.

$$M_c = H_{10} + 5\log[\Delta] + 10\log[r] \quad (1.5)$$

$$M_c = H_0 + 5\log[\Delta] + 2.5n\log[r] \quad (1.6)$$

In (1.5), M_c is the measured comet magnitude made visually and corrected to a standard aperture of 6.8 cm, H_{10} is the normalized magnitude which is the brightness in stellar magnitudes that the comet has if it is 1.0 au from both the Earth and Sun, Δ is the comet-Earth distance, and r is the comet-Sun distance. Both r and Δ are in astronomical units. In (1.6), 2.5n is an adjustable parameter and H_0 is the normalized magnitude for a given value of 2.5n, and the other quantities are the same as in (1.5). The term 2.5n in (1.6) is called the pre-exponential factor. Equation (1.5) has only one adjustable parameter (H_{10}) whereas (1.6) has two adjustable parameters (H_0 and 2.5n). An example of the evaluation of H_0 and 2.5n is presented in Chap. 3.

If one uses a photoelectric photometer or a CCD camera with a filter transformed to the Johnson system then (1.5) and (1.6) should be re-written as:

$$V_c = V_{10} + 5\log[\Delta] + 10\log[r] \quad (1.7)$$

$$V_c = V_0 + 5\log[\Delta] + 2.5n\log[r] \quad (1.8)$$

In these equations, V_c is the V filter brightness of the coma in stellar magnitudes, V_{10} and V_0 are the normalized magnitudes based on V_c values. The Δ, n and r terms are the same as in the original (1.5) and (1.6). It is important for people to distinguish between brightness measurements with the eye versus those made with calibrated filters.

Figure 1.36 shows the distribution of H_{10} values for 350 comets (not including numbered periodic comets) appearing between 1801 and 1959. The average H_{10} value for these comets is 7.01. The distribution of H_{10} values appears to match a standard bell curve with a standard deviation of around two stellar magnitudes. Essentially the data in this Figure are consistent with the average comet shining at magnitude 7.01 when it is 1.0 au from the Earth and Sun.

Figure 1.37 shows the distribution of normalized magnitude, H_0, values based on (1.6), and Fig. 1.38 shows the distribution of pre-exponential factor, 2.5n values. Both figures are based on H_0 and 2.5n values of 127 numbered Short-period comets. The average H_0 value is 9.70 with a standard deviation of about two stellar magnitudes. The average value of the pre-exponential factor, 2.5n (1.6) is 13.9, with a standard deviation of about five stellar magnitudes. This value is somewhat larger than the assumed value of $2.5n = 10$ or $n = 4$ in (1.5).

Figure 1.39 shows the distribution of normalized magnitude (H_0) values for certain Long-period comets, and Fig. 1.40 shows the distribution of pre-exponential factor (2.5n) values for certain Long-period comets. Both figures are based on values from 181 comets. The average H_0 value for these comets is 7.49 stellar magnitudes, and the average value of the pre-exponential factor is 9.16.

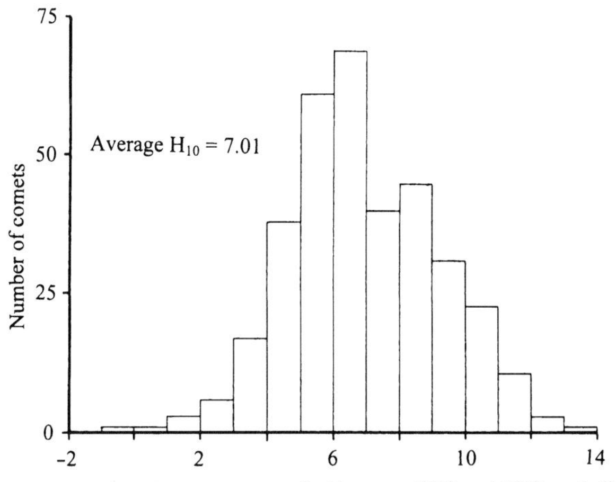

Fig. 1.36. This graph shows the distribution of H_{10} values (1.5) for 350 comets studied between 1801 and 1959. Numbered periodic comets are not included. Data are from *Cometography*, Volumes II through IV ©2003, 2007 and 2009 by Gary Kronk. These values have not been corrected for aperture (credit: Richard W. Schmude Jr.).

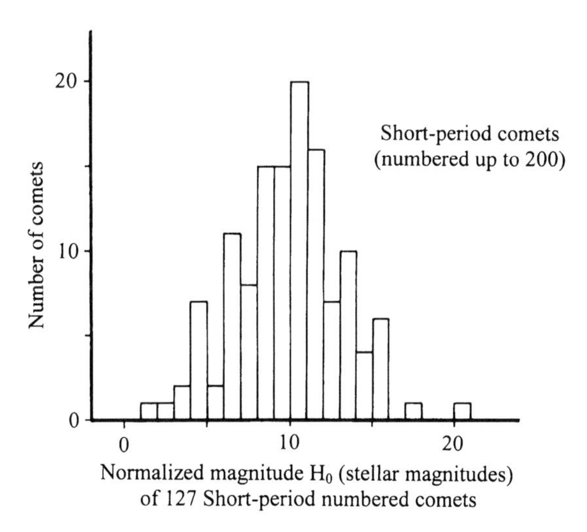

Fig. 1.37. This graph shows the distribution of normalized magnitude (H_0) values for 127 numbered Short-period comets. Most of the values are from the *International Comet Quarterly*, Vol. 28, No. 4a and Vol. 2, No. 4a. Some of the data came from Jonathan Shanklin's website, *The Comet's Tail* newsletter (written by Jonathan Shanklin) and the author's analysis of a few comets. All of the values are believed to have been corrected to a standard aperture of 6.8 cm (credit: Richard W. Schmude Jr.).

Table 1.7 summarizes the average values of H_0, 2.5n and H_{10} for numerous short- and long-period comets. One difference between these two groups of comets is that the H_0 and H_{10} values are about two to three magnitudes fainter for the short-period comets than long-period comets. This may be due to short-period comets

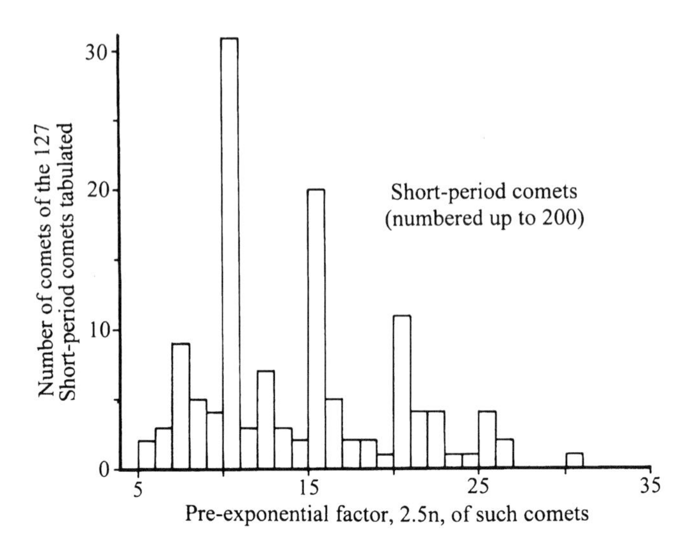

Fig. 1.38. This graph shows the distribution of pre-exponential factor (2.5n) values for 127 numbered Short-period comets. Most of the values are from the *International Comet Quarterly*, Vol. 28, No. 4a and Vol. 2, No. 4a. Some of the data came from Jonathan Shanklin's website, *The Comet's Tail* newsletter (written by Jonathan Shanklin) and the author's analysis of a few comets. All of the values are believed to have been corrected to a standard aperture of 6.8 cm (credit: Richard W. Schmude Jr.).

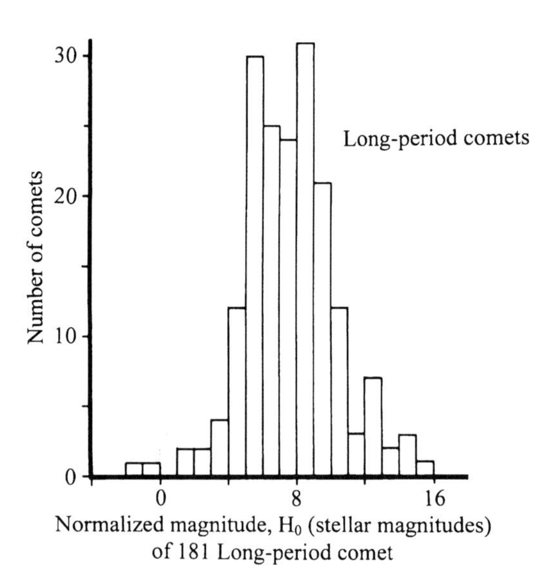

Fig. 1.39. This graph shows the distribution of normalized magnitude (H_0) values for 181 Long-period comets. Most of the values are from the *International Comet Quarterly*, Vol. 28, No. 4a and Vol. 2, No. 4a. Some of the data also came from Jonathan Shanklin's website, *The Comet's Tail* newsletter (written by Jonathan Shanklin), the *Journal of the Association of Lunar and Planetary Observers* and the author's analysis of a few comets. All of the values are believed to have been corrected to a standard aperture of 6.8 cm (credit: Richard W. Schmude Jr.).

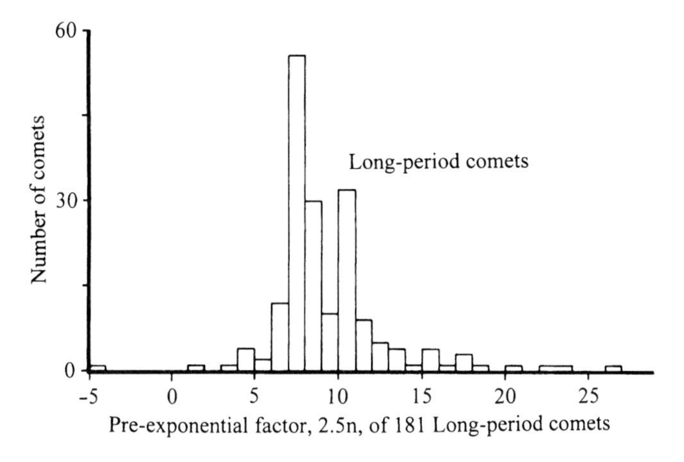

Fig. 1.40. This graph shows the distribution of pre-exponential factor (2.5n) values for 181 Long-period comets. Most of the values are from the *International Comet Quarterly*, Vol. 28, No. 4a and Vol. 2, No. 4a. Some of the data also came from Jonathan Shanklin's website, *The Comet's Tail* newsletter (written by Jonathan Shanklin), the *Journal of the Association of Lunar and Planetary Observers* and the author's analysis of a few comets. All of the values are believed to have been corrected to a standard aperture of 6.8 cm (credit: Richard W. Schmude Jr.).

Table 1.7. Average values of H_{10}, H_0 and 2.5n in (1.5) and (1.6) for large groups of Short- and Long-period comets. These values have been computed by the writer using values compiled from several sources including the *International Comet Quarterly*, *The Journal of the Association of Lunar and Planetary Observers*, Jonathan Shanklin's Website, *The Comet's Tail* newsletter (written by Jonathan Shanklin) and the author's analysis of a few comets

Type	H_0	2.5n	H_{10}	Number of comets
Numbered Short-period comets	9.70	13.9	–	127
Long-period comets	7.49	9.16	–	181
All comets listed in the two rows above (Short- and Long-period comets)	8.40	11.1	–	308
Numbered periodic comets (Short-period comets)	–	–	10.19	68
Comets with a C/ designation in Cometography Vols. II–IV (Long-period comets)	–	–	7.01	350
All comets in the two rows above (Short- and Long-period comets)	–	–	7.53	418

having lost most of their volatile substances due to their many past close encounters with the Sun. There is also a sizable difference in the 2.5n values for short-period comets and long-period comets. This difference may be due to the short-period comets having less highly-volatile substances compared to those with long periods.

Can the H_0 value tell us anything about the size of the nucleus? This is a difficult question to answer. A preliminary discussion is presented below. In this regard, note first that Table 1.8 lists characteristics of a few short-period cometary nuclei. These characteristics are based on spacecraft images. The comet with the largest nucleus in the table, 1P/Halley, has the smallest H_0 value, which is what should be expected. After all, the largest nucleus should have the largest coma and, hence, reflect the largest amount of light. The comet with the smallest nucleus, 81P/Wild 2,

Table 1.8. Characteristics of the nuclei of four comets. The sources for the areas of the nuclei are listed in Tables 2.2, 2.5, 2.12 and 2.15. The author computed the H_0 values from data reported in the *International Astronomical Union Circulars* and a few issues of the *International Comet Quarterly*

Comet	Area of nucleus (km^2)	Log (area in km^2)	H_0
1P/Halley	500	2.70	4.1[a]
19P/Borrelly	90	1.95	6.7
81P/Wild 2	46	1.66	6.9
9P/Tempel 1	119	2.08	5.5

[a]Average of pre- and post-perihelion values in 1985–1986

Table 1.9. Values of the area, equivalent diameter and the logarithm of the area of nuclei of Short-period comets based on (1.9) in the text. The author computed the values in this table using (1.9) in the text

H_0	Logarithm of the nucleus area (area is in km^2)	Area (km^2)	Equivalent diameter (km)
−2	4.64	44,000	118
2	3.34	2,200	26
6	2.03	110	5.9
10	0.73	5.3	1.3
14	−0.58	0.26	0.29
18	−1.89	0.013	0.064

has the largest H_0 value and, hence, its coma reflects the smallest amount of light of the four comets listed.

Furthermore, I have computed the logarithm of the areas of the cometary nuclei listed in Table 1.8. The abbreviation for logarithm is log. The log[area in km^2] values and the H_0 values in Table 1.8 were fitted to a linear least squares equation. The result:

$$\text{Log}\left(\text{nucleus area in km}^2\right) = 3.99 - 0.327\text{H}_0 \quad \textit{Short-period comets} \quad (1.9)$$

The data follow this (1.9) well. The correlation coefficient, r, for (1.9) equals 0.96. This equation was used to generate approximate areas and diameters of nuclei of Short-period comets with different values of H_0. These results are summarized in Table 1.9 and may be useful in answering the question posed.

Finally, it is believed that (1.9) and Table 1.9 should be considered as a very preliminary answer to the question posed. There are several factors that one should consider when relating H_0 values to the size of cometary nuclei. Three of these are (1) the fraction of the surface of a nucleus which is active, (2) the amount of dust released during the release of volatile substances and (3) the particle size distribution of dust released. It is believed that the largest factor for consideration is the fraction of the surface of the nucleus which is active. For example, only about 10% of the surfaces of both 1P/Halley and 19P/Borrelly were active when space probes flew past them. If 50% of the surface had been active for, say, 1P/Halley in 1986 instead of 10%, it would have been much brighter, and, hence, the H_0 value would have been different.

Parts of a Comet

Comets often have at least four visible parts, namely, the central condensation, coma, dust tail and the gas tail. A fifth, and unseen part, is the nucleus. In many cases, the gas and/or dust tail will also be invisible. The gas tail is sometimes called the ion tail or the plasma tail. In this book, the term gas tail is used. The four parts are shown in Fig. 1.1 and each is discussed below.

Central Condensation of the Nucleus

The central condensation is the brightest part of the coma. It often lies near the center of the coma, but it can also be off to one side. The central condensation contains the nucleus along with possibly jets and dense clouds of gas and dust. It should not be confused with the solid nucleus. The central condensation can be a few hundred to a few thousand kilometers across. The nucleus of an average size comet, on the other hand, is usually only a few kilometers across. It is doubtful that anybody has ever seen a comet's nucleus.

Cometary nuclei contain most of a comet's mass. This mass may be anywhere from a few meters to a several kilometers across. The nucleus is believed to resemble an "icy dirtball." What this means is that it contains mostly dirt but with some ices that sublime. This ice is mixed with darker, non-volatile material. Some of the water may also be locked up in hydrated compounds. A hydrated compound contains water molecules that are chemically bonded to salt ions. Epsom salt is an example of a hydrated compound. Over 40% of the weight of Epsom salt is water. We know that the nuclei studied by space probes have low albedos and, hence, must have dark material. The nuclei listed in Table 1.8 all have average albedos of 0.06 or less; this is similar to the albedo of coal. The ices on Triton, Neptune's moon, have an average albedo of 0.72. Therefore ices on the nuclei (listed in Table 1.8) must have dark contaminants. Much of the ice may also be buried underneath a thin layer of dark material.

When a nucleus is far from the Sun, it is cold, and, hence, its volatiles do not become gases. As a result, little or no gas and dust leaves it. Essentially, a cold nucleus does not have a coma. As the nucleus approaches the Sun, gases begin to leave. This activity kicks up dust. Both the expelled gas and much of the dust leaves the nucleus and becomes part of the coma.

Why doesn't the gas and dust in the coma simply fall back to the nucleus like it would on the Earth? The answer is that a comet's nucleus is much smaller than the Earth, and, hence, its gravity is weaker. A nucleus also has a much lower escape speed than the Earth.

In this regard, the escape speed is the speed needed to escape the gravitational pull of an object. The escape speed of the Earth, for example, is about 11,000 m/s or 25,000 miles/h. Since gas atoms and molecules in our atmosphere move at speeds of a few hundred meters per second, they do not and could not escape from Earth. In other words, the gases in our atmosphere do not reach Earth's escape speed. The escape speed, V_e, is defined as:

$$V_e = \left[(2 \times G \times M)/d\right]^{0.5} \quad (1.10)$$

In this equation, M is the mass of the nucleus or planet, d is the distance to the center of the massive object and G is the gravitational constant which is 6.67×10^{-11} $m^3/(kg\,s^2)$. The Earth's escape speed is high because its mass is high. As one gets farther from the Earth, its escape speed drops. The escape speed of a comet is much lower than Earth because of its smaller mass. A spherical nucleus with a diameter of 4 km and a density of 0.5 g/cm^3 has an escape speed of about one meter/second or about 2 miles/h. Therefore, dust leaving at a speed greater than one meter/second will escape the gravitational pull of this nucleus. Slow moving dust particles, with speeds of less than one meter/second on the other hand, will fall back to this nucleus.

What is the strength and structure of a comet's nucleus? A few recent observations and calculations have shed some light on this question. Space probe data are consistent with a few nuclei having densities well below that of water ice. Nuclei are believed to have some empty space inside and this will weaken them. Astronomers have also examined the forces that broke comet Shoemaker-Levy 9 (C/1993 F2) apart. One group estimates that this nucleus had a tensile strength of between 100 and 10,000 N/m^2, or 0.015–1.5 $pounds/in.^2$. This is similar to that of a dirt clod, but is much less than that of glass (5,000–30,000 $pounds/in.^2$) or low-carbon steel (22,000 $pounds/in.^2$).

Figure 1.41 shows what a comet's nucleus might look like. There are areas of both volatile and non-volatile materials. The volatile material may contain water, carbon monoxide or other substances. The volatile material may be concentrated in ice patches as illustrated in the figure. When ice sublimes, dust is often released. This dust is probably some combination of ice crystals and non-volatile material. It may continue moving outwards from the nucleus or it may fall back, depending

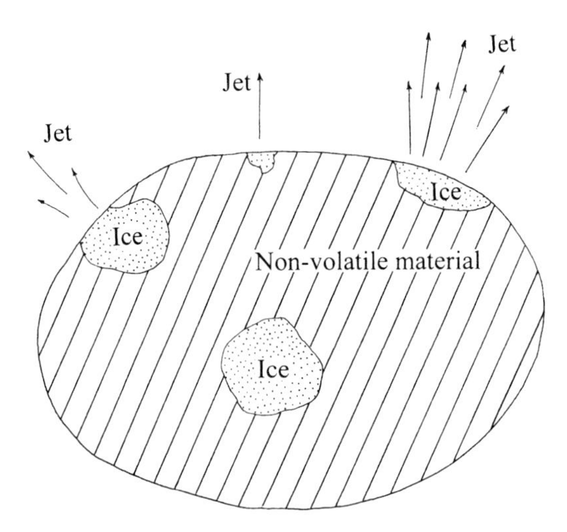

Fig. 1.41. A drawing of a comet's nucleus. Ice facing the Sun will sublime and move away from the nucleus. The non-volatile material will not sublime unless the nucleus makes a very close approach to the Sun (credit: Richard W. Schmude Jr.).

on whether it is moving faster than the escape speed of the nucleus. Many escaping ice grains may themselves sublime, releasing more gas and dust.

At what Sun distance does the coma develop? The answer to this question depends on the chemical composition of the surface of the nucleus. Significant amounts of nitrogen sublime at very low temperatures ~50° Kelvin which, in turn, can kick up dust. Much higher temperatures, however, are needed before large amounts of water sublime, and very high temperatures are needed before significant amounts of silicate materials sublime. Generally speaking, the substances on cometary nuclei are broken down into two categories – volatiles (or ices) and non-volatiles. Volatile substances are those that evaporate rapidly at room temperature (25°C or 77°F). Examples of volatile substances include nitrogen, carbon monoxide, carbon dioxide, ammonia, methane, ethane and water. Examples of non-volatile materials include iron, sodium, silicate materials and carbonaceous-chondrite materials. Volatile substances have higher vapor pressures at room temperature than non-volatile substances. Because of this, they sublime quicker, resulting in thicker comas.

A substance's vapor pressure describes its equilibrium pressure at a given temperature. If, for example, water is placed in a closed container, it would start to evaporate. After a short time, there would be gaseous water in the air above the liquid water. Some of the gaseous water would collide with the liquid water below and rejoin it. This is called condensation. The more water that evaporates, the faster that water from the gaseous phase returns to the liquid phase. At some point, the rate of evaporation will equal the rate of condensation. When this occurs, equilibrium is reached. The equilibrium vapor pressure is the pressure that the gaseous water exerts under equilibrium conditions.

As applied to comets, the higher a substance's vapor pressure, the faster it will sublime. Table 1.10 lists a few substances and their equilibrium vapor pressures at four different temperatures (50, 100, 150 and 200°K). Nitrogen has a higher vapor pressure than the other substances. Water, on the other hand, has a lower vapor pressure compared to other volatile substances. Hence, a nucleus with lots of pure nitrogen ice will develop a coma at a greater distance from the Sun than a nucleus that has just pure water ice. We know, for example, that comet Hale-Bopp (C/1995

Table 1.10. Equilibrium vapor pressure for various substances that may be found on the nuclei of comets. All vapor pressures are measured in pascals; 1 Pa = 0.00001 atm or 0.000147 pounds/in.[2]. The writer used the Clausius–Clapeyron equation in *Physical Chemistry, 7th edition* ©2002 Peter Atkins and Julio de Paula. Values of the heats of sublimation used in the Clausius–Clapeyron equation are from *Handbook of Chemistry and Physics, 89th edition* ©2008, D. R. Lide – editor-in-chief

Substance	Vapor pressure in pascals at 50°K	Vapor pressure in pascals at 100°K	Vapor pressure in pascals at 150°K	Vapor pressure in pascals at 200°K
Water (H_2O)	<0.001	<0.001	<0.001	0.1
Ammonia (NH_3)	<0.001	<0.001	20	>1,000
Carbon dioxide (CO_2)	<0.001	0.02	800	>1,000
Ethane (C_2H_6)	<0.001	10	>1,000	>1,000
Methane (CH_4)	0.3	>1,000	>1,000	>1,000
Carbon monoxide (CO)	100	>1,000	>1,000	>1,000
Nitrogen (N_2)	400	>1,000	>1,000	>1,000

O1) began developing a coma when it was about 18 au from the Sun. This is evidence that highly volatile ices like nitrogen, carbon monoxide and methane were present.

Volatile substances may exist as pure ices, as components of a mixture and as part of a single phase. The data in Table 1.10 applies only to pure substances. A brief description of the situations when a volatile substance is part of a mixture and when it is part of a single phase is presented below.

If a substance is part of a mixture, its vapor pressure will be different than if it is a pure substance. Two or more substances which are admixed are not bonded tightly to one another. A mixture can have a varying composition. Salt and water, for example, form a mixture. One can have slightly salty water or very salty water. The ices on Pluto are believed to be mixtures, and, hence, ices on cometary nuclei may also be mixtures. Predicting the vapor pressure of a specific component in a mixture is complicated, and a discussion of this topic is beyond the scope of this book. Needless to say, it is believed that mixtures of volatile substances may be common on cometary nuclei.

For example, let's say that a cometary nucleus has an ice layer that is composed of 99.87% water, 0.10% ammonia and 0.03% carbon dioxide by mass. Let's assume further that this mixture is an ideal solution, which means that the vapor pressure of each substance obeys Roult's Law. Let's assume also that the nucleus approaches the Sun and the mixture warms up to a temperature of 150°K. The equilibrium vapor pressures for water, ammonia and carbon dioxide in this mixture at 150°K are $<10^{-3}$, 0.02 and 0.1 Pa, respectively. These vapor pressures are sufficient for the development of a coma. What is significant here is that the vapor pressures of ammonia and carbon dioxide are much lower than what they are at 150°K in pure ices (20 and 8,00 Pa, respectively). As a result, the more volatile substances (ammonia and carbon dioxide ice) will survive for a much longer time if they are part of a mixture.

When a compound is bonded to a second compound then different chemical laws apply. For example, Epsom salt ($MgSO_4 \cdot 7H_2O$) contains seven water molecules that are bonded to each $MgSO_4$ unit. Once $MgSO_4 \cdot 7H_2O$ is heated, the H_2O molecules begin leaving the $MgSO_4$. Many become gaseous water. We know that some of the water in the soil of Mars is chemically bonded to salts and this may also be the case for comets. The rate of water release depends on both the strength of the bonds connected to the water molecules and the temperature. There are, of course, compounds where other volatile species are bonded.

Coma

Once enough gas and dust gathers above the nucleus, a coma develops. Comas may range in size from less than 1,000 km across to over 1,000,000 km across. Astronomers using modern equipment are able to detect faint comas. This is one reason why they have discovered so many comets since 1990. Table 1.11 lists the diameters of ten comas. In many cases, a coma will change in size as it passes the Sun. For example, the average diameter of Comet Hale-Bopp's (C/1995 O1) coma was 2,200,000 km in mid-1996, but was only 1,200,000 km

Table 1.11. Coma diameters for selected comets

Comet	Coma diameter in km (year or date)
1P/Halley	700,000 (1986)[a]
9P/Tempel 1	80,000 (2005)[b]
17P/Holmes	540,000 (Oct. 29, 2007)[c]
17P/Holmes	1,600,000 (Nov. 9, 2007)[c]
19P/Borrelly	120,000 (1981, 1987–1988, 1994–1995 and 2008)[b]
23P/Brorsen Metcalf	220,000 (Aug. 7, 1989)[d]
81P/Wild 2	68,000 (1990–1991 and 1997)[b]
Lulin (C/2007 N3)	280,000 (Feb. 17, 2009)[c]
Hyakutake (C/1996 B2)	320,000 (Mar. 25, 1996)[c]
Hale-Bopp (C/1995 01)	1,200,000 (early 1997)[e]

[a] Computed from photographs published in *The International Halley Watch Atlas of Large-Scale Phenomena* ©1992 J. C. Brandt et al.
[b] Computed from values published in the *International Comet Quarterly*
[c] Computed from the writer's personal visual estimates
[d] From Machholz (1995)
[e] From Schmude et al. (1999)

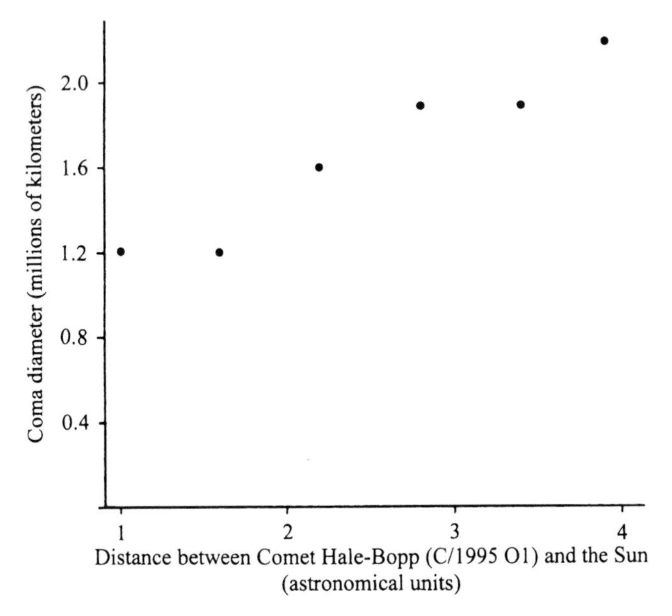

Fig. 1.42. This graph shows the coma diameter of Comet Hale-Bopp (C/1995 01) at different distances from the Sun (credit: Richard W. Schmude Jr.).

by early 1997 as the comet reached perihelion. See Fig. 1.42. The comas of three other comets, Bradfield (C/1987 P1), Wilson (C/1986 P1) and 23P/Brorsen-Metcalf all grew smaller at perihelion. This change may be due to the stronger force of the solar wind near the Sun and/or to the higher amounts of solar radiation near the Sun.

Dust particles in the coma can range in size from less than 0.01 μm (or 0.0000004 in.) to over 1.0 cm (or 0.39 in.). Larger particles are influenced more by their momentum than the gravity of both the Sun and the comet nucleus. Small particles, however, have little momentum and are more heavily influenced by pressure exerted by solar radiation and the solar wind than by gravity. What this means is that if large and small particles are released from the nucleus at the same time, they will separate eventually.

Dust, gas and ions (ions are atoms or molecules with a net electrical charge) will also follow different paths because each of these particles is affected differently by the different forces in outer space. Magnetic fields, for example, will affect the trajectory of ions but not dust or neutral atoms. Both neutral atoms and ions have almost no momentum and hence will be influenced strongly by pressure exerted by both solar radiation and the solar wind. Gravity and momentum, on the other hand, will play some role in the trajectory of dust particles, but will have almost no influence on gas and ions. The net result is that dust, ions and gas will separate. Hence, comets often have two distinct tails – the gas and the dust tails.

I measured the angle between the gas and dust tails of Comet Hale-Bopp in images made during March and April of 1997. The average angle was 29°. This shows that the dust moved in a different direction than the ions and gas.

Dust Tail

Once a coma develops, some of the material can leave to form a dust tail. The particles making up the dust tail are not gravitationally bound to the nucleus. The dust tail is often flat and curved. It is in approximately the same orbital plane as the comet, and, hence, it will appear almost straight as Earth passes through the comet's orbital plane. The curvature of the dust tail probably depends on several factors, including the comet-Sun distance, the dust particle size distribution, the strength of the solar wind and the rotation of the nucleus. The dust tail generally does not have the same shape as the comet's orbit.

What determines the length and shape of the dust tail? This is a difficult question to answer. This is because lighting, sky conditions, the type of instrument and how the observer perceives the tail have a large influence on the perceived tail length and shape. Members of the Association of Lunar and Planetary Observers, for example, reported 20 tail length measurements for Comet Bradfield (C/1987 P1) within a few days of October 12, 1987. The average reported "tail length" was 3.7 million kilometers (km). The standard deviation was 2 million km. The high standard deviation is due undoubtedly to differences in lighting, and the other factors mentioned above. For similar reasons, it is difficult to measure the width and shape of a dust tail.

Theoretical studies offer one way of exploring the length and shapes of cometary dust tails. In some cases, the dust tail is broken down into both synchrones and syndynes. According to Beatty and Bryant (2007), a synchrone contains dust particles given off at the same time. A syndyne is a region in the tail where

Fig. 1.43. An image of Comet McNaught (C/2006 P1) made on January 22, 2007, at the Mudgee Observatory, ©Grahame Kelaher. The synchrones are the nearly vertical features in the dust tail near the coma at left (credit: Grahame Kelaher).

Fig. 1.44. This is an image of Comet Lulin made on January 8, 2009, by Paul Montfield. The tail is to the right of the coma, and the antitail is to the left of the coma. The image is 0.5° wide (credit: Paul Montfield).

particles have nearly the same size. A combination of solar gravity and radiation pressure will push many dust particles into syndynes. The smaller the dust particle, the larger will be the role that radiation pressure plays on its trajectory. An analysis of Comet McNaught (C/2006 P1) by Marco Fulle suggests that the smaller the dust particle, the faster it is pushed from the nucleus. Synchrones were visible in this comet. See Fig. 1.43.

When the Earth passes through the orbital plane of a comet, an antitail is often visible. This is a feature that is pointed towards the Sun. It is made up of large particles that are moving in front of the nucleus. Figure 1.44 shows Comet Lulin (C/2007 N3) and its antitail.

Gas Tail

The fourth part of a comet is its gas tail. It contains both gas and ions. The gas is made up of neutral atoms, molecules and free radicals. A free radical is either one atom, or two or more atoms bonded together, that are highly reactive due to an unstable electron configuration. The electric charge on an ion is due to an unequal number of protons and electrons in an atom or molecule. For example, CO^+ (carbon monoxide ion) has a + sign to indicate that it has a net charge of +1. This ion contains a carbon and an oxygen atom that are bonded together. It has 14 protons (six for carbon and eight for oxygen), but has only 13 electrons. Since there is one more proton than electron, there is a net charge of +1.

A few ions found in gas tails include H_2O^+, OH^+, CO^+, CO_2^+, CH^+, N_2^+ and C_2^+. These species are not necessarily the ones coming off the nucleus. Instead, some of the gases coming off the nucleus are ionized through one or more processes. In many cases the parent molecule is broken into pieces which we view as ions in the gas tail. For example, a water molecule can be broken into an OH^+ ion, an H atom and an electron. The gas tail often has a bluish color in images and photographs. However, this tail generally is too faint to reveal its true color to the visual observer.

The gas tail points within a few degrees of the anti-sunward direction as seen from the nucleus. For this reason, it is possible to compute the length of this tail from angular-size measurements along with parameters associated with the comet's position. This calculation is explained in Chap. 4.

The Sun has a strong magnetic field which extends into outer space. This magnetic field is broken into two regions having two different polarities. The heliospheric current sheet separates these two regions. At times, this sheet lies close to the solar equator but, at other times, it shifts position because the Sun's magnetic field shifts every 11 years. When a comet passes through the heliospheric current sheet its gas tail often breaks. The older tail becomes separated from the comet and a new gas tail develops. Comet Hyakutake (C/1996 B2) crossed the heliospheric current sheet on March 25, 1996, and its gas tail underwent a large break.

Brightness Changes of Comets over Time

Do comets fade with time? The answer is undoubtedly yes. The fading, however, may be gradual, or it may come in spurts. For example, a comet may have a large patch ice that sublimes when near the Sun. This ice may become depleted over a

200-year period, causing the comet to fade. This fading may continue until a new ice patch is uncovered and, upon such occurrence, the comet would brighten and may remain bright for the next 400 years. After this patch becomes depleted, the comet will once again fade. Therefore, comets may not follow predictable trends.

Astronomers have measured the loss rate of water and other gases for several comets. A typical loss rate for water is 3×10^{28} molecules per second. If this loss rate is sustained for 3 months, 2.3×10^{35} H_2O molecules would be lost which would be equivalent to 7×10^9 kg. On a spherical nucleus that has a diameter of 4 km, this would be equivalent to a layer that is 140 cm (or 55 in.) deep if spread over 10% of the surface of the nucleus. Due to this fact, a nucleus can produce a coma for dozens of trips around the Sun, provided that its volatile-rich deposits are at least 100-m deep and the perihelion distance is greater than about 0.6 au.

There are other factors which affect comet brightness which need consideration. Brightness is affected by comet-Sun and comet-Earth distances. For this reason, a normalized magnitude, such as H_{10}, must be considered instead of the comet's actual brightness. This is because the normalized magnitude is the brightness of a comet when it is 1.0 au from both the Earth and Sun. As a result, different distances do not change the H_{10} value. See (1.5) above. A second factor is the uncertainty in the brightness measurements. Measurements made with the eye have an uncertainty of 0.5 magnitudes, whereas those made with a photometer or a CCD camera may have an uncertainty of 0.01 magnitudes. Naturally, brightness measurements with small uncertainties are more likely to reveal long-term changes.

Lots of data over a wide range of comet-Sun distances are needed to compute H_0 values. This data is often not available. For this reason, I concentrated on H_{10} values. In most cases, my aim has been to look for brightness changes over the last 100 years.

Common sense dictates that comets showing the first sign of fading should be those that have short orbital periods. Accordingly, I have examined several comets with short orbital periods, i.e. those with orbital periods less than 200 years. One problem with older H_{10} values is that they are not corrected to a standard aperture of 6.8 cm. To make matters more complicated, astronomers in the 19th and early 20th centuries made most of their visual brightness estimates with *refractors*. Since the 1960s, astronomers have made most of their brightness estimates with Newtonian *reflectors*. Newtonian reflectors have a much smaller aperture correction factor than refractors of the same diameter. For example, a 0.25 m (10 in.) refractor has an aperture correction factor of −1.2 stellar magnitudes, whereas the same-sized Newtonian reflector has a correction factor of only −0.3 stellar magnitudes. As a result, it is believed that all H_{10} comet values, with the exception of 1P/Halley, should be corrected to a standard aperture of 6.8 cm. Since most of the H_{10} values for Comet 1P/Halley are based on data collected before the invention of the telescope, the H_{10} values based on unaided-eye estimates deviates only slightly form one apparition to the next.

I have considered nine periodic comets in the analysis. One of those, 2P/Encke, has been studied already, and its results are quoted without modification. With corrected brightness, the estimates of comets 4P/Faye, 6P/d'Arrest, 7P/Pons-Winnecke, 8P/Tuttle, 10P/Tempel 2, 19P/Borrelly and 81P/Wild 2 are corrected to a standard aperture of 6.8 cm. If the aperture was not obvious for a brightness estimate it was not used. I did not use photographic magnitudes because photographic plates are more sensitive to blue light than human eyes. Once normalized magnitudes, H_{10}, were computed I then computed linear equations relating the H_{10} values to the year. Figures 1.45–1.48 show the results for the comets 6P/d'Arrest, 7P/Pons-Winnecke, 19P/Borrelly and 81P/Wild 2.

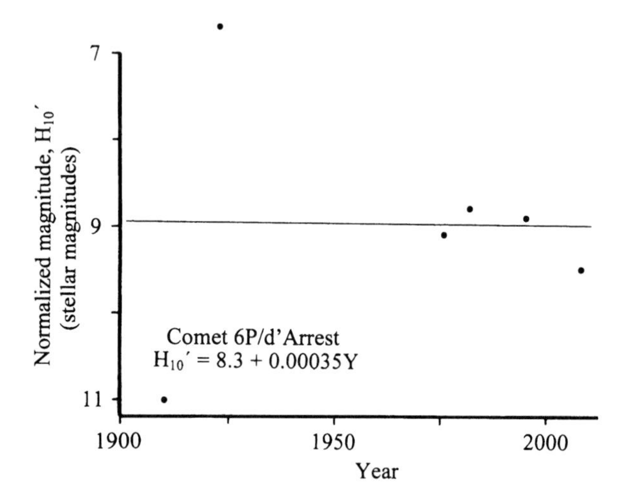

Fig. 1.45. This graph shows a linear fit to the H_{10}' values of Comet 6P/d'Arrest between 1910 and 2008. Data before 1933 are based on brightness estimates reported in *Cometography* vol. III ©2007 by Gary Kronk which have been corrected to a standard aperture of 6.8 cm. Data after 1974 are from an analysis by the author of measurements reported in IAU Circulars. The author corrected these data to a standard aperture of 6.8 cm. In the equation, Y is the year (credit: Richard W. Schmude Jr.).

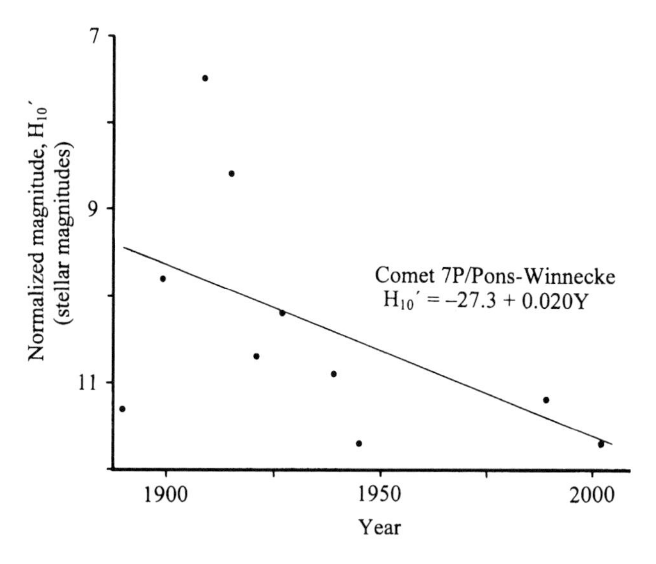

Fig. 1.46. This graph shows a linear fit to the H_{10}' values of Comet 7P/Pons-Winnecke measured between 1892 and 2002. Data before 1959 are based on brightness estimates reported in *Physical Characteristics of Comets* ©1964 by S. K. Vsekhsvyatskii and *Cometography* vol. II and III © 2003, 2007 by Gary Kronk which have been corrected to a standard aperture of 6.8 cm. Data after 1974 are from an analysis by the author of measurements reported in IAU Circulars and are corrected by him to a standard aperture of 6.8 cm. In the equation, Y is the year (credit: Richard W. Schmude Jr.).

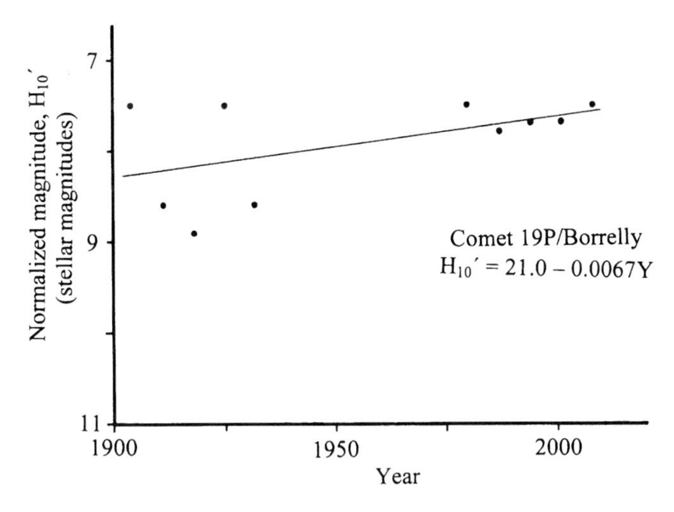

Fig. 1.47. This graph shows a linear fit to the H_{10}' values of Comet 19P/Borrelly measured between 1904 and 2008. Data before 1934 are from *Cometography* vol. III ©2007 by Kronk. Data after 1974 are from an analysis by the author of measurements reported in IAU Circulars and corrected by him to a standard aperture of 6.8 cm. In the equation, Y is the year (credit: Richard W. Schmude Jr.).

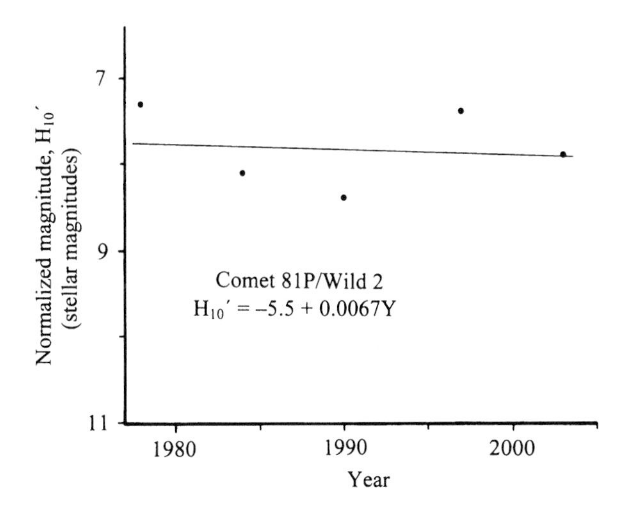

Fig. 1.48. This graph shows a linear fit to the H_{10}' values of Comet 81P/Wild 2 measured since 1978. All data are from an analysis by the writer of measurements reported in IAU Circulars. The author corrected the data to a standard aperture of 6.8 cm. In the equation, Y is the year (credit: Richard W. Schmude Jr.).

In each of these figures, a linear equation is given which relates the normalized magnitude to the year – Y. In the case of comets 4P/Faye, 7P/Pons-Winnecke and 81P/Wild 2, the normalized magnitude value increases with time. This means that these comets grew fainter over the time increment covered by the data. The normalized magnitude of Comet 7P/Pons-Winnecke, for example, dimed at an average rate of 0.02 stellar magnitudes/year during the twentieth century. Comets 10P/Tempel 2 and 19P/Borrelly, on the other hand, grew brighter during most of the twentieth century. The change in the brightness of Comet 6P/d'Arrest, for example, is so small that it is considered negligible.

Table 1.12 summarizes the relationship between the normalized magnitude, H_{10}', and year for nine periodic comets. Four of the comets show little or no change in their H_{10}' value over time, that is, the brightness change is less than 0.001 magnitudes/year. Three of them have a dimming trend in their H_{10}' values, while two have a brightening trend. When the results are averaged, these comets dimmed at a rate of 0.003 magnitudes/year. This rate however is well below the standard deviation for the nine brightness change values of 0.009 magnitudes/year.

Table 1.13 shows the data in Table 1.12 in terms of the comet's orbital period. When taken together, the average change in the normalized magnitude is 0.026 stellar magnitude (or about 2%) per orbit for all nine comets.

Table 1.12. A summary of changes in the H_{10} values and brightness for nine periodic comets over time. The author computed the H_{10} values from data in the International Astronomical Union Circulars, the *International Comet Quarterly* and *Cometography*, Volumes I–III ©1999, 2003 and 2007 by Gary Kronk

Comet	Magnitude equation[a]	Brightness change (magnitude/year)	Time interval	Comment
1P/Halley	$H_{10} = 4.1 + 0.00015\ Y$	<0.001	239BC–1986	Little or no change with time
2P/Encke	No change	~0	1832–1987	Little or no change with time
4P/Faye	$H_{10} = 29.8 + 0.0019\ Y$	0.0019	1910–1991	Dims with time
6P/d'Arrest	$H_{10} = 8.3 + 0.00035\ Y$	<0.001	1851–2008	Little or no change with time
7P/Pons-Winnecke	$H_{10} = -27.3 + 0.020\ Y$	0.020	1892–2002	Dims with time
8P/Tuttle	$H_{10} = 9.1 - 0.0004\ Y$	<0.001	1912–2007	Little or no change with time
10P/Tempel 2	$H_{10} = 25.7 - 0.0088\ Y$	−0.0088	1899–1990	Brightens with time
19P/Borrelly	$H_{10} = 21.0 - 0.0067\ Y$	−0.0067	1910–2008	Brightens with time
81P/Wild 2	$H_{10} = -5.5 + 0.0067\ Y$	0.0067	1978–2003	Dims with time

Average 0.003

[a]In all equations, Y is the year

Table 1.13. A summary of changes in the brightness per orbit for nine periodic comets over time. The author computed the H_{10} values from data in the International Astronomical Union Circulars, the *International Comet Quarterly* and *Cometography*, Volumes I–III ©1999, 2003 and 2007 by Gary Kronk.

Comet	Average period (years)	Brightness change (magnitudes/orbit)
1P/Halley	77.5	0.012
2P/Encke	3.30	~0
4P/Faye	7.36	0.014
6P/d'Arrest	6.47	0.002
7P/Pons-Winnecke	6.05	0.12
8P/Tuttle	13.6	0.005
10P/Tempel 2	5.28	−0.046
19P/Borrelly	6.87	−0.046
81P/Wild 2	6.30	0.042

Average 0.026

Since the average normalized magnitude value has a slight increase in time, it is a sign of dimming. The scatter in the data, however, prevents me from making a reliable conclusion for these comets. Meisel and Morris (1982) examined 11 periodic comets and concluded that there is little or no dimming taking place over time.

One way to address the question of comet dimming in the future is to make brightness estimates with CCD cameras and standard filters. Such measurements should contain the entire coma. One can attain an accuracy of 0.01 stellar magnitudes with CCD measurements. Highly accurate magnitude measurements made over a few decades will serve as a powerful tool in the search for long-term brightness changes in periodic comets.

Comet Impacts in the Near Past

Throughout history, there have been many comet and asteroid impacts on Earth and the Moon. The Moon's cratered surface is a testament to this. One large impact on Earth may have triggered the demise of the dinosaurs 65 million years ago. This chapter concludes by describing three impact events that have happened since 1900. These are the 1908 Tunguska Impact in the Siberian region of Russia, the 1994 Shoemaker-Levy 9 Impact on Jupiter, and an unexpected July 2009 impact on Jupiter.

In the Tunguska Impact, many astronomers believe that a piece of a comet, perhaps the size of a large restaurant, exploded a few miles above the ground. The explosion took place at about 61° N, 100° E or about 3,800 km or 2,350 miles due east of St. Petersburg, Russia. The explosion leveled trees over an area about the size of the State of Rhode Island. The explosion was as strong as several hundred atomic bombs and kicked up a large amount of dust. While this impact was very powerful, a more powerful one took place in July 1994, when the Comet Shoemaker-Levy 9 crashed into Jupiter.

In July 1992, Comet Shoemaker-Levy 9 apparently broke apart when it made a close pass to Jupiter. The fragments continued on a similar path, and about 20 large ones crashed into Jupiter in July 1994. These collisions created large amounts of infrared radiation and lots of dark material. The infrared radiation came from the tremendous amount of heat that was given off during the collisions. I have observed several of the impact spots soon after they formed and estimated an albedo of 0.15 for the darkest portions of the spots. The impact areas started off as dark spots with diameters of a few thousand kilometers. After a few days, they spread out. See Fig. 1.49. By early 1995, the spots had evolved into diffuse dark blobs. The impact area remained darker than average in mid-1995, but returned to a nearly normal appearance by 1996.

These impacts took place near 47° S, which lies in Jupiter's South South South temperate current. John Rogers of the British Astronomical Association reports an average drift rate of −10.9°/30 days for the impact spots in 1994 (see Rogers 1996). This value is near the average drift rate of the South South South temperate current, which is −8.3 (standard deviation = 5.7°) degrees per 30 days.

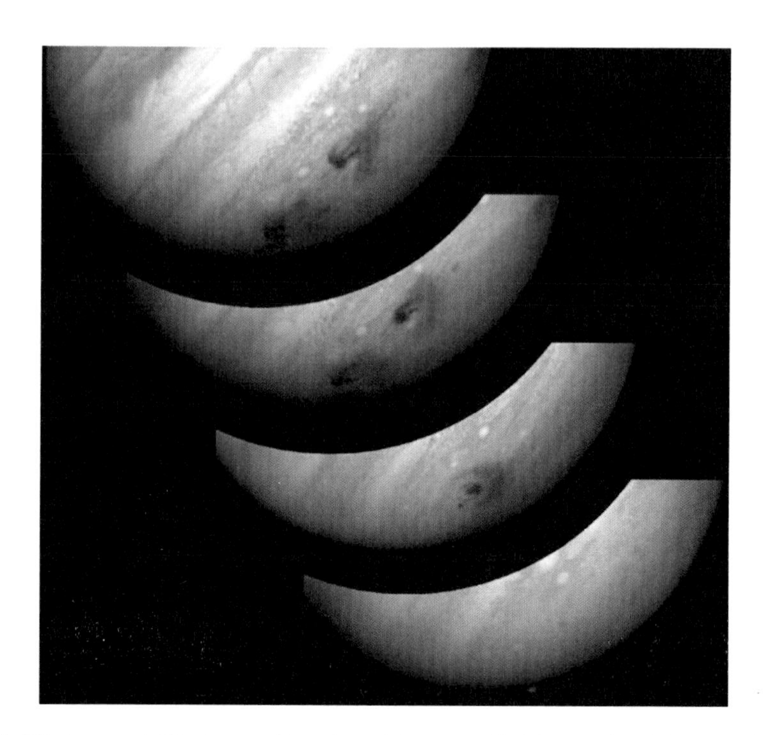

Fig. 1.49. A sequence of visible light images of Jupiter showing dark impact spots. The dark spots are the result of the impact of two fragments of Comet Shoemaker-Levy 9 in July 1994. The *bottom right image* shows Jupiter just after impact; the impact plume is the small bump at the bottom, the dark impact spot has not rotated into view. The next three frames show the development of the dark impact spot between July 18 and July 23, 1994. Note that additional dark impact spots are visible in the two upper left images (credit: JPL/NASA/STScI).

John Westfall of the Association of Lunar and Planetary Observers carried out photoelectric photometry measurements of Jupiter between June 7 and August 7, 1994. He used filters that were close to the Johnson B and V system. According to his measurements, Jupiter did not get much brighter or dimmer in green and blue light as a result of the impacts. Astronomers were aware that Comet Shoemaker-Levy 9 would crash into Jupiter months before impact. A second Jupiter impact, however, occurred without warning on July 19, 2009.

On July 19, 2009 Anthony Wesley, an amateur astronomer from Australia discovered a dark spot in Jupiter's southern hemisphere. See Fig. 1.50. A few hours after this discovery, professional astronomers imaged this spot in infrared light. They found that a large amount of infrared light was near this spot. This meant that large amounts of heat were given off. The obvious conclusion is that a large object struck Jupiter. Astronomer Don Parker took an image of this dark spot 5 days after its discovery. See Fig. 1.51.

We have an idea of a few characteristics of the impact spot and the object that created it. On July 19, the impact spot was located at a system II longitude of 208° W and at a zenographic latitude of 56.5° ± 0.5° S. The darkest part of this spot had a width of 6.2° of longitude. This translates as 4,200 km or 2,600 miles. The north-south dimension on that date was 2,800 km or 1,700 miles. This spot had an

Fig. 1.50. This image shows the dark spot discovered by Anthony Wesley. Anthony Wesley took this image on July 19, 2009, at 16:52 UT from Murrumbateman, Australia. The system II central meridian of this image is 244° W. The dark spot is near Jupiter's edge at about the 11:30 position (credit: Anthony Wesley).

Fig. 1.51. Don Parker took this image of Jupiter and the dark impact spot on July 24, 2009, at 6:36:18 UT. The dark spot is near Jupiter's edge at about the 11:30 position. He took this image from Coral Gables, Florida. Don used a 0.4-m or 16-in. Newtonian reflector along with a SKYnix 2-0 camera to take this image. The system II longitude of this image is 264° W (credit: Don Parker).

approximate area of 9 million km^2 which is roughly the area of the United States of America. Between July 19 and July 26, this spot moved at an average rate of two-thirds of a degree west per day. It should be pointed out that more data over a longer time interval are needed in order to evaluate an accurate drift rate. One professional astronomer states that the impacting object was "several football fields across".

Chapter 2

Comets 9P/Tempel 1, 1P/Halley, 19P/Borrelly, and 81P/Wild 2

In this chapter, comets 9P/Tempel 1, 1P/Halley, 19P/Borrelly and 81P/Wild 2 are described in detail. These analyses are possible because we have sent spacecraft to each of them. Furthermore, both professional and amateur astronomers have observed them during several of their apparitions. Finally, humans have witnessed every return of Comet 1P/Halley since 239 BC.

These comets are periodic ones. Three of them belong to the Jupiter Family and, hence, their orbital periods are short. Orbital characteristics of these comets are summarized in Table 2.1.

I have broken this chapter into four parts. These four parts describe our current knowledge of Comets 9P/Tempel 1, 1P/Halley, 19P/Borrelly and 81P/Wild 2.

R. Schmude, *Comets and How to Observe Them*, Astronomers' Observing Guides,
DOI 10.1007/978-1-4419-5790-0_2, © Springer Science+Business Media, LLC 2010

Table 2.1. A few orbital characteristics of Comets 1P/Halley, 9P/Tempel 1, 19P/Borrelly and 81P/Wild 2. All values are from *Catalogue of Cometary Orbits 2008, 17th edition* ©2008 by Brian G. Marsden and Gareth V. Williams.

Comet-year	Perihelion distance (au)	Orbital inclination (°)	Orbital period (years)
1P/Halley-1986	0.5871	162.2	76.0
9P/Tempel 1-2005	1.506	10.5	5.52
19P/Borrelly-2008	1.355	30.3	6.85
81P/Wild 2-2003	1.590	3.2	6.40

Part A: Comet 9P/Tempel 1

Introduction

On April 3, 1867, Ernst W. L. Tempel discovered a new comet in the Constellation Libra. This comet was probably close to tenth magnitude at the time of discovery. Bruhns used position measurements, and calculated the first elliptical orbit for this comet. This individual computed an orbital period of 5.74 years. Others confirmed the short period from the 1867 position measurements. E. J. M. Stephan recovered this comet on April 4, 1873. Since it was the ninth periodic comet to be recognized, the first two symbols in its name are 9P. Since this is the first of two numbered periodic comets with the name "Tempel", it has a 1 after its name. The full name is 9P/Tempel 1. Astronomers observed this comet in 1867, 1873 and 1879. After 1879 it made three close passes to Jupiter which affected its orbit. As a result, 9P/Tempel 1 was lost until Elizabeth Roemer photographed it on June 8, 1967. Astronomers have detected it on each subsequent apparition through 2005.

The orbital period of 9P/Tempel 1 is currently 5.52 years. As a result, its apparitions alternate between being favorable and unfavorable. Figure 2.1 illustrates the positions of the Earth, Sun and 9P/Tempel 1 in its 2005 and 1999–2000 apparitions. At the turn of twenty-first century, the comet was almost directly behind the Sun (as seen from Earth). As a result, it appeared close to the Sun and was not well observed. The situation, however, was different 5.5 years later. As shown in Fig. 2.1, it was at a much more favorable location in July 2005. As a result, it was well observed. Therefore, we do not have much data for 9P/Tempel 1 during the unfavorable apparitions of 1966–1967, 1977–1978, 1988–1989 and 1999–2000.

In this part, visual observations of this comet are discussed. Thereafter, an overview of the Deep Impact spacecraft mission is presented. This will reveal what we know about the nucleus of this comet. Part A continues with summaries of the Deep Impact spacecraft projectile event, along with the coma and tail of 9P/Tempel 1. It ends with a discussion of important events of this comet in the second decade of the twenty-first century.

Visual Observations

This section describes the brightness and normalized magnitude of this comet during its 1983, 1994 and 2005 apparitions. The size of its coma and the degree of condensation (DC) value follows.

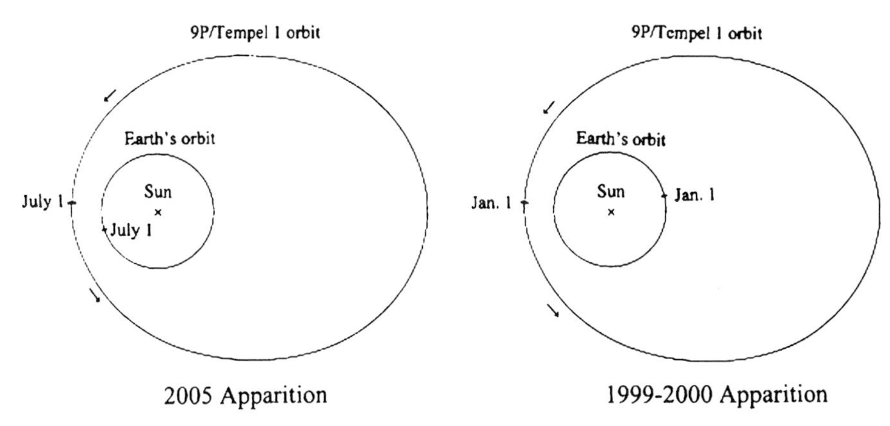

Fig. 2.1. The positions of the Sun, Earth and Comet 9P/Tempel 1 are shown in 2005 (*left*) and in 1999–2000 (*right*). The positions of the Earth and 9P/Tempel 1 are shown by *tick marks*. This comet is close to the opposite direction from the Sun at perihelion in a favorable apparition like 2005, and hence it is seen easily from Earth. The comet, however, is nearly behind the Sun at perihelion during an unfavorable apparition like 1999–2000 and, hence, it is difficult to see from Earth (credit: Richard W. Schmude, Jr.).

Astronomers estimated the brightness of 9P/Tempel 1 in 1983, 1994 and 2005. These measurements have been converted into H_{10}' values according to (2.1). This equation is:

$$H_{10}' = M_c - 5\log[\Delta] - 10\log[r]. \qquad (2.1)$$

In it, M_c is the measured brightness of the comet in stellar magnitudes corrected to a standard aperture of 6.8 cm, Δ is the comet-Earth distance and r is the comet-Sun distance. Both Δ and r are in astronomical units (au). The H_{10}' value is the normalized magnitude and represents how bright the comet would be at a distance of 1.0 au from both the Earth and the Sun. The difference between H_{10}' and H_{10} is that H_{10}' values are determined from individual brightness values, whereas H_{10} is the Y-intercept in a graph of $M_c - 5\log[\Delta]$ versus log[r] graph where the slope is forced to equal 10. There is only one value of H_{10} per apparition, whereas there are as many H_{10}' values as there are brightness estimates. Otherwise, H_{10} and H_{10}' are the same.

The value of computing the normalized magnitude, H_{10}', is that the effects of the changing comet-Earth and comet-Sun distances are eliminated. Figure 2.2 shows how the H_{10}' value changed with the dates for the 1983, 1994 and 2005 apparitions. Two trends are apparent in this figure. The first one is that the H_{10}' value becomes smaller as Comet 9P/Tempel 1 approaches perihelion. The steady decline of H_{10}' values near perihelion repeats in all of these apparitions. This is evidence that the comet brightens rapidly as it approaches the Sun.

The second trend is that the H_{10}' values are lower in 1983 than in 1994 and 2005. This means that the comet reflected more light in 1983 than in 2005. The higher the magnitude value of an object, the dimmer that object will be. Is the dimming trend in Fig. 2.2 real? The best way to answer this question is to resort to (1.6) and determine both the normalized magnitude, H_0 and the pre-exponential factor, 2.5n.

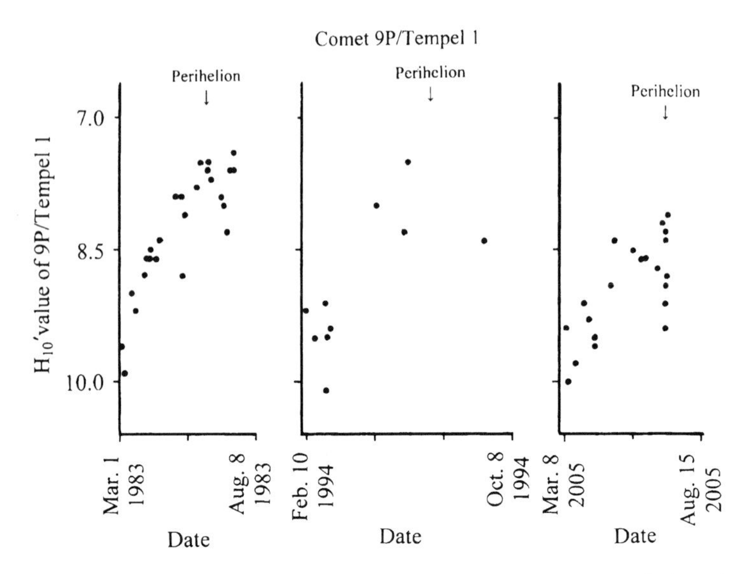

Fig. 2.2. Values of H_{10}' for Comet 9P/Tempel 1 are plotted against varying dates in 1983, 1994 and 2005. All values are based on (2.1) and brightness estimates reported in the International Astronomical Union Circulars. All brightness estimates were corrected to a standard aperture of 6.8 cm using the equations in Chap. 3 (credit: Richard W. Schmude, Jr.).

For the 1983, 1994 and 2005 apparitions, the brightness in stellar magnitudes of 9P/Tempel 1 follow (2.2)–(2.4). These equations are:

$$M_c = 5.11 + 5\log[\Delta] + 24.8\log[r] \quad 1983\ \textit{apparition.} \tag{2.2}$$

$$M_c = 5.42 + 5\log[\Delta] + 24.0\log[r] \quad 1994\ \textit{apparition.} \tag{2.3}$$

$$M_c = 5.92 + 5\log[\Delta] + 23.7\log[r] \quad 2005\ \textit{apparition.} \tag{2.4}$$

In these equations, M_c, Δ and r are the same as in (2.1). Equations (2.2)–(2.4) are based on data published in the International Astronomical Union Circulars. All brightness measurements for aperture have been corrected. Equations (2.2)–(2.4) show that 9P/Tempel 1 brightens rapidly as it approaches the Sun. In fact, this comet brightens by 1.1 stellar magnitudes when its distance to the Sun drops by 10%. A comet that follows (1.5) brightens by only 0.46 stellar magnitudes when its distance to the Sun drops by 10%. Therefore, (1.5) underestimates the magnitude change caused by a decreasing comet-Sun distance for Comet 9P/Tempel 1. This underestimation explains the sharp drop in H_{10}' values as the comet approached perihelion in 1983, 1994 and 2005.

Equations (2.2)–(2.4) also show a fairly consistent trend in both the normalized magnitude values, H_0 (5.11, 5.42 and 5.92), and the pre-exponential factors (24.8, 24.0 and 23.7). The consistency of these three equations points to a similar process of coma development in the three apparitions.

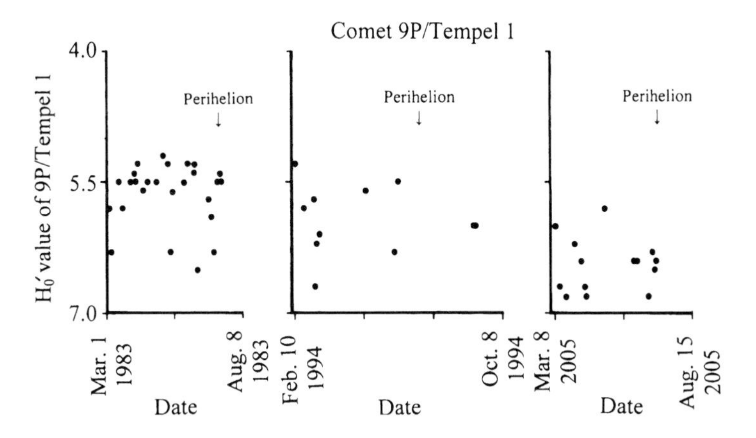

Fig. 2.3. Values of H_0' for Comet 9P/Tempel 1 are plotted against varying dates in 1983, 1994 and 2005. All values are based on (2.5)–(2.7) and brightness estimates reported in the International Astronomical Union Circulars. All brightness estimates were corrected to a standard aperture of 6.8 cm using the equations in Chap. 3 (credit: Richard W. Schmude, Jr.).

Equations (2.5)–(2.7) are similar to (2.2)–(2.4). In (2.5)–(2.7) H_0', instead of M_c, is isolated. These equations are:

$$H_0' = M_c - 5\log[\Delta] - 24.8\log[r] \quad 1983\ \textit{apparition.} \tag{2.5}$$

$$H_0' = M_c - 5\log[\Delta] - 24.0\log[r] \quad 1994\ \textit{apparition.} \tag{2.6}$$

$$H_0' = M_c - 5\log[\Delta] - 23.7\log[r] \quad 2005\ \textit{apparition.} \tag{2.7}$$

In these equations M_c, Δ and r are the same as in (2.1).

Equations (2.5)–(2.7) were used to compute normalized magnitude values, H_0', for 9P/Tempel 1 for the 1983, 1994 and 2005 apparitions. The H_{10}' values are shown in Fig. 2.3. The trend of decreasing normalized magnitudes near perihelion is no longer present because (2.5)–(2.7) are more accurate models for the brightness of 9P/Tempel 1 than (2.1).

The lower H_0' values in 1983 compared to 2005 are consistent with this comet reflecting more light in 1983 than 22 years later. The difference in H_0' values is clearer in Fig. 2.4. This figure shows how the 1983 brightness estimates (dots) compare to (2.7) (curve – 2005 trend). It is evident that 9P/Tempel 1 reflected more light in 1983 than in 2005. In fact, the data in Fig. 2.4 are consistent with the normalized magnitude being 0.5–0.6 stellar magnitudes brighter in 1983 than in 2005. This means that 9P/Tempel 1 reflected 60–70% more light in 1983 than in 2005.

Figure 2.5 shows graphs plotted against time of the coma radius, DC (degree of condensation) value and the normalized magnitude, H_0' value. The DC value describes how diffuse or condensed the coma appears; it is described further in Chap. 3. Average values were computed for 30-day time intervals for the 1983, 1994 and 2005 apparitions, except that the July 4 and 5 values in 2005 were not included in the averages because of the Deep Impact spacecraft projectile event.

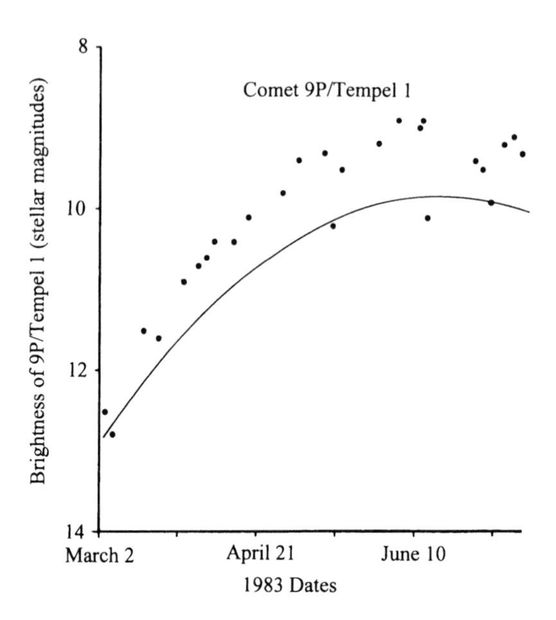

Fig. 2.4. This graph shows a comparison between the brightness of Comet 9P/Tempel 1 in 1983 (*dots*) and the (2.7) *curve* – 2005 trend. This comet was 0.5–0.6 stellar magnitudes brighter in 1983 than in 2005. As stellar magnitude drops, the brightness increases (credit: Richard W. Schmude, Jr.).

There are at least two trends in Fig. 2.5. One is that the H_0' values are higher in 2005 than in 1983 and 1994. As mentioned earlier, this points to the comet reflecting less light in 2005 than in 1983. In spite of this dimming, the coma radius near perihelion remained nearly the same in the three apparitions. Therefore, the comet reflected less light in 2005, but its coma did not shrink. This may appear to be a discrepancy. One explanation for this may lie in the DC value. The DC value just before perihelion dropped from about four in 1983 to less than three in 2005. This meant that the central portion of the coma was brighter in 1983 than in 2005. It will be interesting to see if this trend continues in 2016.

Is the dimming trend in Figs. 2.2–2.5 consistent with measurements made by professional astronomers? Yes. One astronomer reported that 9P/Tempel 1 produced less water and dust in 2005 than in 1983. In a second study, data from the International Ultraviolet Explorer were consistent with this comet producing less OH gas in 2005 than in 1983. We know that pure OH does not exist as an icy compound. Instead, OH is most likely a by-product of H_2O, and hence a low OH production rate points to a low H_2O production rate. The low amount of H_2O in 2005 is consistent with less material ejected into the coma. This will lead to a dimmer absolute magnitude.

Has comet 9P/Tempel 1 changed between 1867 and 2005? S. K. Vsekhsvyatskii reports normalized magnitudes, H_{10}, in 1867 (8.4), 1873 (9.2) and 1879 (10.4). These values are fainter than those from 1983 (8.3), 1994 (8.9) and 2005 (9.0). Vsekhsvyatskii did not correct the brightness measurements to a standard aperture of 6.8 cm, whereas the 1983–2005 values were corrected to the standard aperture. This can explain why the nineteenth century values are larger than those between 1983 and 2005. While I believe that the normalized magnitude, H_{10} has remained nearly the same between 1867 and 2005.

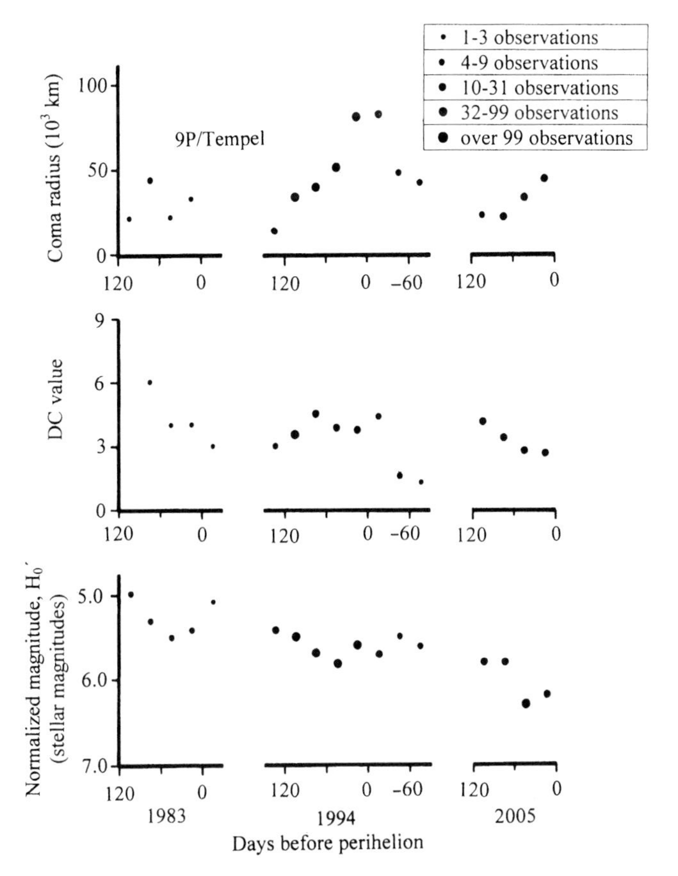

Fig. 2.5. These are graphs of the coma radius, DC value (degree of condensation value) and normalized magnitudes, H_0', of Comet 9P/Tempel 1 for the 1983, 1994 and 2005 apparitions. Data are from the International Astronomical Union Circulars and from the *International Comet Quarterly*. The normalized magnitudes are based on brightness estimates which were corrected to a standard aperture of 6.8 cm (credit: Richard W. Schmude, Jr.).

The coma fluctuates in size from one apparition to the next. The respective average coma radii for 1867, 1873, 1879, 1983, 1994 and 2005 are 40,000 km, 20,000 km, 20,000 km, 30,000 km 50,000 km and 40,000 km, respectively. During all six apparitions, coma size estimates fluctuated and, as a result, the figures represent estimated values valid to just one significant figure.

One characteristic of interest is the enhancement factor. This is the brightness difference, in stellar magnitudes, between the comet and the bare nucleus at full phase. The nucleus at full phase has a solar phase angle of 0°. The enhancement factor is zero when the comet has no coma or tail. When a coma develops, the enhancement factor rises above zero. As the coma gets larger, the enhancement factor increases. The enhancement factor is often highest when the comet is near perihelion.

The enhancement factor of 9P/Tempel 1 near perihelion was 5.9 ± 0.3 stellar magnitudes in 2005. The enhancement factor was a little higher in the 1983 and 1994 apparitions.

How does Comet 9P/Tempel 1 compare to other comets? The average H_0 value for it is equivalent to a stellar magnitude of 5.5 which is much lower than the corresponding value for the 127 Short period comets considered in chapter 1 ($H_0 = 9.70$). Therefore, 9P/Tempel 1 reflects over 40 times more light than a typical short-period comet. The pre-exponential factor for 9P/Tempel 1, 24.2, is well above the corresponding average based on 127 numbered Short-period comets. This means that 9P/Tempel 1 brightens at a much faster rate as it approaches the Sun than most of the numbered Short-period comets summarized in Table 1.7. The nucleus of 9P/Tempel 1 has a mean radius of 3 km. This is larger than most of the computed radii of other Jupiter Family comets. The normalized magnitude, V(1,0) of the nucleus of 9P/Temple 1 is brighter than most of the nuclei of other Jupiter family comets. Therefore, this concludes that 9P/Tempel 1 is large for a Jupiter Family comet.

Overview of the 2005 Deep Impact Mission

The Deep Impact flyby spacecraft imaged the nucleus of 9P/Tempel 1 before, during and after impact. This spacecraft was a fly-by mission. This means that it flew past the nucleus without orbiting it. It also released a projectile that crashed into the nucleus. This projectile had an approximate mass of 370 kg, and it struck the nucleus at a speed of 10 km/s (or 23,000 miles/h). Since the projectile was moving at this speed with respect to the nucleus, it deposited 2×10^{10} Joules of kinetic energy. For comparison, a fully-loaded 100-car freight train with a mass of 5 million kg (or 5,500 tons) moving at a speed of 63 miles/h (or 100 km/h) has a kinetic energy of only 2×10^{9} Joules, or about one-tenth that of the Deep Impact projectile.

The Deep Impact projectile struck the nucleus on July 4, 2005, at 5:44:36 Universal Time (UT). Since the nucleus was 0.894 au from the Earth, it took the light from the impact 7 min and 26 s to reach the Earth. Earth-based astronomers first detected this event within 3 s of 5:52:02 UT. The projectile struck an area at latitude 30–35°. The soil temperature at the impact site was about 310°K or 98°F.

Our knowledge of Comet 9P/Tempel 1 is based on data collected from Earth-based observation and from spacecraft including the Deep Impact mission and projectile impact. These data cover a broad range of wavelengths of electromagnetic radiation. As it turns out, different wavelengths yield different information. Therefore, when the data from several different experiments are combined, a better understanding of the comet occurs. Figure 2.6 gives a sample of the instruments, spacecraft/observatories and wavelengths that astronomers used to study 9P/Tempel 1. As can be seen, astronomers carried out measurements using X-ray, ultraviolet, visible, infrared and microwave wavelengths.

Nucleus: Introduction and General Characteristics

The Deep Impact probe recorded close-up images of the nucleus of 9P/Tempel 1 through different color filters. As a result, we have images of about 25% of the nucleus at a resolution of 10 m or better. This means that we can see features as small as 10 m (or 33 ft) in some images. Table 2.2 lists some physical and photometric characteristics of the nucleus.

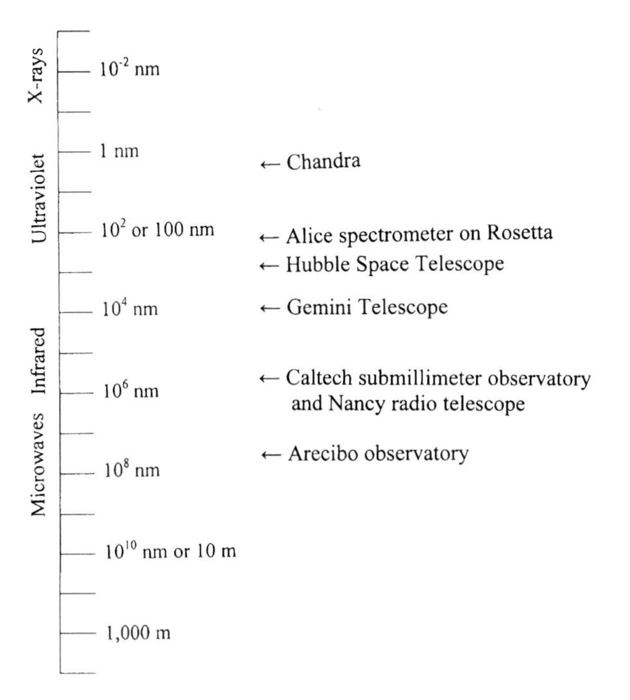

Fig. 2.6. This drawing shows some of the different wavelengths used by different spacecraft, telescopes and observatories in studies of Comet 9P/Tempel 1 during projectile impact. Astronomers used microwave, infrared, visible, ultraviolet and X-ray wavelengths in their studies. Note that visible light is a very narrow interval between the infrared and the ultraviolet portions in the figure. These studies are from papers published in "Deep Impact at Comet Tempel 1 – a special issue of *Icarus*" vol. 191, No. 2s (credit: Richard W. Schmude, Jr.).

Table 2.2. Physical and photometric constants of the nucleus of Comet 9P/Tempel 1

Characteristic	Value	Source
Radii	3.9×2.8 km	Lamy et al. (2007a)
Radius (mean value)	3.0 ± 0.1 km	Thomas et al. (2007)
Density	0.4 g/cm^3	Richardson et al. (2007)
Surface area	119 km^2	Thomas et al. (2007)
Mass	4.5×10^{13} kg	Richardson et al. (2007)
Average acceleration due to gravity (surface)	~0.0003 m/s^2	Thomas et al. (2007) and Richardson et al. (2007)
Rotation rate	41–42 h	Bensch et al (2007) and Lamy et al (2007a, b)
Volume	~110 km^3	The Author computed this value from the mass and average density
Escape speed	~1 m/s	The Author computed this value from the mass and mean radius
Geometric albedo (red light)	0.072 ± 0.016	Fernández et al. (2007b)
Geometric albedo (green light)	0.056 ± 0.007	Li et al. (2007)
R–J	1.46 ± 0.13	Hergenrother et al. (2007)
B–V	0.84 ± 0.01	Li et al. (2007)
V–R	0.50 ± 0.01	Li et al. (2007)
R–I	0.49 ± 0.02	Li et al. (2007)
V(1, 0)	15.2 ± 0.2	Ferrín (2007)
Solar phase angle coefficient	0.055 ± 0.007 mag./degree	Li et al. (2007)

In this section, some of the general characteristics of the nucleus are described. This is followed by a description of some of its geological features along with some ways that erosion occurs on it. Discussions on the sublimation rate, the surface temperature and coma development are set forth. The Section ends with an overview of what we know about the top few meters of the nucleus.

The nucleus has an irregular shape because its gravity is too weak to pull itself into a spherical shape. See Fig. 2.7. The Earth, on the other hand, has a nearly spherical shape because its gravity is strong enough to pull itself into that shape. Since the nucleus has an irregular shape, its brightness changes as it rotates. In one study, the R-filter brightness measurements changed by 39%, or almost 0.4 stellar magnitudes due to rotation.

Two important terms used in geology which can be applied to describe the nucleus are porosity and permeability. The porosity of a soil or rock sample is the percentage of the sample that is void of material or open to material. A piece of granite without any cracks has a very low, perhaps zero, porosity, whereas pumice has a high porosity. Closely related to porosity is permeability. Permeability describes how well a soil or rock sample transmits a substance through it such as a liquid or a gas. Material with a high permeability, for a specific substance, allows it to pass through without much resistance.

Consider, for example, a crate of lemons, with each lemon constituting a separate piece of material. The gap between each lemon constitutes a void which is open to other material and represents porosity. Consider further that when another substance, such as water is poured into the crate of lemons, it flows

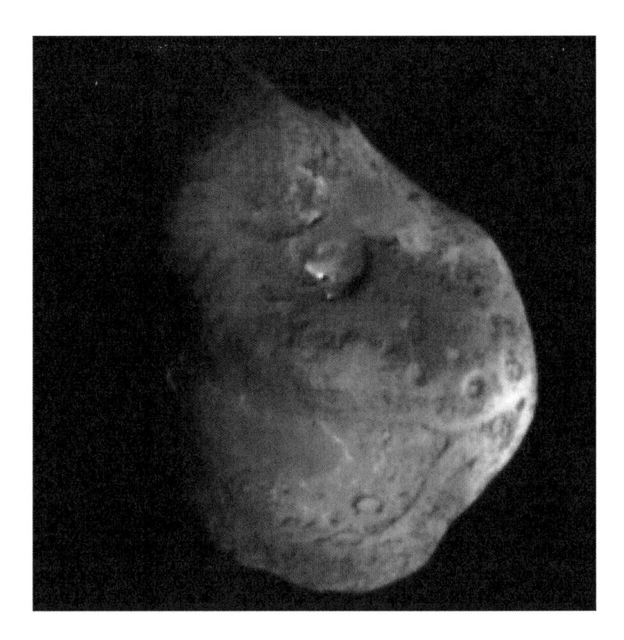

Fig. 2.7. This shows an image of the nucleus of Comet 9P/Tempel 1. The Deep Impact spacecraft took this image on July 4, 2005 just before impact of its projectile. North is in the upper right direction. The image is 9.6 km across (credit: NASA/JPL-Caltech/UMD).

around each lemon and into the voids and passes through the crate. This represents permeability. In order for the permeability to be high, the voids must be interconnected and be relatively large. Large voids are needed because the atoms or molecules making up the material which are moving through the rock or soil sample are often attracted to the surfaces of the grains of the material. This attraction will impede the flow. Therefore, as the hole size gets larger, the effects of the attraction between the material moving through the sample and the grain surfaces are reduced. Therefore, just because a material has a high porosity does not mean that it has a high permeability. Furthermore, a given material may have a high permeability for some substances but a low permeability for others. For example, a silicon rich rock may have a lower permeability for water vapor than for gaseous ethane (C_2H_6).

One group analyzed the trajectory of the impact ejected after the Deep Impact event and was able to determine an approximate mass and density of the nucleus. The density (0.4 g/cm^3) is much less than that of water ice (~0.9 g/cm^3) and, hence, this has lead astronomers to believe that the nucleus has a high porosity. The acceleration due to gravity varies across the surface but is close to 0.0003 m/s^2. For a comparison, the acceleration due to gravity on the Earth is 9.8 m/s^2. Since the weight of an object is proportional to the gravitational acceleration, this means that things only weigh $0.0003 \div 9.8$ or about 0.00003 times as much on the nucleus of 9P/Tempel 1 as on Earth. A 220 lb (100 kg) astronaut on Earth would weigh only 0.007 lb or 3 g on the nucleus of this comet.

The nucleus has an albedo of about 0.05 in green light and an albedo of about 0.07 in red light. This means that it only reflects a few percent of the visible light falling on it. This is similar to the nuclei of a few other comets. The low albedo of 9P/Tempel 1 means that most of the sunlight is absorbed by the nucleus. This will cause its surface temperature to rise.

The color of the nucleus is different from sunlight. The Sun's B–V value is 0.65, whereas the corresponding value for the nucleus of 9P/Tempel 1 is 0.84. This means that the B filter magnitude of the comet is 0.84 stellar magnitudes dimmer than the V filter magnitude. The corresponding difference for the Sun is only 0.65 stellar magnitudes and, hence, means that the Sun is bluer than the nucleus.

How do the colors of 9P/Tempel 1 compare to the nuclei of other short-period comets? Thanks to studies carried out with the Hubble Space Telescope and other advanced equipment, one group of astronomers has compiled a table of color indexes of over 40 nuclei of short-period comets. (The color index is a quantitative way of expressing the color of an object.) The average B–V, V–R and R–I values for the nuclei of the short-period comets studied – with their standard deviations in parentheses – are 0.87 (0.19), 0.50 (0.12) and 0.46 (0.10), respectively. These values are close to those of Comet 9P/Tempel 1. See Table 2.2. Therefore, the nucleus of Comet 9P/Tempel 1 has a color that is similar to that of the nuclei of other short-period comets.

The areas imaged by the Deep Impact flyby spacecraft had nearly the same color. This should not be surprising since the nucleus is probably covered with a layer of dust. This is the dust that breaks lose from the surface but then falls back.

Geological Features and Erosion

The nucleus contains several geological features. These include rimless depressions, nearly filled depressions with raised rims, and depressions with rims, smooth plains, and scarps. These features are shown in Figs. 2.7 and 2.8. Erosion has played undoubtedly an important role in the development of these features. Some of the ways that erosion takes place on the nucleus are described, followed by a description of some of the features on the nucleus.

The dominant form of erosion on the nucleus is the release of volatile substances. These substances may be released either through sublimation or through the decomposition of compounds having volatile molecules bonded to them like hydrated salts. These processes will lead to the removal of both volatile and non-volatile material. One group of astronomers estimates that 1.3×10^9 kg of water left the nucleus between June and September of 2005. If this water came from 10% of the surface of the nucleus, this amount would correspond to an ice layer 12 cm or almost five inches thick. When volatile substances leave the nucleus, other materials, including non-volatile dust grains, are also released.

A second way that erosion takes place is through meteoroid impacts. Since the nucleus is not surrounded by a thick layer of gas, fast-moving meteoroids can strike its surface without being slowed down. Based on one study, approximately 10^{10} meteors, with a stellar magnitude of +10 or brighter, strike the Earth every-day. This corresponds to a rate of about 20 very tiny meteors per square kilometer strike each day. Since the nucleus of 9P/Tempel 1 has an area of 119 km^2 this

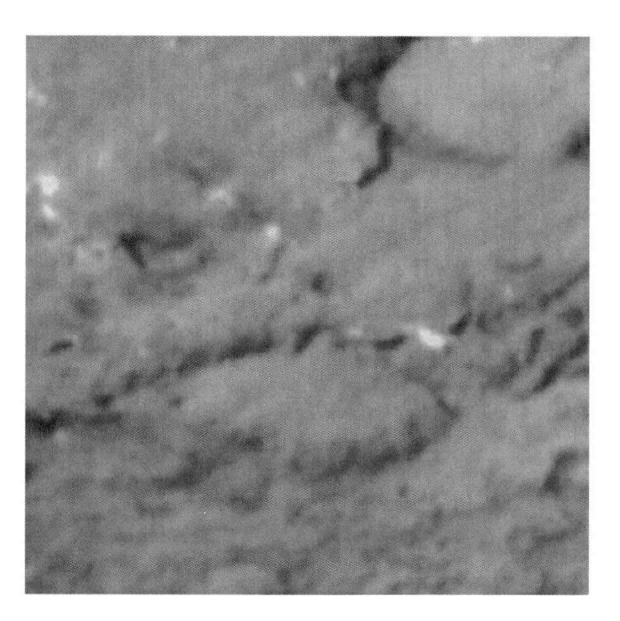

Fig. 2.8. The Deep Impact spacecraft took this close-up image in 2005 of the nucleus of Comet 9P/Tempel 1 just before projectile impact. North is in the *upper right* direction. This image is 0.83 km across. Scarps and a nearly circular depression are visible. The depression appears to have an *elliptical shape* here because of foreshortening (credit: NASA/JPL-Caltech/UMD).

means that over 2,000 meteoroids strike it each day if the impact rate is the same as it is for the Earth. Most of these objects would be tiny, but nevertheless, even small objects can cause some erosion if they are moving fast.

A third way that erosion takes place is the interaction between high-energy radiation and material on the surface. High-energy radiation includes cosmic rays, gamma rays, x-rays and ultraviolet radiation from our Sun and, when applicable, other stars. High energy radiation is able to break chemical bonds. For example, ultraviolet radiation can break a water molecule up into H and OH. The H (atomic hydrogen) and OH would leave as gases.

Undoubtedly other types of erosion have helped shape the surface features on the nucleus. In the next few paragraphs, some of the features on the nucleus are presented. The Deep Impact probe imaged just some of the surface.

Three types of circular features on 9P/Tempel 1 may be impact craters. One of these features has a raised rim and it resembles a mesa. I have called this feature the nearly filled depression with a raised rim. See Fig. 2.9a. Perhaps this is an impact crater that was later filled up with dust. We know that the nucleus is covered with dust that has fallen back. Two other features that resemble craters are the rimless depression and the depression with rims. See Fig. 2.9b and c. The rimless depression has flat floor and resembles a shallow sinkhole on Earth.

The largest rimless depression on the nucleus is about 250 m (800 ft) across. One group of astronomers suggests that this feature may be an eroded impact crater. The depression with a rim appears to have a flat floor. The largest depression with a rim on the nucleus is around 400 m (1,300 ft) across. Fresh impact craters on the Moon, which are less than a few kilometers across, have a "bowl-shaped" interior. See Fig. 2.9d. The depression with rims on 9P/Tempel 1 is different from a similar-sized impact crater on the Moon because of the difference in floor shape.

One group of astronomers counted a few dozen circular depressions and identified them on an image. This count reveals that smaller depressions are more numerous than larger ones. This is consistent with impact craters. I believe that there are several impact craters on the nucleus of 9P/Tempel 1. Erosion, however, has modified them.

There are at least three smooth areas on the nucleus. They lack circular depressions and other irregularities. Two of them have areas of 1 or 2 km^2. The nature

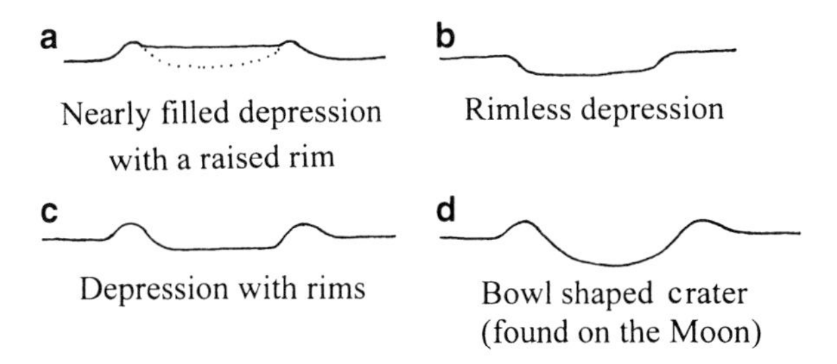

Fig. 2.9. (**a–c**) show side views of different types of circular depressions on Comet 9P/Tempel 1. (**d**) Shows the side view of a typical bowl-shaped crater on the Moon. Note that the circular depressions in (**a–c**) are different from a *typical bowl-shaped crater* on the Moon (credit: Richard W. Schmude, Jr.).

of these features is poorly understood. One group of astronomers suggests that one of the smooth areas is a flow feature.

Scarps are cliff-like features. There are several of them on the nucleus. Two scarps follow a circular path and may be the remnants of eroded impact craters. Two of these circular features have diameters of 0.9 and 1.1 km (0.6–0.7 miles). Another scarp runs for at least 4 km. This scarp is probably a few meters high. One group of astronomers suggests that it surrounds an area that has been stripped of material. Perhaps this area had a higher percentage of volatiles which led to greater rates of sublimation.

Factors which Affect the Release of Volatile Substances

How is it that some areas can be stripped preferentially of material? The answer depends on at least five factors, namely, the nature and amount of the volatile substances, the orientation of the spin axis, the slope of the surface, the thermal characteristics of the surface material and permeability. These are described below.

The nature and amount of the volatile substances will affect how quickly material is released into the coma. If pure ices are considered the vapor pressure would be the dominant factor. Nitrogen with its high vapor pressure sublimes much faster than water. Hence, an area with lots of nitrogen ice will lose material faster than one with little or no nitrogen ice. Furthermore, areas with lots of volatile material will lose material quicker than other areas with less volatile material.

A second factor which affects the release rate of volatiles is the orientation of the spin axis. Earth's spin axis is pointed near the North Star which is at least 65° from the Sun as seen from the Earth. Consequently, Earth's Polar Regions are usually very cold and, hence, little sunlight reaches them. The spin axis for 9P/Tempel 1 is pointed near the star Tau-Draco which is in the constellation of Draco. However, the axis might change. One group of astronomers reports that the nucleus may undergo some chaotic movement. The orientation of the rotational axis will determine which areas receive the most sunlight. This, in turn, will affect how quickly volatiles are released.

A third factor to consider is the slope of the surface. Have you ever noticed that the part of a roof facing south during a sunny day usually loses its frost or snow layer before that part of its roof facing north? This occurs in most of the northern hemisphere because the part facing south receives more sunlight than the part facing north. The nucleus of 9P/Tempel 1 has an irregular shape and, hence, different parts are at different slopes. A region may be sloped in such a way that it receives more direct solar energy than other areas. As a result, this region will have higher temperatures. This will lead to a higher release rate of volatile substances.

A fourth factor for consideration is the thermal characteristics of the surface material. One thermal characteristic is heat conductivity. This quantity describes how well a substance transmits heat. Insulation, for example, impedes the flow of heat and, hence, is a poor conductor of heat. Large areas of insulating material will impede the flow of heat to deeper layers. This, in turn, will mean a slower rate of release of volatile substances. Astronomers believe that the porous surface of 9P/Tempel 1 does not conduct heat well and, hence, it is similar to a layer of insulation.

A fifth factor for consideration is the permeability of the upper layer of the nucleus. High permeability will enhance the release of gas, whereas low permeability will hinder it. To make matters more complicated, the permeability depends on both the material undergoing sublimation and the nature of the upper layer of the nucleus.

Surface Temperature and Coma Development

The nucleus has a wide range of temperatures at perihelion. The temperature rises as high as 336° ± 7°K or 145° ± 13°F at the sub-solar point. The sub-solar point is the area where the Sun is directly overhead. Areas that are about half-way between the sub-solar point and the terminator reach temperatures of about 300°K or 80°F. Areas at the terminator receive almost no sunlight. Temperatures at the terminator reach a peak temperature of 272° ± 7°K or about 30° ± 13°F. The temperatures on the night half undoubtedly fall well below the freezing point of water even at perihelion. This is because there is almost no atmosphere to trap the heat.

There are three reasons for the high temperatures on the nucleus of 9P/Tempel 1. One reason is that almost all of the sunlight strikes the surface. On Earth, our atmosphere absorbs around 25% of the visible light coming from the Sun. This helps keep temperatures at moderate levels. A second reason is the comet's slow rotation. The 41–42 h rotation period of the nucleus means that there is more time for the surface to heat up. A third reason is the low albedo of the nucleus. Most people in the southern United States know how hot a dark asphalt road can get on a hot and sunny summer day. This is because the dark asphalt absorbs lots of the sunlight falling on it and consequently it heats up. The dark nucleus behaves in a similar way.

The coma of 9P/Tempel 1 does not begin developing until the nucleus is 3.5 au from the Sun. For a comparison, Comet 1P/Halley begins developing a coma at a distance of just over six au, and Comet Hale-Bopp begins developing a coma at a distance of about 18 au from the Sun. Comets 1P/Halley and Hale-Bopp have highly volatile substances like nitrogen, methane and ammonia ice, and that this is the reason why these comets begin developing comas at large distances from the Sun. Comet 9P/Tempel 1, on the other hand, lacks significant quantities of highly-volatile substances near its surface and, hence, it must get closer to the Sun before it develops a coma.

The Top Few Meters

What is the nature of the surface of Comet 9P/Tempel 1? Much of the surface is made up of dust and non-volatile material. One group of astronomers has coined the phrase 'icy dirtball' to describe the nucleus of 9P/Tempel 1 instead of the more familiar term "dirty snowball". This group used the "icy dirtball" term because of the high dust-to-gas ratio in the coma, especially after the impact. The

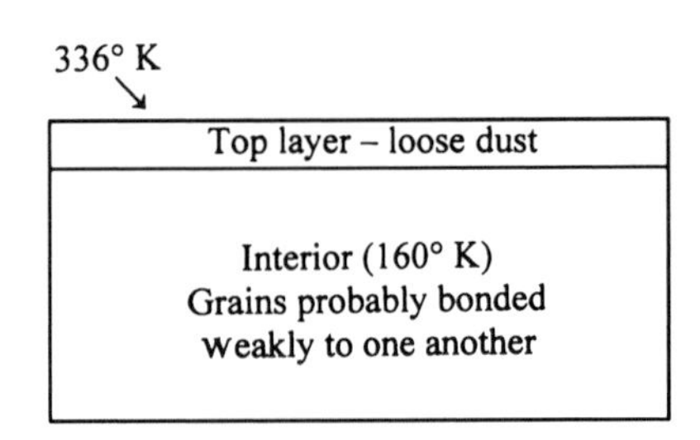

Fig. 2.10. This drawing shows a cross-section view of the top few meters on the nucleus of Comet 9P/Tempel 1. The top layer of the nucleus contains loose dust. The interior contains grains which probably are bound weakly to one another. While the surface temperatures may reach 336°K, the deeper layers probably remain very cold, perhaps only 160°K even when the comet is at perihelion (credit: Richard W. Schmude, Jr.).

low bulk density of the nucleus implies a fluffy surface. One group of astronomers reports that the surface has porosity of at least 50% and that it may be as high as 90%. Part of the reason for the fluffy surface is the low gravity of the nucleus. Dust weighs very little and, hence, it is not compacted as much by its weight, like it is on the Earth.

Figure 2.10 shows a cross-section view of two different layers of the nucleus near the sub-solar point. The top layer consists of loose dust which probably is not bonded tightly to other grains. Below this layer lies the interior. Particles in the interior are probably bounded weakly to one another. We know that the interior has a high porosity because of the low average density of the nucleus. The composition of the volatile substances in the top meter of the interior is similar to what it is 20 m (66 ft) below. One difference between the top layer and the interior is the temperature. Surface temperatures can get quite hot as discussed earlier; however, the temperatures even a few meters below the surface are probably very cold. One group of astronomers carried out several thermal modeling studies of the nucleus and found that the temperature of the interior is around 160°K (−172°F). This is due to the insulating effect of the top layers.

A second difference between the top layer and deeper layers of the interior is the crystalline state of materials. Crystalline materials are those with a regular arrangement of atoms. Layers below the surface are more likely to have a crystalline structure than those at the surface. This is due probably to the fact that surface materials are exposed to more high-energy radiation than deeper layers. As mentioned earlier, high-energy radiation can break chemical bonds, which can lead to the destruction of crystalline structure.

The material at the surface is made up mostly of non-volatile material like silicon and carbon-rich materials. Many carbon-rich materials are very dark, and hence have low albedos. The low albedo of 9P/Tempel 1 is consistent with large amounts of carbon material.

Deep Impact Event

The Deep Impact Mission was one of several missions in NASA's Discovery Program. (Other missions in the Discovery Program included several relatively low-cost missions to Mars, comets and other celestial bodies.) The Deep Impact Mission was highly-successful. One goal of it was to learn more about the differences between the material on the surface and the material several meters below the surface of a target object such as a comet. A second goal was to learn more about the physical structure of the top few meters of a cometary nucleus. During the Deep Impact (DI) event on July 4, 2005, over a dozen instruments both on Earth and in outer space collected data of this event. Amateur astronomers also watched the DI event and collected valuable data before, during and after it. This section discusses the physical changes which took place on the nucleus and the chemical changes which occurred in the coma which resulted from the DI event.

Astronomers were not sure what to expect as the fast moving projectile struck the comet. A few had witnessed previous impacts, including lunar meteors, the collision of comet Shoemaker-Levy 9 (D/1993 F2) into Jupiter and the controlled crash of the Lunar Prospector spacecraft into the Moon. However, all of these impacts involved a large body being impacted – the Moon or Jupiter. The DI event was different in that it did not involve a large body. The nucleus of 9P/Tempel 1 is small compared to the Moon or Jupiter, and, hence, it has low gravity. What would happen to the nucleus and the ejecta? Astronomers awaited the results of the DI experiment.

From outer space the impact appeared different from that viewed from Earth. When the projectile struck the nucleus, the nucleus started to brighten in ultraviolet, visible and infrared wavelengths. The nucleus brightened gradually over several minutes. This is consistent with observations made by amateur astronomers. The Deep Impact flyby spacecraft also imaged a flash of light in visible wavelengths less than a second after the impact. This flash lasted for less than a second and was not observed from Earth. The initial flash was probably due to some of the projectile's kinetic energy being transformed into light. A bright plume of gas and dust quickly followed the flash. Much of the material in the plume reflected sunlight. At this stage, the plume produced a shadow. The luminous plume reached a size of 700 m (2,300 ft). It also moved at a rate of about 1,000 m/s in the direction away from where the projectile came.

From Earth the impact of July 4, 2005, did not produce an observable flash of visible light through moderate sized (0.2–0.6 m) telescopes. Professional astronomers, who used much larger instruments, also did not detect a flash of light right after impact. A couple of people saw the central condensation brighten starting at about 6:00 UT or about 8 min after impact. This brightening continued until the comet sank into the western sky. The light from the coma became more concentrated after impact; or, in other words, the degree of condensation (DC) rose. The DC value is described in Chap. 3. According to two amateur astronomers, the comet brightened by several tenths of a magnitude, or increased in brightness by about a factor of two between 5:52 and 6:10 UT. This brightening was gradual.

Several individuals also imaged this comet during the impact. Their images show clearly the comet becoming brighter minutes after the impact. Two of these images are in the October, 2005 issue of *Sky and Telescope* magazine.

What did the Deep Impact (DI) event tell us about the nature of the surface of 9P/Tempel 1? The best way to answer this question is to examine similar experiments on Earth. One group of scientists carried out a series of experiments similar to that of the DI event. They fired 0.30 g projectiles moving at velocities of up to 5.5 km/s (12,300 miles/h) at a variety of materials. During their experiments, they imaged the impacts using high-speed cameras. From these experiments, this group learned that the image flash depends on the porosity and the chemical composition of the target material. One consequence of these studies is that volatile materials, like water ice, lower the brightness and duration of the impact flash.

The impact flash resulting from the Deep Impact experiment is consistent with the nucleus having both a porous surface and a large quantity of volatile material. In addition, we know that both silicate and carbon-rich dust are present on the nucleus. Finally, by studying the trajectory of the dust leaving the impact site, one group of astronomers was able to measure the acceleration due to gravity along with the mass and average density of the nucleus.

Some of the kinetic energy of the impacting projectile was released as visible light. One group of astronomers reports a luminous efficiency of 7×10^{-5}. This value is lower than what was expected form projectile experiments carried out in labs. The lower-than-expected value may be due to the presence of volatile substances on Comet 9P/Tempel 1 while some of the projectile's kinetic energy was released as infrared light, most of it was released as kinetic energy to the nucleus.

Approximately 1–10 million kilograms (2–20 million pounds) of water ice was released after the impact. Different groups of astronomers come up with different amounts, but most estimates are close to or within the range just stated. One group also reports that a substantial amount of carbon monoxide (CO) was released after the impact. A large amount of dust was also released and one group reported that the impact was very dusty. Much of the dust left the nucleus at speeds of about 100–200 m/s (220–450 miles/h). The gases left at even higher speeds.

Shortly after the impact, a dust plume developed. See Fig. 2.11. Initially, this plume was in the southwest direction. After a couple of days, it was clear that radiation pressure was pushing most of the dust back into the tail towards the east. There is some evidence that the many forces near the nucleus separated the dust particles according to size. No new jets developed after the impact. The mean particle size in the coma fell after the impact. This may be due to the release of a large number of small particles from the impact.

The dust kicked up by the impact contained both silicate and carbon based material. Some of the dust had crystalline olivine and crystalline pyroxene in it, and there was an enhancement of crystalline material in the coma after the impact. This lead one group of astronomers to conclude that the layers a few meters below the surface were more likely to be crystalline than material at the surface.

Shortly after the impact, the percentages of different gaseous species remained nearly the same. This is evidence that the surface and deeper layers have a similar chemical composition.

What were the long-term effects of the impact on the comet? The impact undoubtedly created some sort of crater on the nucleus. The crater, however, was poorly resolved. There were three reasons for this, namely, the camera was out of focus, the debris from the impact blocked the camera view and the probe flew past the nucleus at a high speed. This last reason meant that there was not much

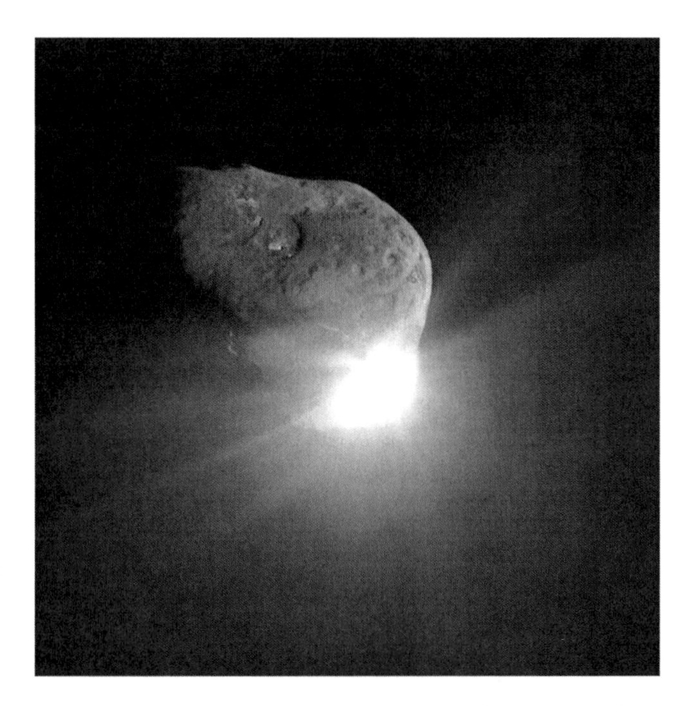

Fig. 2.11. On July 4, 2005, the Deep Impact spacecraft took this image of the nucleus of Comet Tempel 1 just after impact. The *fan-shaped rays* are caused by sunlight hitting ejecta from the impact (credit: NASA/JPL-Caltech/UMD).

time to record an image. Many astronomers believe that the final crater was around 100 m or 330 ft across.

We have, however, a much clearer picture of how the impact affected the comet's coma and tail. The impact caused a temporary surge in the amount of gases and dust in the coma. Some of this extra material ended up in the gas and dust tails of the comet. The higher-than-normal amounts of gas and dust in the coma, however, did not persist. The average normalized magnitude between July 31 and Aug. 6, 2005 was $H_0' = 6.3$, which is essentially the same as the corresponding average between March and early July of 2005 ($H_0' = 6.1$). Therefore, the coma was pretty much back to normal by about July 9, 2005. We also know that the impact did not lead to any new observable jets.

Coma and Tail

Table 2.3 lists a few photometric constants of the coma of Comet 9P/Tempel 1. Most of these constants are based on dozens of measurements made mostly by amateur astronomers.

Several people estimated the size of the coma. Values ranged from less than 1 arc-min up to several arc-minutes. The coma diameter fluctuated during the 1983, 1994 and 2005 apparitions, and 30-day average values are summarized in Fig. 2.5.

Table 2.3. Photometric values of the coma of Comet 9P/Tempel 1

Characteristic	Value	Source
Normalized magnitude, H_0 (before and after perihelion)	5.5 ± 0.3^a	This is based on the 1983, 1994 and 2005 apparitions
Pre-exponential factor, 2.5n (before and after perihelion)	24.2 ± 2.4^a	This is based on the 1983, 1994 and 2005 apparitions
Solar phase angle coefficient, c_v	<0.02 mag./degree	The Author computed this upper limit
Coma radius at a comet-Sun distance of 1.6 au	40,000 km[a]	This is based on the 2005 apparition
Degree of condensation or DC value	3.8[a]	This is based on the 1983, 1994 and 2005 apparitions
Enhancement Factor at perihelion	5.9 ± 0.3^b	Computed from data between June 23, 2005 and July 3, 2005
Color index, B-R, of the inner 9,500 km of the coma	1.1	Walker et al. (2007)
Color index, R-I, of the inner 9,500 km of the coma	0.4	Walker et al. (2007)

[a] These values are based on the Author's calculations. He based his calculations from data published in the International Astronomical Union Circulars and/or several issues of the *International Comet Quarterly*

[b] The Author calculated this value from photometric data of the nucleus and of the comet. Photometric data of the nucleus is from Ferrín (2007). Photometric data of the comet is from the Author's calculations from data published in the International Astronomical Union Circulars and/or several issues of the *International Comet Quarterly*

Over a dozen species including C_2, C_3, CN, CH, NH_2, OH, NH, O, HCN, CH_3OH, CO_2, H_2S and H_2O were in the coma before the impact. Some of these molecules, like H_2O and H_2S, probably came directly form the nucleus without chemical modification. Others like NH and OH are fragments of larger molecules like NH_3 and H_2O, respectively.

Comet 9P/Tempel 1 had several jets just before impact. One group of astronomers reports that these jets appeared to come from isolated areas on the nucleus like vents or cracks. In some cases, the jets originated near ice patches on the surface. One group of astronomers studied images of the nucleus over a few days and concluded that at least two jets rotated with the nucleus. The largest jets rose at least 200 km above the nucleus.

Radio astronomers monitored the abundance of HCN in the coma during mid-2005. They found that the abundance fluctuated within a period of 1.7 days. This is close to the rotation period of the nucleus. The fluctuation of HCN is probably due to a single source on the nucleus.

I found that the H_0' values computed from (2.5) to (2.7) did not increase or decrease with the solar phase angle. Therefore, I selected an upper limit to the solar phase angle coefficient of the coma to be 0.02 magnitudes/degree.

The B-R and R-I color indexes of the inner coma (within 9,500 km of the nucleus) are based on images made 5 days before impact. The corresponding values for the Sun are $B - R = 1.19$ and $R - I = 0.34$ according to *Allen's Astrophysical Quantities*, fourth edition, ©2000, Arthur Cox – editor. Thus the inner coma has a color similar to that of the Sun. The central 2,500 km of the coma was redder than the remainder of the inner coma. This color difference may be due to sublimation of dust and/or ice grains in the coma.

The production rates of several gas species (OH, NH, CN, C_2 and C_3) were highest about 4–8 weeks before perihelion in the 1983, 1994 and 2005 apparitions. This is unusual since one would expect production rates to be highest at peak temperatures and these temperatures should be reached just after perihelion. One possible explanation for the high sublimation rates before perihelion is that a thin layer of dust with lots of volatile substances fell back to the nucleus about 1–2 years after perihelion. This material would sublime as soon as temperatures started to rise.

Comet 9P/Tempel 1 underwent several brightness increases in 2005. These increases are called "outbursts." One outburst took place on June 14. It caused the comet to brighten by about 50% in visible light. This brightening was due at least partially to extra dust in the coma. There may have also been an extra amount of water which was released. A second outburst took place on June 22. During this outburst, an additional amount of water was released. In fact, more water was released from this outburst than from the human-made outburst 12 days later. A third outburst took place on July 2. One group of astronomers found that seven of the ten outbursts occurred on the side of the nucleus near where the July 4 impact took place.

This comet has a short tail. It is not clear whether this was a dust or a gas tail. The average tail lengths in 1972, 1983, 1994 and 2005 were 0.03°, 0.04°, 0.06° and 0.02°, respectively. The average position angle of the tail in April, 1994 was 224° with a standard deviation of 20°. The position angle is measured with respect to north. The respective position angles of north, east, south and west are 0°, 90°, 180° and 270°.

Projection of Comet 9P/Tempel 1 in 2013–2019

Comet 9P/Tempel 1 will be at a favorable position for viewing in 2016. Table 2.4 lists predicted right ascension and declination values along with other of the characteristics. The comet will be well placed in the evening during June and July for people viewing from the Northern Hemisphere. The situation reverses in August and September. During these months, the comet will be in the southern part of the sky and, hence, it will be well placed for people in the Southern Hemisphere.

Table 2.4. Predicted positions, angular diameter values of the coma and brightness values for Comet 9P/Tempel 1 during mid-2016. The right ascension and declination values are from the Horizons On-Line Ephemeris System. The Author computed the angular diameters of the coma and brightness values from predicted distances and standard equations

Date (2016)	Right ascension	Declination	Angular diameter of the coma (arc-min)	Brightness (stellar magnitudes)
May 1	11 h 40 min 29s	18° 46′ 12″	2.3	11.6
May 16	11 h 42 min 21 s	15° 41′ 17″	2.2	11.3
May 31	11 h 52 min 45 s	11° 38′ 59″	2.1	11.1
June 15	12 h 10 min 25 s	06° 55′ 06″	2.0	10.9
June 30	12 h 34 min 05 s	01° 42′ 33″	1.9	10.7
July 15	13 h 02 min 45 s	−03° 46′ 37″	1.8	10.7
July 30	13 h 35 min 46 s	−09° 18′ 45″	1.7	10.8
Aug. 14	14 h 12 m 43 s	−14° 38′ 44″	1.6	10.9
Aug. 29	14 h 53 min 16 s	−19° 30′ 13″	1.5	11.2
Sept. 13	15 h 36 min 58 s	−23° 37′ 52″	1.4	11.5

Measurements of the opposition surge of the nucleus will be of great value. The opposition surge is a brightening of the nucleus at solar phase angles of less than 5°. This surge can yield information on the characteristics of the dust layer at the surface. The three best dates to measure the opposition surge will be Dec. 3, 2013, Nov. 19, 2018, and Dec. 15, 2019. The respective solar phase angle values on these dates will be 0.13°, 0.64° and 0.67°, respectively. The bare nucleus will be close to 21st magnitude on these dates and will be located in the constellation Taurus.

The best time to image an antitail will be between May 24, 2016, and June 5, 2016. During this time, the Earth will be within 1° of the orbital plane of the comet. According to the Horizons On-Line Ephemeris System, the Earth will cross the orbital plane on May 29, 2016, at 21:19 UT.

Part B: Comet 1P/Halley

Introduction

From early times to the twentieth century, people from many parts of the world have observed Comet 1P/Halley. The earliest recorded sightings date from at least 239 BC. The Chinese have consistently watched this comet for about 2,000 years. During 1986, astronomers on the seven Continents studied this comet as part of the International Halley Watch. The former Union of Soviet Socialist Republics, Japan, the United States and some countries in Europe were involved in the sending of six space probes to this comet in 1986. Comet 1P/Halley is truly a comet of international interest!

Sir Edmund (or Edmond) Halley was the first to compute an orbit for this comet. Having done so, he predicted that this comet would return in 1758 or 1759. Johann G. Palitzsch, a farmer and an amateur astronomer, confirmed the reappearance of it on Dec. 25, 1758. It was widely studied during the first half of 1759 and subsequently became known as Halley's Comet. This comet has the name 1P/Halley and this name is used throughout this Book.

Amateur and professional astronomers studied 1P/Halley during its 1985–1986 apparitions. Most of what we know about it is based on what we learned in the 1980s. There is an excellent book showing hundreds of images of 1P/Halley taken in the 1980s and early 1990s. See *The International Halley Watch Atlas of Large-Scale Phenomena*. This book is part of an archive of photographs of 1P/Halley. NASA, in conjunction with dozens of astronomers based on all Continents helped put this book together.

Where will Comet 1P/Halley be in its orbit during the early and mid-twenty-first century? Figure 2.12 shows the orbit of 1P/Halley along with a few of its predicted future positions from 2010 to 2061. Due to Kepler's Second Law of planetary motion, this comet moves faster when it is near the Sun and moves slower when it is far away from it. In late 2023, the comet will be at aphelion–its farthest point from the Sun. At this time, its solar distance will be 35 au. The point "R_{on}" is the location of 1P/Halley when a coma first develops (visually). Similarly, R_{off} is

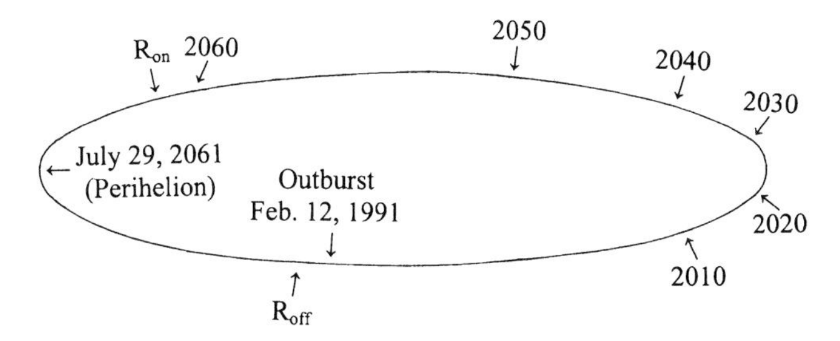

Fig. 2.12. This drawing shows the projected positions of Comet 1P/Halley on Jan. 1 of 2010, 2020, 2030, 2040, 2050 and 2060. The drawing shows also the locations of 1P/Halley when its coma develops (R_{on}) and dissipates (R_{off}), and when it had a huge outburst on February 12, 1991. This comet will reach perihelion on July 29, 2061. Horizons On-Line Ephemeris System were used to compute all positions (credit: Richard W. Schmude, Jr.).

the point where the coma is no longer consistently present (visually). Both R_{on} and R_{off} in Fig. 2.12 are based on what the comet did in the 1980s. According to the Horizons On-Line Ephemeris System, 1P/Halley will reach R_{on} on March 11, 2060. It will reach R_{off} on August 28, 2065. It will be interesting to see if the coma forms and dissipates, respectively, on these dates.

In this part, I will discuss visual observations of the comet, including its brightness history and light curve. This is followed by a discussion of the comet's nucleus, coma and tail. Part B ends with a brief discussion of the next apparition of the comet in 2061.

Visual Observations

Figure 2.13 is a graph of visual brightness estimates of 1P/Halley plotted against varying dates. The comet brightened rapidly in late 1985 and reached a peak brightness of about second to third magnitude in March, 1986. Part of the reason why the comet was brighter in March than at perihelion (February 9) is that it was closer to the Earth in March. After March, the comet dimmed rapidly dropping to 11th magnitude in July. For a brief period, 1P/Halley could be seen with the unaided eye, and people reported brightness estimates of this comet with the unaided eye between late February and early May of 1986. Due to its southerly declination during most of 1986, it was best seen from the Southern Hemisphere.

It is difficult to report an average H_{10}' value for the 1986 apparition because the H_{10} value grew progressively smaller from July 1985 to June 1987. See Fig. 2.14 (top). The average H_{10}' value within 4 months of perihelion (October 9, 1985 to June 9, 1986) is 4.1 with a standard deviation of 0.7 stellar magnitudes. If all data in Fig. 2.14 are considered (July 27, 1985 to June 15, 1987), the average H_{10}' value would be 3.8, with a standard deviation of 1.36 stellar magnitudes. Finally, if we considered only those brightness estimates made with the unaided eye, corrected to a standard aperture of 6.8 cm, the average H_{10}' value would be 3.6. The corresponding value in 1910 was 4.3, meaning that 1P/Halley reflected more light in 1986 than in 1910. As it turns out, 1P/Halley was brighter in 1910 than in 1986 because in the earlier apparition it was closer to the Earth at the time of perihelion.

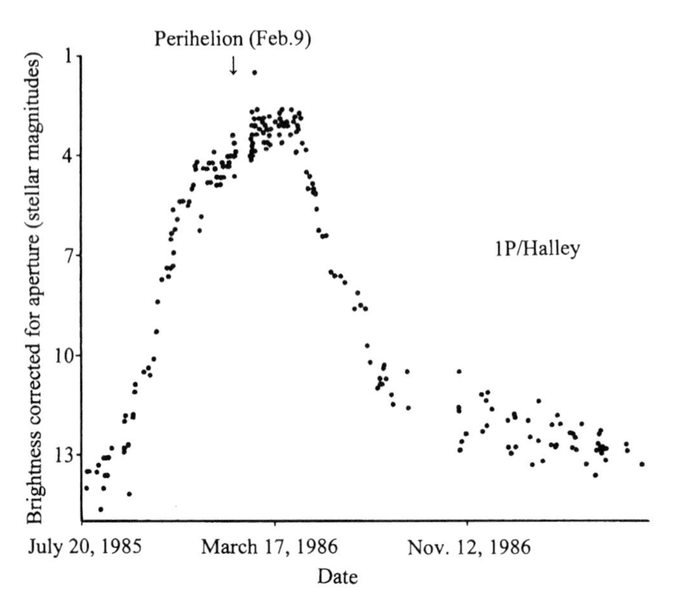

Fig. 2.13. This graph shows the brightness estimates of Comet 1P/Halley, in stellar magnitudes, against varying dates in 1985–1986. All brightness estimates are from the International Astronomical Union Circulars. In all cases, the brightness estimates were corrected to a standard aperture of 6.8 cm using the equations in Chap. 3. Note that this comet brightened rapidly in late 1985 and it reached a peak brightness of about second or third magnitude in March of 1986 (credit: Richard W. Schmude, Jr.).

Should a comet dim as its solar phase angle increases? If it behaves like the planets Venus, Jupiter, Uranus and Neptune, it should dim as its solar phase angle increases. I have prepared two graphs in Fig. 2.14 to determine how the solar phase angle affects the brightness of 1P/Halley. The bottom half of Fig. 2.14 shows the solar phase angle of 1P/Halley versus varying dates and the top half show values of H_{10}' plotted versus such dates. I placed these two graphs together so that they could be compared. Comet 1P/Halley underwent large increases in its solar phase angle in late November of 1985 and in late February of 1986. If this comet behaves like the gaseous planet Jupiter, its H_{10}' value should have gotten larger at these two times. As it turned out, there was little change. Any brightness change was 0.2 magnitudes or less. Accordingly, I compute an upper limit of 0.2 magnitudes $\div$ 40 degrees $=$ 0.005 magnitudes/degree for the solar phase angle coefficient of 1P/Halley when it has a bright coma. For a comparison, the respective solar phase angle coefficients for Venus, Jupiter, Uranus and Neptune based on my compilations are 0.0063, 0.007, 0.0011 and 0.0015 magnitudes/degree. In all cases these solar phase angle coefficients are for green light. All of these planets have a thick, gaseous atmosphere. Hence, their solar phase angle coefficients are low and are dominated by their atmospheres. In a similar way, the gaseous coma of 1P/Haley will dominate the value of the solar phase angle coefficient. As it turns out, the upper limit of the solar phase angle coefficient for 1P/Halley is comparable to those of the planets with thick atmospheres.

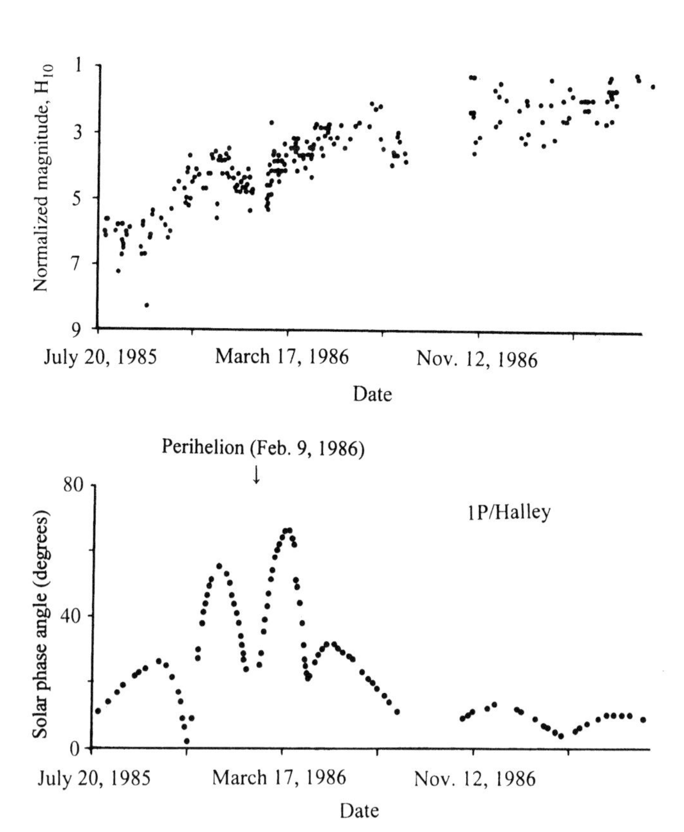

Fig. 2.14. The *top graph* shows a graph of the normalized magnitude, H_{10}', of Comet 1P/Halley in stellar magnitudes versus varying dates. Equation (2.1) was used to compute the H_{10}' values. In all cases, the brightness estimates came from the International Astronomical Union Circulars. These estimates were corrected to a standard aperture of 6.8 cm before computing H_{10}' values. The *bottom graph* shows a graph of the solar phase angle of 1P/Halley versus varying dates. Horizons On-Line Ephemeris System were used to compute all solar phase angles (credit: Richard W. Schmude, Jr.).

One reason for the changing H_{10}' values in Fig. 2.14 (top) is that the pre-exponential factor does not equal 10 for 1P/Halley. The H_{10}' values are based on 2.5n = 10. The declining H_{10}' values are evidence that 2.5n does not equal 10 and, hence, (1.5) is not an adequate model for the brightness of 1P/Halley. Accordingly, (1.6) is used to model the brightness of this comet.

Figure 2.15 shows the values of M_c – 5 log[Δ] for Comet 1P/Halley plotted against various values of log[r]. This graph contains data before the time of the comet's perihelion passage. M_c, Δ and r are defined in (2.1). Figure 2.16 is similar to Fig. 2.15 except that it covers the period after perihelion. In both graphs, there is a moderate amount of scatter in the data because all data are based on visual brightness estimates which have uncertainties of about 0.5 magnitudes. The slope of the data in Fig. 2.15 will equal the pre-exponential factor, 2.5n, and the y-intercept of the data will equal the normalized magnitude, H_0.

The data in Fig. 2.15 do not follow a single straight line but instead follow two lines. This means that either H_0 or 2.5n changed. The break occurs at log r ~ 0.15

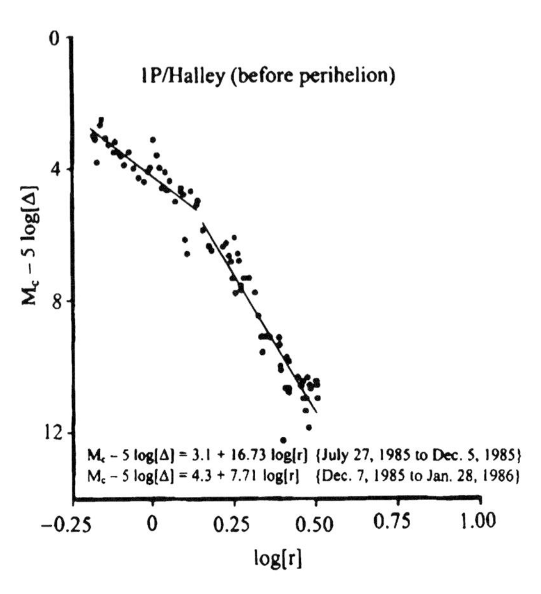

Fig. 2.15. A graph of $M_c - 5\log[\Delta]$ plotted against log[r] for all brightness estimates of Comet 1P/Halley reported in the International Astronomical Union Circulars before perihelion (February 9, 1986). Values of M_c, Δ and r are the same as in (1.6). Note that there is a break in the slope near log[r] = 0.15. The data follow the two equations near the *bottom* of the graph (credit: Richard W. Schmude, Jr.).

or r ~ 1.4 au. Green and Morris (1987) report a similar break in their plot of $M_c - 5\log[\Delta]$ versus log[r]. Based on the data in Fig. 2.15, the brightness of 1P/Halley follows $M_c - 5\log[\Delta] = 3.1 + 16.73\log[r]$ for log[r] values of between 0.15 and 0.51 and $M_c - 5\log[\Delta] = 4.3 + 7.71\log[r]$ for log[r] values of between −0.19 and 0.15. These two equations are re-arranged so that the brightness, M_c, is isolated. Then:

$$M_c = 3.1 + 5\log[\Delta] + 16.73\log[r] \quad \{0.15 < \log[r] < 0.51\}. \tag{2.8}$$

$$M_c = 4.3 + 5\log[\Delta] + 7.71\log[r] \quad \{-0.19 < \log[r] < 0.15\}. \tag{2.9}$$

In these equations, M_c, Δ and r are the same as in (2.1).

There is a sizable difference between (2.8) and (2.9). The biggest difference is the value of the pre-exponent factor. In (2.8), 2.5n = 16.73 or n = 6.692 and in (2.9), 2.5n = 7.71 or n = 3.084. This means that 1P/Halley brightened much more rapidly as it approached the Sun when log[r] > 0.15 than when it was closer to the Sun.

Unlike the data in Fig. 2.15, the data in Fig. 2.16 follows a single straight line. Apparently there was no break near log[r] = 0.15. The brightness of 1P/Halley after perihelion follows the equation $M_c - 5\log[\Delta] = 3.9 + 7.60\log[r]$. Once again, this equation is re-arranged so that the brightness, M_c is isolated as shown:

$$M_c = 3.9 + 5\log[\Delta] + 7.60\log[r] \quad \{-0.22 < \log[r] < 0.78\}. \tag{2.10}$$

Once again, M_c, Δ and r are the same as in (2.1). The normalized magnitude after perihelion is $H_0 = 3.9$ which is brighter than the value of $H_0 = 4.3$ just before

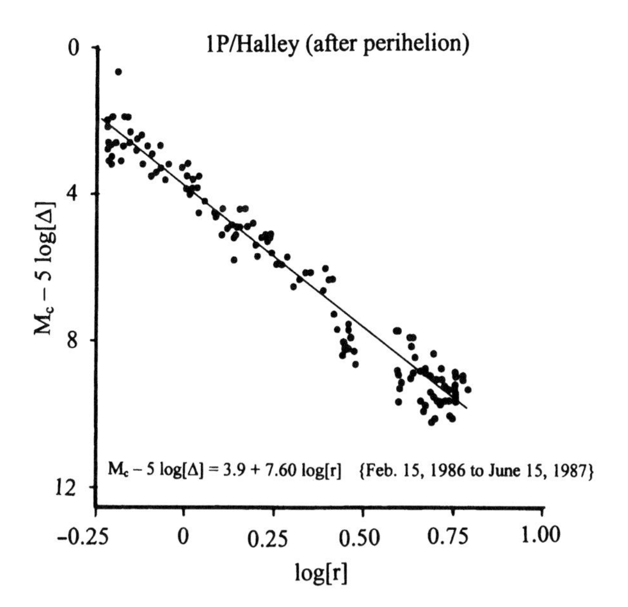

Fig. 2.16. This is a graph of $M_c - 5 \log[\Delta]$ plotted against log[r] for brightness estimates of Comet 1P/Halley reported in the International Astronomical Union Circulars after perihelion (Feb. 9, 1986). Estimates made with telescopes having diameters greater than 0.61 m were not used since I believe that the aperture-correction equations in Chap. 3 do not apply to these large instruments. Values of M_c, Δ and r are the same as in (1.6). The data follow the equation near the *bottom* of the graph (credit: Richard W. Schmude, Jr.).

perihelion. In spite of this difference, the values of n just before and after perihelion are virtually identical; that is n = 3.084 just before perihelion and n = 3.04 just after perihelion.

The enhancement factor for 1P/Halley reached 10.5 ± 0.2 stellar magnitudes. This means that the coma was ~16,000 times brighter than its bare nucleus. This value is higher than that for 9P/Tempel 1, 19P/Borrelly and 81P/Wild 2 because this comet got much closer to the Sun than the other three. The enhancement factor of 1P/Halley reached a peak value of ~10.6 stellar magnitudes about 3 weeks after perihelion.

Did 1P/Halley undergo sudden brightness changes in 1985–1986, and, if so, did they show up in amateur brightness data? Outbursts did occur and, in some cases, they showed up in amateur brightness data. A group of Japanese astronomers detected a strong outburst between 9:07 and 12:29 UT on December 12, 1985. They reported the development of a jet of material during this time. Green and Morris (1987) report that 1P/Halley brightened by about 0.3 magnitudes at about the same time as this outburst. According to Fig. 2.13 there was also a steeper-than-expected brightening near Dec. 12.

A group of American and Australian astronomers reported a second brightening of the comet along with a rise in the amount of various gases in the coma on March 24–25, 1986. Based on amateur data, Green and Morris reported a brightening of about 0.3 magnitudes at about the same time. Many amateur astronomers reported that 1P/Halley dimmed in early April 1986. Much of this dimming was due probably to the tail undergoing two disconnection events on April 1 and April 6, 1986. During this time, Green and Morris reported two brightness drops on April 4 and 8. A sharp brightness drop at this time is also evident in Fig. 2.13.

Against this background of sudden brightness changes, a second question comes to the fore: Has this comet reflected more light during some apparitions compared to others? The answer is "yes". Figure 2.17 shows the normalized magnitude of Comet 1P/Halley based largely on estimates computed by Kronk since 239 BC. The 1910 and 1986 values are based on unaided eye brightness estimates only. In spite of the large scatter in the data, I believe the comet reflected more light than normal around the year 300 AD and less light than normal around the year 1300 AD.

One final question is: Is this comet reflecting less light over time? To answer this I have fitted the normalized magnitude values in Fig. 2.17 as a function of the year. The data follow the following equation

$$H_{10}' = 4.05 + 0.00015\,Y. \tag{2.11}$$

In this equation, Y is the year where the year is positive if it is followed by AD and is negative if it is followed by BC. This equation is consistent with the normalized magnitude growing 0.00015 magnitudes fainter each year. The large scatter in the data, however, casts substantial uncertainty on this result. Therefore, this comet is not changing much with time.

Does 1P/Halley reflect more or less light than a typical Short-period comet? The answer calls for an examination of normalized magnitude values. The average value of H_{10} based on the 30 apparitions between 239 BC and 1986 AD is 4.2,

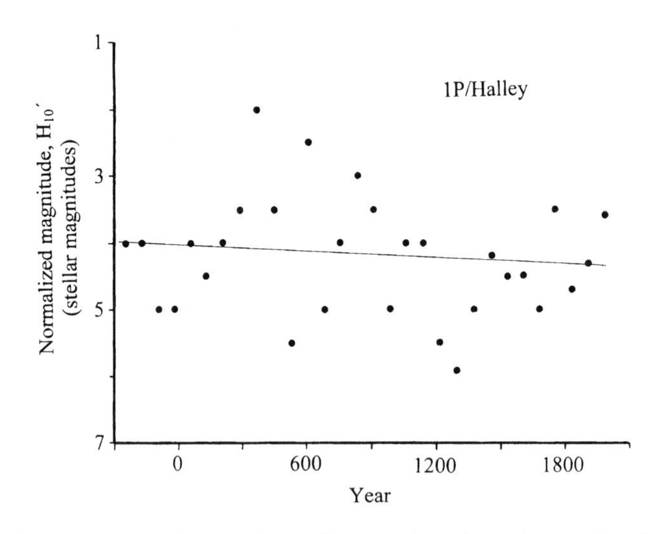

Fig. 2.17. This is a graph of the normalized magnitude, H_{10}', of Comet 1P/Halley in relation to the year. In almost all cases, the data are recomputed brightness values listed in *Cometography*, Vols. I, II and III by Kronk (1999, 2003, 2007). The 1910 and 1986 data are based on visual estimates of 1P/Halley by various people as reported in the International Astronomical Union Circulars and in *Atlas of Comet Halley 1910 II* ©1986 by B. Donn, J. Rahe and J. C. Brandt (credit: Richard W. Schmude, Jr.).

with a standard deviation of 0.8 stellar magnitudes. The average H_{10} value for 68 short-period comets is 10.19 and, hence, Comet 1P/Halley reflects more light than the typical comet. Similarly, the H_0 value of 1P/Halley just before and after perihelion (4.3 and 3.9) are much lower than the corresponding value for 127 numbered short-period comets ($H_0 = 9.70$). The fact that the normalized magnitude of 1P/Halley is much brighter than a typical short-period comet is undoubtedly one reason why it has been faithfully watched for over 2,000 years. The pre-exponential factor, 2.5n, for 1P/Halley after perihelion is about half of the average value of 127 numbered short-period comets summarized in Table 1.7. This means that 1P/Halley does not brighten as fast as a typical short-period comet when it, 1P/Halley, approaches the Sun.

Unless there is a large outburst of gas and dust, 1P/Halley will remain faint until 2,061. The brightness of the nucleus, in stellar magnitudes, follows (2.12). Thus:

$$M_N = 15.2 + 5\log[\Delta \times r] + 0.046 \times \alpha. \quad (2.12)$$

In this equation, M_N is the brightness of the nucleus in stellar magnitudes, α and r are the same as in (2.1) and α is the solar phase angle in degrees. Since the nucleus has an irregular shape, its brightness can fluctuate by several tenths of a magnitude. The nucleus probably will remain fainter than 27th magnitude until the early 2050s.

Figure 2.18 shows the location of 1P/Halley between 2010 and 2060. The comet's right ascension and declination does not change much during this time because of its huge distance from the Sun. During much of this time, the nucleus will be at opposition in January or early February. The nucleus, however, will

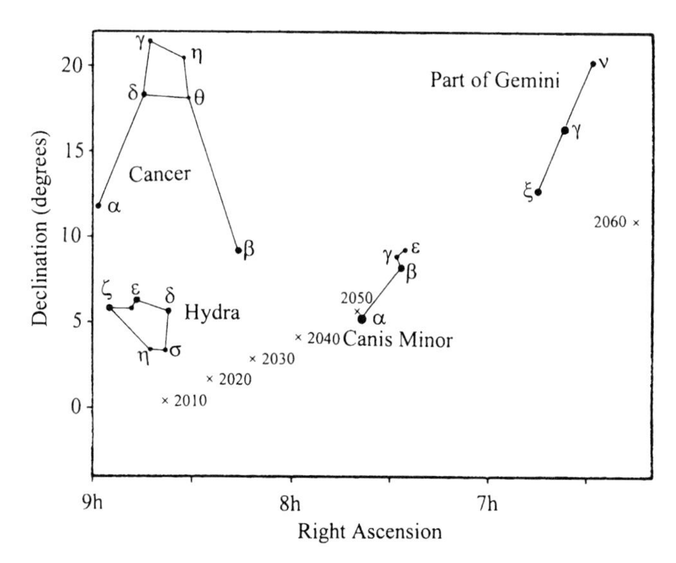

Fig. 2.18. This drawing shows positions of Comet 1P/Halley in the sky on January 1 of 2010, 2020, 2030, 2040, 2050 and 2060. These data are from the Horizons On-Line Ephemeris System. During much of the first half of the twenty-first century, this comet will be in the constellations of Hydra and Canis Minor (credit: Richard W. Schmude, Jr.).

brighten only by a few tenths of a magnitude at opposition because of its great distance from the Sun. The position will change only a little during the course of a year because this comet is so far away from the Sun.

Astronomers continue to image the nucleus of 1P/Halley. They are using these images to measure positions and to look for outbursts. If the instrumental sensitivity improves, astronomers would be able also to use images to better understand the complex rotation of its nucleus.

Overview of Spacecraft Studies

A total of three space probes, Vega 1 (the former Union of Soviet Socialist Republics or USSR), Vega 2 (USSR) and Giotto (Europe) made close flybys of the nucleus while three others Suisei (Japan), Sakigake (Japan) and the International Cometary Explorer (United States of America or USA and Europe) made more distant flybys. Many other countries helped by receiving the radio signals from these probes as they made their flybys. Several probes in other parts of our Solar System also collected data on Comet 1P/Halley during early 1986.

The two Vega spacecraft moved at speeds of 78 km/s (175,000 miles/h) relative to the nucleus of 1P/Halley. Giotto moved at a speed of 69 km/s (154,000 miles/h) relative to the nucleus. These high speeds meant that measurements and images had to be made quickly. Dust collisions with the spacecraft were also a serious problem, and some instruments on these spacecraft were damaged as a result of high-speed dust collisions. To make matters worse, dust also damaged the solar panels on the Vega spacecraft. These panels were the source of power for the many instruments on those spacecraft. We also know that radio signals from Giotto stopped reaching the Earth just before its closest approach to the nucleus. The most likely explanation is that one or more dust particles collided with Giotto causing it to wobble. (As discussed later, a dust particle can be the size of a large raindrop.) This wobble caused the antenna to point away from the Earth, thus explaining the loss of signal.

This section contains three parts, namely, "Nucleus", "Coma" and "Tail". A discussion on each part follows.

Nucleus

The three flyby spacecraft Vega 1, Vega 2 and Giotto collected images of the nucleus near peak activity. As a result, jets from the nucleus obscured much of the view. To make matters worse, much of the nucleus was in darkness and, hence, only part of it was imaged. Damage caused by dust also hindered such efforts. Because of these limitations, our knowledge of the mass and density of the nucleus is uncertain. Table 2.5 lists some physical and photometric constants of the nucleus.

Like Comet 9P/Tempel 1, the nucleus of 1P/Halley has an irregular shape. It appears to have the shape of a peanut. Since we do not have high resolution images of its entire surface, its three dimensional shape is somewhat uncertain.

Table 2.5. Physical and photometric constants of the nucleus of Comet 1P/Halley

Characteristic	Value	Source
Radii	$7.6 \times 3.6 \times 3.6$ km	Merényi et al. (1990)
Radius (mean value)	4.6 km	Computed from radii by the Author
Density	0.4 g/cm^3 (estimated)	Author's estimate from limits in Peale (1989)
Surface area	$\sim$500 km^2	Keller et al. (1986)
Estimated mass	$1–2 \times 10^{14}$ kg	Author's estimate
Acceleration due to gravity	0.0005 m/s^2	Computed from mean radius and assumed mass by the Author
Rotation rate	Complex; there is some evidence for a 2.2 and a 7.4 day period	Peale (1989)
Volume	400 km^2	Peale (1989) and Merényi et al. (1990)
Escape speed	$\sim$2 m/s	Computed from the mean radius and the estimated mass by the Author
Geometric albedo (visible and near infrared wavelengths)	0.04 ± 0.02	Sagdeev et al. (1986b)
V(1, 0) in visible wavelengths	14.1	Ferrín (2007)
B-V	0.65	Lamy and Toth (2009)
V-R	0.35	Lamy and Toth (2009)
R-I	0.28	Lamy and Toth (2009)
Solar phase angle coefficient	0.046 mag./degree	Ferrín (2007)

One of the biggest surprises of the spacecraft data is the discovery of the complex rotation of the nucleus. Apparently the nucleus does not have a single axis of rotation. Should this be a surprising result? I believe that the answer is "No." In this regard, the mass of the entire nucleus is around 1.6×10^{14} kg; the nucleus may lose around 2×10^{10} kg of material during each trip to the inner Solar System, and since 239 BC, the nucleus has made 30 trips to the inner Solar System. Therefore, it has lost about 6×10^{11} kg of material over the last 2,200-plus years. This is about 0.4% of the total mass of the nucleus. Such a loss will change the moment of inertia and will introduce also non-gravitational forces to the rotating nucleus. These two factors can cause the rotational axis to shift. Comets that pass close to the Sun will lose mass so fast that their nuclei will probably not rotate around a single axis. Comets that remain farther from the Sun stand a better chance of having a fixed axis of rotation.

Astronomers are only able to identify features on the nucleus of 1P/Halley larger than about 500 m or about 1,600 ft. This is due to a combination of the limited resolutions of the Vega and Giotto images along with the large amount of gas and dust in front of the nucleus. From close-up images, we know that the terminator, the portion separating day from night, is jagged. A jagged terminator implies lots of topographical relief. In fact, one group of astronomers believes that the topographical relief is about 1 km (or about 3,300 ft).

The nucleus has several circular features which resemble impact craters. It is not clear whether these are due to impacts, ground collapse or from jets. One feature is 2.1 km across and is 0.1 km deep. A fresh 2.1 km impact crater on the Moon would have a depth of around 0.4 km. Hence, the 2.1 km feature is much shallower than a similar-sized, fresh-impact crater on the Moon. A second circular feature, ~4 km across, is also present, along with several smaller circular features.

What is the surface like? In the regions imaged, there were no large areas of high albedos. Therefore, any pure ice on the surface is mixed with darker material. When the nucleus is near perihelion, the surface temperature can reach 330°K (134°F) on the day side. Therefore, any pure ice on the surface would sublime quickly. There is probably a range of surface temperatures at perihelion due to differences in the Sun's altitude, slope of the surface, and the orientation of the spin axis. The surface may be covered with a layer of carbon-rich dust which would be a poor conductor of heat. This could lead to a large temperature gradient between the surface of the nucleus and a meter below it.

Even though the temperature on the surface of the nucleus reaches 330°K near perihelion, it drops to less than 100°K or −280°F for most of the time. As a result, the temperature in the interior is probably close to 100°K because there would not be enough time for the heat to reach the interior due to the short time that the nucleus is close to the Sun.

What would an astronaut see if he or she were standing on the daylight side of the nucleus of 1P/Halley? First of all, near perihelion, our astronaut would notice only small amounts of dust rising as a result of the release of volatile substances. The dust would probably be barely noticeable and would be thin enough that the astronaut would be able to see through it. It would be thinner than a dust cloud kicked up by a small wind on Earth. Essentially, our astronaut would have no trouble seeing the brightest stars at night because the Vega spacecraft were able to image the nucleus in spite of having to peer through over 8,000 km of the dust and gases making up the inner coma. After taking a few steps, our astronaut would realize how weak the gravity is on the nucleus. Our 150 lb (68 kg) astronaut would only weigh a fraction of an ounce – about the weight of a United States five-cent coin or a British one-penny coin on Earth. With this low gravity, our astronaut would have to be careful not to jump or run lest he/she depart into space. Because of the low gravity of the nucleus, the astronaut would be able to walk on the fluffy surface without sinking.

Does the nucleus undergo outbursts when it is far from the Sun? Yes it does. Astronomers at the European Southern Observatory photographed a huge outburst on February 12, 1991. The comet brightened from about magnitude 25 to about magnitude 19 during this time. Essentially, it brightened by a factor of over 100! The cloud of gas and dust from this event was 230,000 km long and 140,000 km wide on that date. The average coma radius between February 12 and March 24 was near 100,000 km, and the coma was about 100 times brighter than the bare nucleus. Between April 3 and 13, 1991, the coma radius had dropped to 50,000 km and the coma was about 20 times brighter than the bare nucleus. During this outburst, the nucleus was over 14 au from the Sun and the daytime temperature of the nucleus was probably around 100°K or −280°F. If the gas and dust traveled at a speed of 700 m/s, the outburst would have occurred early on February 11, 1991.

Coma

Comet 1P/Halley begins to develop a coma when it is around 6 au from the Sun. The coma grows larger and thicker as the comet approaches the Sun. The coma can reach a radius exceeding 350,000 km or over 200,000 miles near perihelion.

The coma of 1P/Halley is much larger than that of 9P/Tempel 1. This is due probably to the larger size of the nucleus of 1P/Halley. Table 2.6 lists physical and photometric constants of the coma near perihelion.

Figure 2.19 shows half-month averages of the coma radii during 1985–1986. These radii are based on measurements from photographs. The coma kept a nearly constant size between late February and June of 1986. This is different from

Table 2.6. Physical and photometric constants of the coma of Comet 1P/Halley near perihelion

Characteristic	Value	Method
Normalized magnitude, H_{10}	4.1 ± 0.2[a]	Average value within 4 months of perihelion (Feb. 9, 1986)
Normalized magnitude, H_0 (just before perihelion in 1986)	4.3 ± 0.3[a]	Dec. 7, 1985 to Jan. 28, 1986
Pre-exponential factor, 2.5n (just before perihelion in 1986)	7.71 ± 0.8[a]	Dec. 7, 1985 to Jan. 28, 1986
Normalized magnitude, H_0 (after perihelion in 1986)	3.9 ± 0.3[a]	Feb. 15, 1986 to June 15, 1987
Pre-exponential factor, 2.5n (after perihelion in 1986)	7.60 ± 0.8[a]	Feb. 15, 1986 to June 15, 1987
Solar phase angle coefficient	<0.005 mag./degree	Author's estimate of computed normalized magnitude values
Coma radius at a comet-Sun distance of 1.6 au	350,000 km	Author's analysis of 1985–1986 photographs published in Brandt et al. (1992)
Degree of condensation or DC value	5.5	[b]
Enhancement Factor at perihelion (stellar magnitudes)	10.5 ± 0.2	[c]

[a] The Author computed all of these values from an analysis of brightness measurements published in the International Astronomical Union Circulars
[b] The Author computed this value from data published in the *International Comet Quarterly*
[c] The Author calculated this value from photometric data of the nucleus and of the comet. Photometric data of the nucleus is from Ferrín (2007). Photometric data of the comet is from the Author's calculations from data published in the International Astronomical Union Circulars and/or several issues of the *International Comet Quarterly*. All data used was collected between Jan. 26, 1986, and Feb. 24, 1986

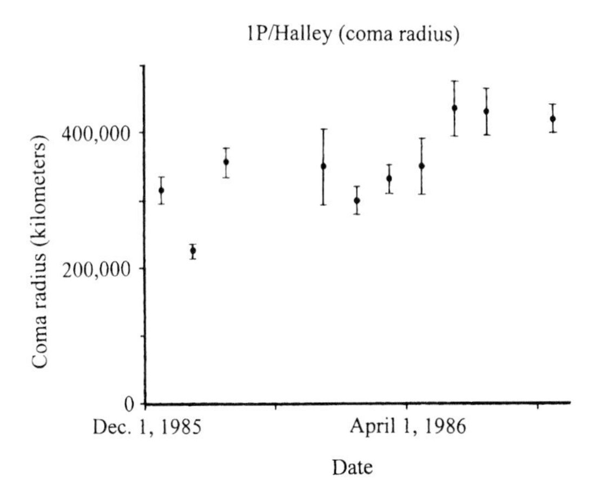

Fig. 2.19. This graph shows the average coma radius of Comet 1P/Halley for different dates in 1985 and 1986. I measured the diameter of the coma of Comet 1P/Halley from photographs published in *The International Halley Watch Atlas of Large-Scale Phenomena*, ©1992 by J. C. Brandt, M. B. Niedner, Jr. and J. Rahe. The coma radii are half-month averages. The error bars equal the standard deviation divided by the square root of the number of data points. (credit: Richard W. Schmude, Jr.).

Comet Hale-Bopp (C/1995 O1) where the coma grew smaller at perihelion. One possible reason why the coma of 1P/Halley did not get smaller at perihelion in 1986 may be due to the existence of frequent outbursts of water vapor which took place at that time.

The average coma radius of all 101 different dates between December 1, 1985 and July 6, 1986, was 350,000 km. This is quite a bit higher than the average radius based on visual estimates of 200,000 km. This discrepancy may be due to the fact that visual observers miss faint parts of the coma. The Vega and Giotto probes detected dust well beyond the visible coma. This is evidence that the coma extends beyond 200,000 km from the nucleus. Photographs, on the other hand, will pick up fainter parts of the coma.

The coma of 1P/Halley was not quite circular in 1985–1986. It was a little longer in the direction of the gas tail. Based on 31 measurements, I determined the longest dimension of the coma to be 1.12 times the length of the dimension that is perpendicular to the gas tail. See Fig. 2.20. The long axis lies along the Sun-gas tail line. This asymmetry may be due to the fact that most jets develop on the Sun-facing side of the nucleus. Hence, these jets would cause more material to be released in the part of the coma facing the Sun. This, in turn, would lead to a longer coma dimension along the Sun-gas tail line.

The coma does not have a uniform color. Essentially some parts reflect more red light than others. Astronomers believe that this non-uniform color is due to a different particle-size distribution of dust in different parts of the coma. I agree with this conclusion. In addition, jets often have a different color than the coma. For example, one jet imaged on March 1, 1986, was redder than the rest of the coma. This can explain also the changing color of different parts of the coma over time.

Astronomers report that in March, 1986, the nucleus released about 3,000–10,000 kg of dust every second. This is about 10–30% of the amount of gas released by the nucleus. Most of the dust was released from the side of the nucleus facing the Sun.

What are the characteristics of the dust particles? According to one study, the mean density of the dust grains is around 0.35 g/cm^3. This is only about one-third of the density of water, but is similar to the bulk density of the nucleus of Comet 9P/Tempel 1. This low density is consistent with dust particles being fluffy. We know that over 10,000 dust grains collided with the ~12,000 lb Giotto spacecraft. In some cases, the collisions were strong enough to move the spacecraft. One group of astronomers was able to measure these slight movements and compute

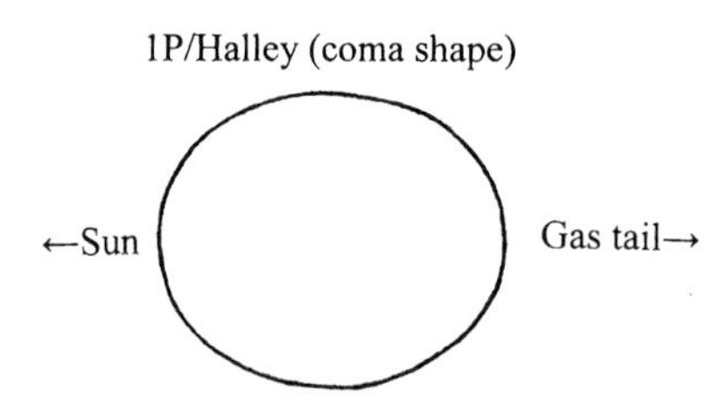

Fig. 2.20. This drawing shows the approximate shape of the coma of Comet 1P/Halley during 1985–1986. The coma had an *oval shape* and was longest in the direction of the gas tail (credit: Richard W. Schmude, Jr.).

approximate masses of the larger impacting grains. The largest dust grain to collide with Giotto had a mass of 40 mg which is about the mass of a fairly large raindrop. There were several other grains with masses of between 1.0 and 30 mg – roughly the mass of a medium-sized raindrop.

What elements were in the dust? Thanks to the mass spectrometers and other instruments onboard the Vega and Giotto spacecraft, this question was answered. The elements carbon, nitrogen, oxygen, magnesium, aluminum, silicon, sulfur, calcium and iron were present. Of these nine elements, oxygen appears to be the most abundant, followed by carbon and magnesium. The oxygen is bonded undoubtedly to other elements like silicon and iron. Much of the carbon is bonded probably to other elements as well. One group of astronomers reported spectroscopic evidence for the presence of carbon–hydrogen bonds. This, along with the large amount of carbon, is consistent with the presence of material similar to what is found in carbonaceous chondrite meteorites.

Astronomers determine the distribution of dust in the coma by studying its color and brightness intensity. If dust flows in a uniform behavior away from the nucleus, its density should fall with the square of the distance of it from the nucleus. For example, the dust density at a distance of 1,000 km from the nucleus should be four times the density at a distance of 2,000 km. To make matters more complicated, one must also consider depth – the third dimension. Taking depth into account, the dust signal should drop in proportion to its distance from the nucleus because the dust signal incorporates depth. Therefore, the dust signal at a nucleus distance of 2,000 km should only be half the signal at a distance of 1,000 km from the nucleus.

However, in March of 1986, the dust signal did not drop off linearly with increasing distance from the nucleus. This meant that there was a non-uniform flow of dust through part of the coma. This may have been due to the acceleration of dust particles caused by gas escaping from the coma or the pressure of solar wind. In addition, sudden outbursts of dust could cause the dust signal to deviate from the expected behavior.

The Vega and Giotto spacecraft measured the density and composition of dust as they traveled through the coma. These spacecraft measured dust grains with masses of at least 10^{-17} g. Both the Vega and Giotto spacecraft detected dust more than 200,000 km from the nucleus. Closer in, all three spacecraft measured the number of dust particles per unit volume. At a distance of about 9,000 km from the nucleus, Vega 1 measured a density of three dust particles per cubic meter on March 6. (An average size refrigerator has a volume of about 1 m^3.) Three days later, Vega 2 measured a density of only 0.1 dust particle per cubic meter at a nucleus distance of 8,000 km. These two values differ by a factor of 30. This difference is consistent with the fact that dust does not flow in a uniform behavior but instead comes in spurts.

The peak mass of the coma of 1P/Halley was calculated to be around one to two million metric tons. This figure includes the mass of all gases and dust out to a distance of 150,000 km from the nucleus. Beyond this distance, the amount of gas and dust is so low that it will not affect materially the total coma mass. Based on the mass just reported, the nucleus has about 100,000 times the mass as the coma. Therefore, even though the coma can be over 10,000 times brighter than the nucleus, its mass is much smaller than that of the nucleus.

When 1P/Halley is within 1 au of the Sun, water (H_2O) is the most abundant gas in the coma. At this distance, the comet releases water at an average rate of 30,000 kg/s or 66,000 lb/s. Water, like dust, is released in spurts. Earth-based observatories detected outbursts of water on February 19–20, March 20 and March 24, 1986.

What causes water outbursts? One group of astronomers believes that there are four possible causes for the outbursts. When lots of solid water molecules in random orientations (amorphous ice) turn into a solid where the molecules are in a specific orientation (crystalline ice), lots of energy is released. This release of energy can lead to the explosive release of water vapor. This group suggests also that chemical explosions, thermal stress and pockets of compressed gas can lead to an outburst of water.

Table 2.7 lists approximate production rates of different gaseous species in proportion to that of water. At the present time, we do not know whether these species come from the nucleus, from dust grains in the coma or both the nucleus and dust grains. The carbon dioxide (CO_2) and ammonia (NH_3) may come from ices. Other species, like diatomic sulfur (S_2), may be fragments of larger molecules.

Table 2.8 lists different molecular and ionic species which existed in the coma in 1986. These species are broken into three groups, namely, stable molecules, molecular fragments and ionic species. The stable molecules are fairly unreactive. They can exist as ices or as gases. Molecular fragments and ionic species, on the other hand, are very reactive. They can not exist as pure ices. Molecular fragments are fragments of larger molecules. For example, OH is probably a fragment of H_2O. Ionic species are those that have either lost or gained one or more electrons. Ultraviolet light from the Sun or high-energy collisions can knock one or more electrons off of a neutral atom or molecule creating an ion. There are apparently a large number of ions in the coma when 1P/Halley is near perihelion. The Vega 1 spacecraft, for example, detected an ion density of 4,000 ions per cubic centimeter. This value is comparable to the ion density in the ionospheres of the planets Uranus and Neptune.

While jets cover just part of the Sun-facing side of the nucleus of Comet 1P/Halley, most of them form on the afternoon side of the nucleus. One group of astronomers

Table 2.7. Production rates of various gases from the nucleus of Comet 1P/Halley just after perihelion. All production rates are in terms of that for water vapor

Substance	Approximate production rate in terms of that for water	Source
Carbon monoxide, CO	0.05	Combes et al. (1988) and Samarasinha et al. (1994)
Carbon dioxide, CO_2	0.03	Combes et al. (1988) and Krankowsky et al. (1986)
Methane, CH_4	0.02	Krankowsky et al. (1986) and Allen et al. (1987)
Formaldehyde, H_2CO	0.04	Combes et al. (1988)
Ammonia, NH_3	0.01	Krankowsky et al. (1986), Allen et al. (1987), and Magee-Sauer et al. (1989)
Disulfur radical, S_2	0.001	Wallis and Swamy (1987)
Dicarbon radical, C_2	0.0028	Fink (2009)
Azanyl radical, NH_2	0.00276	Fink (2009)
Cyanide, CN	0.00147	Fink (2009)

Table 2.8. Different molecular and ionic species which were in the coma of Comet 1P/Halley in 1986

Stable molecules	Molecular fragments	Ionic species
Carbon monoxide, CO[h]	Methylidyne, CH[a]	C^+[d]
Carbon dioxide, CO_2[h]	Cyanide, CN[a]	$^{12}CH^+$[d]
Naphthalene, $C_{10}H_8$[i]	Dicarbon radical, C_2[a]	CO^+[c]
Phenanthrene, $C_{14}H_{10}$[i]	Tricarbon radical, C_3[a]	CO_2^+[f]
Hydrogen cyanide, HCN[i]	Azanylidene radical, NH[a]	C_2^+[d]
Water vapor, H_2O[h]	Azanyl radical, NH_2[a]	$^{56}Fe^+$[d]
Formaldehyde, H_2CO[h]	Hydroxyl radical, OH[a]	H_2^+[e]
Ammonia, NH_3[i]	Sulfur monoxide, SO[g]	H_2O^+[d]
	Disulfur radical, S_2[g]	H_3O^+[d]
	Carbon, C[b]	Na^+[d]
	Iron, Fe[b]	$^{16}O^+$[d]
	Magnesium, Mg[b]	OH^+[d]
	Oxygen, O[b]	$^{32}S^+$[d]
	Silicon, Si[b]	$^{34}S^+$[d]
	Sodium, Na[b]	

[a] From Moreels et al. (1986)
[b] From Kissel et al. (1986b)
[c] From International Astronomical Union Circular 4183
[d] From Krankowsky et al. (1986)
[e] From Balsiger et al. (1986)
[f] From Korth et al. (1986)
[g] From Wallis and Swamy (1987)
[h] From Combes et al. (1988)
[i] From Crovisier and Encrenaz (2000)
[j] From Magee-Sauer et al. (1989)

reported that one jet started from a 7 km^2 area on the nucleus. A second group of astronomers report that 27% of the total surface area of the nucleus was active. I believe that most of the jets are gentle features which cover large areas, and doubts that there are many jets which exert a force similar to that of a fire hose.

The Vega and Giotto spacecraft detected small magnetic fields in the coma. Peak magnetic fields were around 65–80 nanotesla. This is several hundred times weaker than Earth's magnetic field. One group of astronomers believes that the magnetic field in the coma is due to the interaction between the solar wind and the coma material. The magnetic field also influences the movement of ions and small dust particles with a net electrical charge.

Tail

The first photograph published in *The International Halley Watch Atlas of Large-Scale Phenomena* which shows a definite tail was made on November 9, 1985. At this time, 1P/Halley was 1.5 au from the Sun. For the next 8 months the gas tail was present. In a photograph taken on July 6, 1986, a thin gas tail is visible. At this time, the comet was 2.5 au from the Sun. After mid-1986, the photographic coverage dropped off.

Near perihelion, Comet 1P/Halley has both a dust and a gas tail. One group of astronomers reported that the gas tail underwent 19 disconnection events between December 4, 1985, and April 13, 1986. A disconnection event happens when part of the gas tail breaks off. This group also reported that all 19 events occurred when 1P/Halley was within 30° of the heliospheric current sheet, and that 11 of these events occurred when the comet was within 10° of the heliospheric current sheet. (The heliospheric current sheet is described in Chap. 1.) A second group of astronomers studied two disconnection events. They used data from several spacecraft to determine the nature of the solar wind during these events in April, 1986. They concluded that both disconnection events occurred when 1P/Halley crossed the heliospheric current sheet.

Figure 2.21 shows a graph of the length of the gas tail on different dates between November 29, 1985, and July 6, 1986. The length of the gas tail changed from 0.02 au up to 0.27 au. Much of this change was due to the numerous disconnection events which took place during 1985–1986. The tail lengths do not correlate well with the H_{10}' values plotted in the top graph of Fig. 2.14. That is, the times when the comet had a long gas tail does not correspond to it being brighter or having a lower H_{10}' value. There seems to be some correlation, however, between the average coma size and the gas tail length. During early January and May of 1986 both the coma and the gas tail were larger than their average values.

During 1986, the dust tail was broader than the gas tail. Its color was yellowish-white. During March of 1986, it was about 2–3° long. The brightest part of this tail made an angle of about 8° with the longest portion of the gas tail. This is about one-third of the separation of the two tails of Comet Hale-Bopp when it was at perihelion.

Comet 1P/Halley had a small antitail in May and early June of 1986. I measured both the angular length and the minimum angle between the antitail and the gas

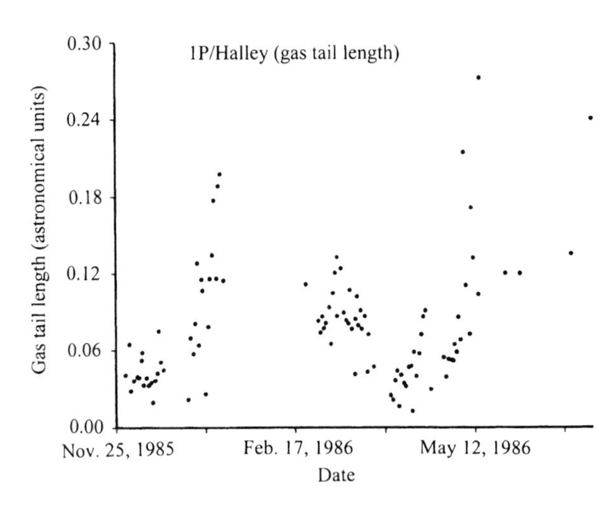

Fig. 2.21. In this graph, the length of the gas tail of Comet 1P/Halley is plotted against varying dates. In all cases, I measured gas tail lengths from both computed the tail lengths from (4.10) and from photographs in the *International Halley Watch Atlas of Large-Scale Phenomena* ©1992 by J. C. Brandt, M. B. Niedner, Jr. and J. Rahe (credit: Richard W. Schmude, Jr.).

Table 2.9. Characteristics of the antitail of Comet 1P/Halley; angle ENP is the angle between the Earth and the orbital plane of that comet. The Author made all measurements from photographs published in *The International Halley Watch Alas of Large-Scale Phenomena* ©1992 by J. C. Brandt, M. B. Niedner and J. Rahe. Values of angle ENP are from the Horizons On-Line Ephemeris System

Decimal date (1986)	Angle between antitail and gas tail (°)	Angular length of antitail (°)	Angle ENP (°)
May 5.11	177	0.1	5
May 6.01	180	0.15	4
May 11.18	179	0.13	2
May 11.41	179	0.17	2
May 11.87	180	0.1	2
May 27.38	180	0.07	1
June 2.03	179	0.05	2

tail on photographs made on several dates in 1986. These results are presented in Table 2.9. In all cases, the antitail was almost 180° from the gas tail. Its angular length was between 0.05° and 0.17°. However, one must be cautious of these dimensions since a longer exposure time may show a longer antitail. In all photographs examined, the antitail was fainter than the gas tail. The Earth moved through the orbital plane of 1P/Halley on May 19, 1986.

Did 1P/Halley change much between 1910 and 1986? As it turns out, we have enough data to answer this question. Comet 1P/Halley did change between its two most recent apparitions. The normalized magnitude, H_{10}, values were 4.3 and 3.6 for the 1910 and 1986 apparitions, respectively. Both values were based solely on unaided-eye brightness estimates. Therefore, this comet reflected about twice as much light in 1986 as in 1910. This is supported by the coma size. The coma radius, based on photographs made 0–30 days after perihelion, was ~270,000 km in 1910 and ~310,000 km in 1986. The larger coma in 1986 would explain the large amount of light that 1P/Halley reflected that year. The average length of the gas tail in 1910 between 0 and 60 days after perihelion was 0.13 au, whereas the corresponding value in 1986 was 0.073 au. The shorter gas tail length in 1986 may be due to the large number of disconnection events. Between 0 and 60 days after the 1986 perihelion date, there were ten disconnection events, whereas the corresponding number for 1910 was two or three. The higher number of disconnection events in 1986 may have been due to the fact that the sunspot number was about twice as high in 1986 as it was in 1910 (Fix 2008).

Projected Apparition of Comet 1P/Halley in 2061

Table 2.10 lists the predicted right ascension, declination and brightness values for Comet 1P/Halley. These predictions are based on the Horizons On-Line Ephemeris System. Non-gravitational forces and planetary perturbations may cause small errors in the predicted positions. Predicted brightness estimates are also listed in Table 2.10. These estimates are based on (2.8)–(2.10).

Comet 1P/Halley will be seen best in the Northern Hemisphere either during the early morning in the eastern sky during late June and early July in 2061 or

Table 2.10. Predicted positions and brightness values for Comet 1P/Halley during 2061. The right ascension and declination values are from the Horizons On-Line Ephemeris System. The Author computed the brightness values from (2.8) to (2.10) in the text

Date (2061)	Right ascension	Declination	Brightness (stellar magnitude)
Feb. 1	03 h 39 min 47 s	13° 34′ 25″	12.9
Mar. 3	03 h 13 min 17 s	14° 21′ 54″	12.1
Apr. 2	03 h 08 min 22 s	16° 05′ 19″	11.0
May 2	03 h 15 min 36 s	18° 32′ 11″	9.3
June 1	03 h 29 min 15 s	21° 54′ 23″	6.9
June 11	03 h 35 min 10 s	23° 26′ 49″	6.2
June 21	03 h 42 min 44 s	25° 26′ 17″	5.4
July 1	03 h 54 min 33 s	28° 16′ 57″	4.4
July 11	04 h 20 min 16 s	32° 56′ 34″	3.1
July 21	05 h 44 min 35 s	40° 50′ 36″	1.7
July 26	07 h 33 min 16 s	42° 36′ 40″	1.1
July 31	09 h 49 min 36 s	33° 51′ 13″	1.0
Aug. 5	11 h 16 min 37 s	19° 39′ 56″	1.1
Aug. 10	11 h 59 min 23 s	09° 22′ 07″	1.8
Aug. 20	12 h 33 min 49 s	−01° 03′ 49″	3.2
Aug. 30	12 h 46 min 59 s	−05° 55′ 06″	4.4
Sept. 29	13 h 04 min 24 s	−12° 36′ 29″	6.7
Oct. 29	13 h 15 min 47 s	−16° 40′ 22″	8.0
Nov. 28	13 h 20 min 42 s	−20° 06′ 59″	8.8

during the early evening in the western sky during mid to late August of 2061. When the comet is brightest, it will be within 20° of the Sun. One feature that people may look for in late July and early August is a brightening due to forward scattered light. In May 1910, J. B. Bullock in Tasmania, Australia, noticed that 1P/Halley became very bright when it was only 2° from the Sun. One person attributed this brightening to forward-scattered light. This should not be surprising since dust particles which are about 0.5 μm across are bright in forward-scattered visible light. The Vega and Giotto probes detected a lot of dust particles near this size. Comet 1P/Halley will get within 20° of the Sun in late July, 2061.

One can look also for an antitail. The Earth will pass through the orbital plane of 1P/Halley on May 20, 2061. One should be able to image the antitail throughout May and early June of 2061.

Part C: Comet 19P/Borrelly

Introduction

On December 28, 1904, A. L. N. Borrelly discovered a new comet. Over the next 2 weeks, many astronomers reported both brightness and position measurements of this comet. G. J. Fayet computed an elliptical orbit for it. He predicted that it would next reach perihelion in December 1911. H. Knox-Shaw recovered this comet on Sept. 20, 1911, and confirmed thereby its periodic orbit. It now has the name 19P/Borrelly which will be used throughout this Book.

Between 1905 and 1911, there were hints that 19P/Borrelly was different than other comets. Several astronomers reported that the central condensation was not centered in the coma. In 1918, astronomers reported that the "nucleus" was elongated. One astronomer reported that the "nucleus" was 50 arc-sec across. We know that the longest dimension of the "nucleus" of 19P/Borrelly was less than 0.03 arc-sec across in late 1918. The reported "nucleus" dimension of 50 arc-sec may have been the result of a large, bright jet. Astronomers now realize that both the long "nucleus" and the fact that it often appears off to one side of the coma is due to the presence of one or more large jets.

The Deep Space 1 spacecraft, developed by NASA, passed about 2,170 km from the nucleus of 19P/Borrelly on September 22, 2001. This spacecraft took images of the nucleus, recorded spectra and collected data on the plasma environment during its flyby. Professional astronomers published several scientific papers on this comet in a 2004 issue of *Icarus*. In addition amateur astronomers across the world studied this comet.

In this Part visual observations of 19P/Borrelly are described, followed by discussions of its nucleus, coma and tail. This Part will end with a discussion of certain predicted events in the second and third decades of the twenty-first century.

Visual Observations

S. K. Vsekhsvyatskii reports normalized magnitudes, H_{10}, of 9.0 (1904), 9.5 (1911), 10.2 (1918), 10.1 (1925) and 9.2 (1932) for 19P/Borrelly. One problem with these early estimates is that they are based on brightness estimates made mostly through 0.2–0.3 m refractors. As mentioned earlier, Vsekhsvyatskii did not correct the brightness estimates to a standard aperture of 6.8 cm. Accordingly, I re-computed normalized magnitudes based on individual brightness estimates which were corrected to an aperture of 6.8 cm. All of his brightness estimates are from Kronk (2007). The respective H_{10} values which he computed for the 1904, 1911, 1918, 1925 and 1932 apparitions are 7.5, 8.6, 8.9, 7.5 and 8.6, respectively.

For the more recent apparitions, I computed H_{10} values of 7.5 (1980), 7.8 (1987–1988), 7.7 (1994–1995), 7.7 (2001) and 7.5 (2008) for 19P/Borrelly. The more recent H_{10} values are a little lower than those re-computed values from the early twentieth century. The average value of H_{10} between 1904 and 1932 is 8.2, and the corresponding average for the 1981–2008 apparitions is 7.6. This is evidence that 19P/Borrelly reflected more light as the twentieth century progressed.

The coma size has fluctuated throughout the twentieth century. The respective average radii for 1904, 1911, 1918, 1925 and 1932 are 30,000 km, 20,000 km, 10,000 km, 20,000 km and 30,000 km, respectively. These radii are smaller than the modern value of 60,000 km. Interestingly, this comet was both faintest and had the smallest coma in 1918. Furthermore, the larger coma during recent apparitions is consistent with that comet reflecting more light. Essentially, a large coma means that more light is reflected and, hence, a lower normalized magnitude value.

During a typical apparition, this comet starts off faint and reaches a maximum brightness near perihelion. In most cases, it reaches perihelion at a time when it is not at its closest distance to the Earth, and unless there is a large outburst, it can not get much brighter than seventh magnitude. During late December of 1987, this comet reached about seventh magnitude which is about as bright as it can get. See Fig. 2.22. This happened because it was closest to the Earth and Sun at almost the same time. During 1987–1988, many people observed 19P/Borrelly because of its nearly optimal position. In other recent apparitions, it was a great deal fainter than seventh magnitude and, as a result, there are fewer observations in other apparitions.

In Fig. 2.23 the H_{10}' values of 19P/Borrelly are plotted against varying dates during the five apparitions between 1980 and 2008. The normalized magnitude, H_{10}', values

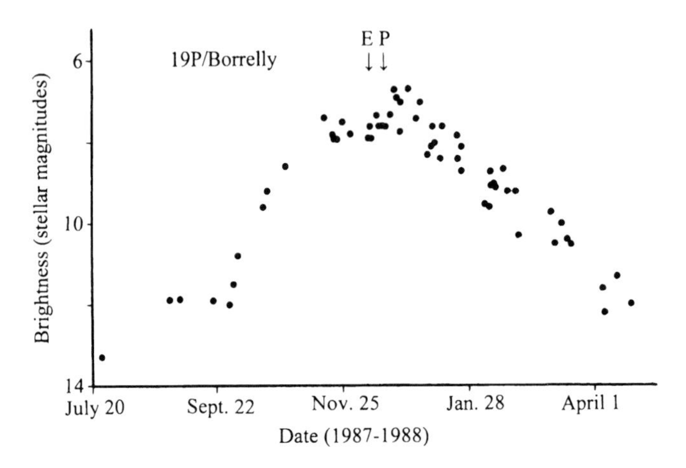

Fig. 2.22. In this graph, the brightness of Comet 19P/Borrelly is plotted against dates in 1987–1988. The time when the comet was closest to the Earth is shown as an E. Similarly the time when it was at perihelion is shown as a P. I corrected these data to a standard aperture of 6.8 cm; the raw magnitude measurements are from the International Astronomical Union circulars (credit: Richard W. Schmude, Jr.).

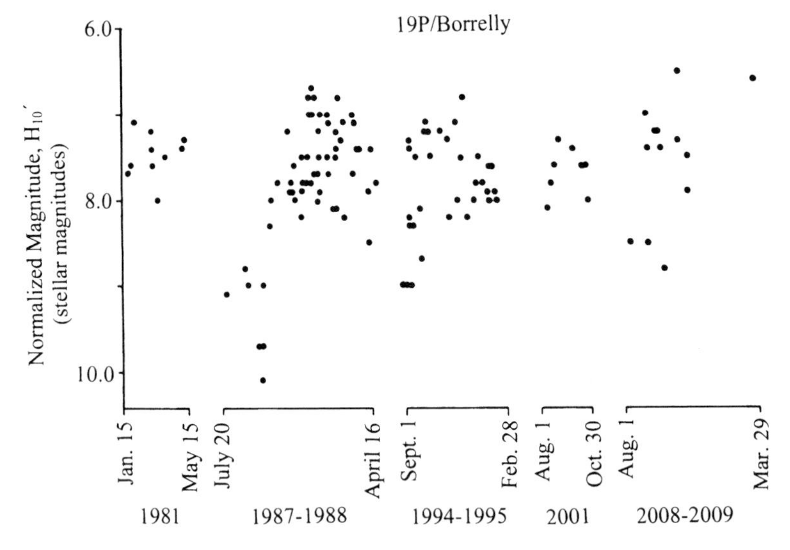

Fig. 2.23. This graph shows the normalized magnitudes, H_{10}', against varying dates for the five apparitions of Comet 19P/Borrelly (1981, 1987–1988, 1994–1995, 2001 and 2008–2009). I corrected these data to a standard aperture of 6.8 cm; the raw magnitude measurements are from the International Astronomical Union circulars (credit: Richard W. Schmude, Jr.).

remained generally within the range of stellar magnitude 7.0–8.4 except during mid-1987. During that time, the H_{10}' values were between a stellar magnitude of 9 and 10. By mid-October of 1987, the comet brightened, and the normalized magnitude changed to a more normal value of magnitude ~8. This represents about a fourfold increase in brightening. Could this have been due to the development of a large jet on the nucleus?

In Fig. 2.24, graphs showing values of 19P/Borrelly's coma radius, DC (degree of condensation) and normalized magnitude, H_{10}' are plotted against varying dates. The DC value describes how diffuse or condensed the coma appears. The DC value is described further in Chap. 3. In all cases, average values were computed for

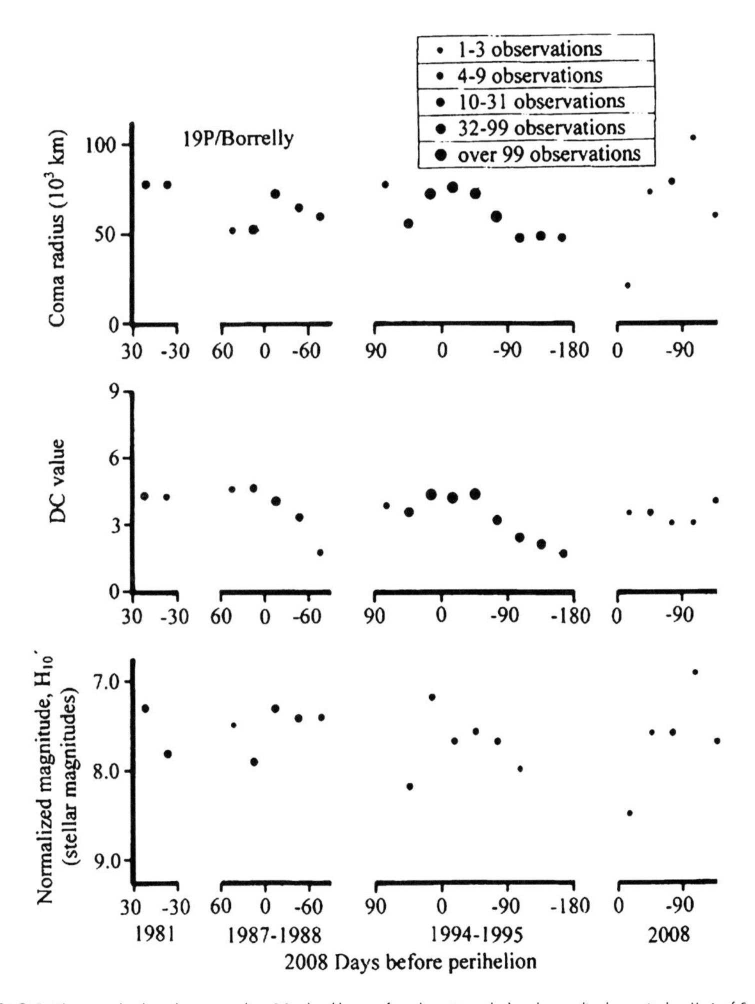

Fig. 2.24. These graphs show the coma radius, DC value (degree of condensation value) and normalized magnitudes, H_{10}', of Comet 19P/Borrelly for its 1981, 1987–1988, 1994–1995 and 2008 apparitions. Data are from the International Astronomical Union Circulars and from the *International Comet Quarterly*. The normalized magnitudes are based on brightness estimates which have been corrected to a standard aperture of 6.8 cm (credit: Richard W. Schmude, Jr.).

30-day time intervals for the 1981, 1987–1988, 1994–1995 and 2008 apparitions. I used the H_{10}' value instead of the H_0' value for 19P/Borrelly because of the large difference in the pre-exponential factor value before and after perihelion.

The most important trend in Fig. 2.24 is that the apparition-average coma radius, DC value and normalized magnitude did not change very much. There were, however, significant changes from 1 month to the next in some apparitions. During the 1987–1988 and 1994–1995 apparitions, the coma size and DC value fell after perihelion.

Figure 2.25 shows a graph of the $M_c - 5\log[\Delta]$ against the log[r] value for the 1987–1988 apparition of 19P/Borrelly. The M_c, Δ and r terms have the same value as in (2.1). Like 1P/Halley, this comet's H_0 and pre-exponential factor values were different before and after perihelion. The data before perihelion are consistent with

$$M_c = 6.35 + 5\log[\Delta] + 21.10\log[r] \quad 1987-1988\ \textit{apparition.} \qquad (2.13)$$

In this equation, M_c is the brightness of 19P/Borrelly in stellar magnitudes and Δ and r are the same as in (2.1). The data after perihelion are consistent with:

$$M_c = 6.9 + 5\log[\Delta] + 13.04\log[r] \quad 1987-1988\ \textit{apparition.} \qquad (2.14)$$

Once again, M_c, Δ and r are the same as in (2.1).

Table 2.11 summarizes normalized magnitudes and pre-exponential factors for 19P/Borrelly for the 1981–2008 apparitions. Weighted averages are computed by weighing apparitional values based on the number of data. Therefore, the values in Table 2.11 have different weights. The weighted average of H_0, in stellar

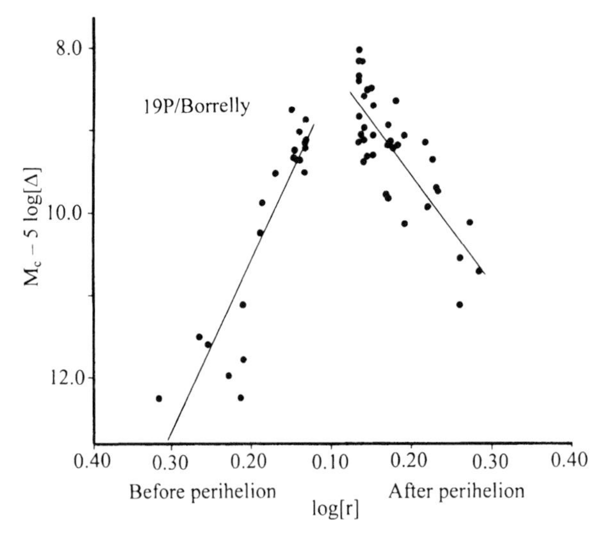

Fig. 2.25. A graph of $M_c - 5\log[\Delta]$ versus log[r] for Comet 19P/Borrelly. The value of r is in astronomical units. The *left half* of the graph summarizes measurements made before the comet reached perihelion and the *right half* summarizes measurements made after the comet reached perihelion (credit: Richard W. Schmude, Jr.).

Table 2.11. A summary of normalized magnitudes and pre-exponential factors for Comet 19P/Borrelly. The Author computed the data from brightness values reported in the International Astronomical Union Circulars and in the *International Comet Quarterly*. He corrected all magnitudes to a standard aperture of 6.8 cm

Apparition	H_{10}	H_0	2.5n
1981 (before perihelion)	7.5	–	–
1981 (after perihelion)	7.5	8.1	5.95
1987 (before perihelion)	8.3	6.4	21.10
1987 (after perihelion)	7.4	6.9	13.0
1994 (before perihelion)	7.8	4.8	28.9
1994 (after perihelion)	7.7	7.0	13.8
2001 (before perihelion)	7.7	–	–
2001 (after perihelion)	7.7	–	–
2008 (after perihelion)[a]	7.5	8.4	6.0
Average	7.6	–	–

[a]The Author is not aware of any data published before perihelion.

magnitudes, is 6.1 ± 0.3 before perihelion and 7.2 ± 0.3 after perihelion. The weighted average values of the pre-exponential factors are 22.4 before perihelion and 11.7 after perihelion.

The normalized magnitudes, H_0, for 19P/Borrelly before and after perihelion are smaller than the corresponding values for 127 numbered short-period comets (9.70). This means that 19P/Borrelly reflects more light than a typical short-period comet. Comet 19P/Borrelly probably reflects more light because its nucleus is larger than that of a typical Jupiter Family comet.

The enhancement factor of 19P/Borrelly jumped from about 6.0 stellar magnitudes in August and September of 1987 to around 7.5 stellar magnitudes during November and December of 1987. This meant that the coma went from being ~250 to ~1,000 times brighter than the nucleus at full phase in 1987. This jump may be due to an outburst which happened in October, 1987.

The enhancement factor for 19P/Borrelly in 1987–1988 is larger than that of 9P/Tempel 1, but is smaller than that of 1P/Halley. Much of this difference is due to the perihelion distance. Comet 19P/Borrelly reached a perihelion distance of 1.36 au in 1987, whereas 9P/Tempel 1 got no closer than 1.49 au from the Sun in 1983–2005, and, currently, 1P/Halley got only about 0.6 au from the Sun in 1986. Naturally one would expect a larger enhancement factor as a comet gets closer to the Sun.

Nucleus

Table 2.12 lists physical and photometric constants of the nucleus of 19P/Borrelly. Much of this data were compiled from the close-up images of the nucleus made by the Deep Space 1 spacecraft. The shape, possible break-up, albedo, surface temperature, rotation, seasons and topography of the nucleus are discussed below.

The nucleus has a shape similar to that of a bowling pin. See Fig. 2.26. It is possible that the nucleus is in two pieces which are held together by gravity. The best images do not show a crack between the narrow and wide ends; however,

Table 2.12. Physical and photometric constants of the nucleus of Comet 19P/Borrelly

Characteristic	Value	Source
Radii	$4.4 \times 2.2 \times 1.5^{a}$ km	Lamy et al. (1998)
Radius (mean value)	~2.4 km	The Author computed the radius using the assumed radii above
Surface area	~90 km^2	Schleicher et al. (2003)
Mass	1.5×10^{13} kg	Author's estimate assuming a density of 0.25 g/cm^3
Average acceleration due to gravity	0.0002 m/s^2	The Author computed this value from the mean radius and the mass
Rotation rate	25 ± 0.5 h	Lamy et al. (1998) and Schleicher et al. (2003)
Direction that the rotational pole points to	RA = 14 h 27 min Dec. = −5° 42 min	Schleicher et al. (2003)
Density	0.18–0.3 g/cm^3	Davidsson and Gutiérrez (2004)
Volume	60 km^3	The Author computed the volume from the mass and density
Obliquity of rotational axis	102.7° ± 0.5°	Schleicher et al. (2003)
Escape speed	~1 m/s	The Author computed the escape speed from the mean radius and the estimated mass
Geometric albedo (wavelength of 500–1,000 nm)	0.029 ± 0.006	Buratti et al. (2004)
V(1, 0) in visible wavelengths	15.9	Tancredi et al. (2006)
J–K	~0.82	Soderblom et al. (2004b)
H–K	~0.43	Soderblom et al. (2004b)
V–R	0.25 ± 0.78	Lamy and Toth (2009)
Solar phase angle coefficient	0.045 ± 0.005 mag./degree	Ferrín (2007)

[a]The third radius was computed by the Author assuming a volume of 60 km^3 and a triaxial ellipsoid shape

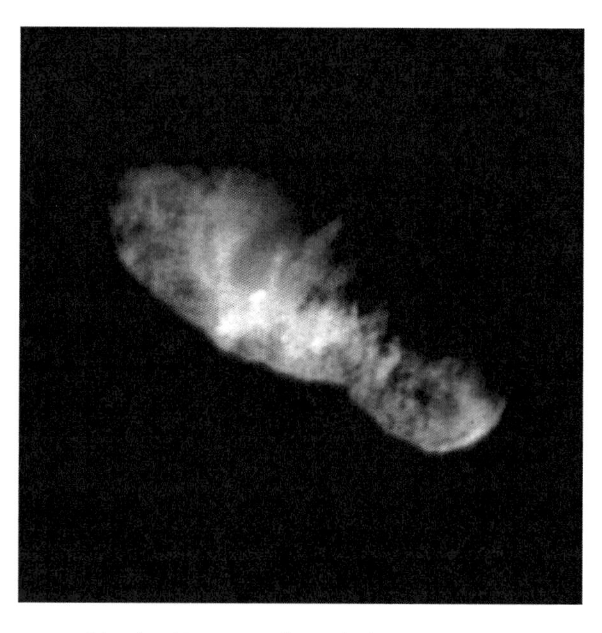

Fig. 2.26. A close-up image of the nucleus of Comet 19P/Borrelly. Note that this object has a shape similar to that of a bowling pin (credit: courtesy NASA/JPL-Caltech.).

dust may have covered one up, or any such crack may be too thin to see in images. One group of astronomers suggested that the nucleus may break apart, perhaps at its thinnest point, and this may have happened already.

In May of 2009, several astronomers reported a bright condensation in the coma of 19P/Borrelly which was a few arc-seconds from the central condensation. This meant that this new condensation was at least several thousand kilometers from the nucleus. One group of astronomers reported that it was about a factor of four dimmer than the central condensation. Did the nucleus break apart, or was the new condensation just a large spurt of dust? Future studies may give an answer.

During the Deep Space 1 encounter, the northern hemisphere of the nucleus was tipped toward the Sun. The sub-solar latitude on the nucleus was at 55°N. This meant that areas 35° from the rotational pole were receiving the maximum amount of sunlight. Based on images and measurements of the northern hemisphere, we have a good idea of its albedo and surface temperature.

Have you ever seen charcoal? The northern hemisphere of 19P/Borrelly is darker than charcoal. It reflects only about 3% of the visible light falling on it. Its nucleus is one of the darkest objects in our Solar System. How did the nucleus get so dark? Carbon is probably responsible for its nearly pitch-black color. Carbon is fairly abundant in our Solar System, and it is a solid at the low to moderate temperatures of the nucleus. The carbon may have come from hydrocarbon compounds which broke apart as a result of ultraviolet light from the Sun. Ultraviolet light is able to break apart carbon–hydrogen bonds, causing hydrogen gas to escape and leave behind a carbon residue. Even when this comet develops a coma, ultraviolet light from the Sun is still able to reach its nucleus. In spite of this, not all areas on the nucleus are equally dark.

Some areas on 19P/Borrelly reflect more visible light than others. In one study, there is a factor-of-three change in reflectivity across its northern hemisphere. That is, some areas reflect three times as much light as others. Some of this change in reflectivity may be due to a difference in the mean particle size. By way of analogy, powdered sugar is brighter than granulated sugar. This is because powdered sugar has a smaller grain size. Astronomers also believe that some parts of the nucleus are darker than others due to a difference in chemical composition. The low albedo of this comet is undoubtedly one reason for its high surface temperature.

The temperature near the sub-solar point rises to 340°K or 152°F when the nucleus is near perihelion. The two reasons for this are the low geometric albedo of the surface, and the fact that the gases in the coma absorb almost no sunlight. The perihelion temperature near the terminator drops to around 300°K or 80°F. Like 9P/Tempel 1 and 1P/Halley, the temperature of the deep interior probably remains very cold. Likewise, the temperatures on the night side of the nucleus are probably very low.

Based on the comet's albedo and temperature of its northern hemisphere, we can conclude that it does not have any large areas of pure water ice on its surface. While some water ice may exist at the surface, it would have to be mixed with dark material to create a low-albedo mixture.

The nucleus rotates around a single axis called the rotational axis. This axis is pointed in the direction listed in Table 2.12. The pole star for the nucleus is sixth magnitude 104 Virgo. The rotational axis is tilted sideways, having an obliquity of

102.7°. The tilt of the axis means that, like Earth, the nucleus of 19P/Borrelly undergoes seasons. Unless the axis undergoes precession, or if the orbit is changed, the nucleus will always be at the same season at perihelion. Very recently, a group of astronomers have tentatively called the pole of the nucleus which pointed near the Sun during perihelion in 2001 the "North Pole". Therefore, the northern spring and summer are the seasons when that comet is close to the Sun and hence the brightest. During the northern fall and winter seasons, the comet is far from the Sun and is very dim. The large tilt of the rotational axis, combined with its orientation, means that the northern hemisphere receives more sunlight than either the equatorial regions or the southern hemisphere. The rotation of the nucleus also affects its brightness as seen from the Earth.

The nucleus undergoes brightness changes due to rotation and its changing solar phase angle. It rotates about once every 25 h. Rotation, along with the bowling-pin shape of the nucleus, means that the brightness fluctuates by over a factor of two. In addition, the nucleus also brightens as the solar phase angle drops. One group of astronomers reported that the nucleus brightens by a factor of 26 in red light when the solar phase angle drops from ~87 to ~7°. This is consistent with the solar phase angle coefficient in Table 2.12. Finally, the opposition surge may affect the brightness of the nucleus.

The opposition surge is a non-linear brightening of an object at a solar phase angle of 0°. Professional astronomers are interested in measuring the value of the opposition surge of the nucleus because it can yield information on the mean particle size of the surface dust along with the compactness of the surface of the nucleus. The opposition surge can be measured only from a graph of the normalized magnitude against the solar phase angle. As mentioned previously, we know already how the normalized magnitude of the nucleus changes with different solar phase angles. We have data of 19P/Borrelly for solar phase angles of between about 2 and 87°. We do not have normalized magnitude values for solar phase angles below 2°. At a solar phase angle of 2°, the nucleus is about 0.1 magnitudes brighter than if it followed the same trend that it followed at larger solar phase angles. If the nucleus has a similar graph of normalized magnitude against solar phase angle, as Uranus' moon Titania, the opposition surge would be about 0.5 stellar magnitudes. It is, however, important to confirm this value.

The albedo and color of the southern hemisphere of the nucleus can reveal important seasonal processes. Since much of this hemisphere does not face the Sun near perihelion, it will remain cold. Hence, water vapor and other gases will condense there. Repeated episodes of condensation should lead to a higher albedo and probably a different color than that of the northern hemisphere. (At the end of this section, I will discuss the best times to observe the southern hemisphere of the nucleus.)

The Deep Space 1 spacecraft obtained a few dozen images of the nucleus at resolutions of between 1 km/pixel and 47 m/pixel. Therefore, the sharpest images reveal features that are about 150 m or 500 ft across. Because the spin axis was pointed near our Sun at the time of encounter, the Deep Space 1 spacecraft was able to image only about half of the surface of the nucleus. Therefore, the discussion of the surface which follows is based on just the half that was imaged. High-resolution images reveal several features on the nucleus which are discussed below.

The nucleus of 19P/Borrelly has mesas, pits, ridges and hills on it. The mesas are smooth areas which rise about 100 m or 330 ft above the surface. The edge of one of the mesas is the source of a large jet centered at the north pole of the nucleus. Over a dozen pits between 200 and 300 m (660–1,000 ft) across are on the nucleus. Many of these pits are on the narrow end of the nucleus. They are probably not impact craters because of their size distribution. Essentially, these pits have nearly the same size. Impact craters, on the other hand, have a different distribution of sizes. In addition, small impact craters greatly outnumber larger ones. A second reason why these pits are probably not impact craters is that many of them lack a raised rim. These pits may have formed as a result of the sublimation of underground ice followed by collapse.

The ridge is a third feature on the nucleus. A ridge is a long and narrow elevated region. It is something like a long, raised hill. The ridges on 19P/Borrelly rise about 200 m or 660 ft above the surrounding terrain and are up to 2 km or just over a mile long. Many of the ridges are concentrated along the narrowest portion of the nucleus. Small hills, rising about 100 m above the surrounding terrain, are also present. Many of them are concentrated on the widest part of the nucleus.

There are no large and obvious impact craters in the area that the Deep Space 1 spacecraft imaged. This may be because the sublimation of ices erases most impact craters after a few dozen trips around our Sun. Since the nucleus is so small, it does not collide with as much space debris as a larger object like the Earth. Impact craters, if any, are most likely to be found in the southern hemisphere – the hemisphere not imaged by the Deep Space 1 spacecraft – because the southern hemisphere faces the Sun only when the nucleus is far away. This means that hardly any volatiles leave the nucleus in that hemisphere and, hence, small impact craters would be preserved better there.

Coma

Table 2.13 lists a few physical and photometric constants of the coma. These constants are based on measurements taken during one or more apparitions between 1981 and 2008. The coma radius near perihelion (60,000 km) is larger than that of 9P/Tempel (40,000 km), but is substantially smaller than that of 1P/Halley (350,000 km). The average degree of condensation (DC) value is 3.8, which is the same as for 9P/Tempel 1, but smaller than that of 1P/Halley.

Thanks to ground-based astronomers who collected both photometric and radio data, we have a good idea of the different chemical species in the coma. Water (H_2O), hydrogen cyanide (HCN), carbon monosulfide (CS), methanol (CH_3OH), carbon monoxide (CO) and hydrogen sulfide (H_2S) are present in the coma. The coma, however, is depleted in both the dicarbon radical (C_2) and the tricarbon radical (C_3). This should not be surprising since about half of the Jupiter family comets are depleted in C_2 and C_3.

One group of astronomers reported that the peak gas production of 19P/Borrelly was around 3.5×10^{28} molecules/s. If we assume that the coma gas is 90% water and the other gases mentioned make up the balance, the nucleus would

Table 2.13. Some physical and photometric values of the coma of Comet 19P/Borrelly

Characteristic	Value	Method
Normalized magnitude, H_{10}	7.6 ± 0.2[a]	This is based on the 1981, 1987–1988, 1994–1995, 2001 and 2008 apparitions
Normalized magnitude, H_0 (before perihelion)	6.1 ± 0.3[a]	This is based on the 1987 and 1994 apparitions
Pre-exponential factor, 2.5n (before perihelion)	22.4 ± 2.2[a]	This is based on the 1987 and 1994 apparitions
Normalized magnitude, H_0 (after perihelion)	7.2 ± 0.3[a]	This is based on the 1987 and 1994 apparitions
Pre-exponential factor, 2.5n (after perihelion)	11.7 ± 1.2[a]	This is based on the 1987 and 1994 apparitions
Coma radius near perihelion	60,000 km	[b]
Degree of condensation or DC value	3.8	[c]
Enhancement Factor at perihelion (stellar magnitudes)	7.7 ± 0.2	[d]

[a] The Author computed all of these values from an analysis of brightness measurements published in the International Astronomical Union Circulars

[b] The Author calculated this value from average coma radii reported in the 1981, 1987–1988, 1994–1995 and 2008 apparitions in *The International Comet Quarterly*

[c] The Author computed this value from data published in the *International Comet Quarterly*, based on the 1981, 1987–1988, 1994–1995 and 2008 apparitions

[d] The Author calculated this value from photometric data of the nucleus and of the comet. Photometric data of the nucleus is from Ferrín (2007). Photometric data of the comet is from the Author's calculations from data published in the International Astronomical Union Circulars and/or several issues of the *International Comet Quarterly*. All data used was collected between December 7, 1987 and January 1, 1988

produce about 1,000 kg/s or 2,200 lb of gas each second when it is near perihelion. This rate is much lower than that of 1P/Halley (30,000 kg/s) but is higher than that for 9P/Tempel 1 (~200 kg/s).

Another group of astronomers suggested that the coma may exert a pressure of 0.03 pascals, or 3×10^{-7} atmospheres, on the surface of the nucleus. The density of gas molecules near the nucleus may approach $10^{13}/cm^3$, which is about the density of the Earth's atmosphere at an altitude of 120 km.

In addition to gas, the coma contains dust. The Deep Space 1 spacecraft moved at a speed of about 16.5 km/s with respect to the comet. As a result, it collided with over a dozen dust grains as it passed through the coma. We know that 19P/Borrelly gave off much less dust than 1P/Halley. This was due probably to the smaller size of the nucleus of 19P/Borrelly and its greater distance from the Sun at perihelion. Much of the dust in the coma of 19P/Borrelly is believed to come from jets.

The most interesting jet, called the Alpha Jet, lies near the north pole of the nucleus. We know this because its orientation did not change much over a 25-h period. See drawings D – F in Fig. 2.27. As the nucleus rotates, one would expect the position angle of the Jet to change as illustrated in the top half of Fig. 2.27. This was not the case though with the Alpha Jet. Instead, this Jet remained in about the same orientation as the nucleus rotated, as shown in the bottom half of Fig. 2.27. The rotational pole does not move during rotation. Therefore, the best explanation for the fixed orientation of the Alpha Jet is that it lies at a rotational pole. It was narrow and extended for over 100 km, or 62 miles, above the nucleus. See Fig. 2.28. One group of astronomers believes that this Jet was observed as far back as 1911. Since the nucleus undergoes seasons, this Jet should be active each time the comet reaches perihelion.

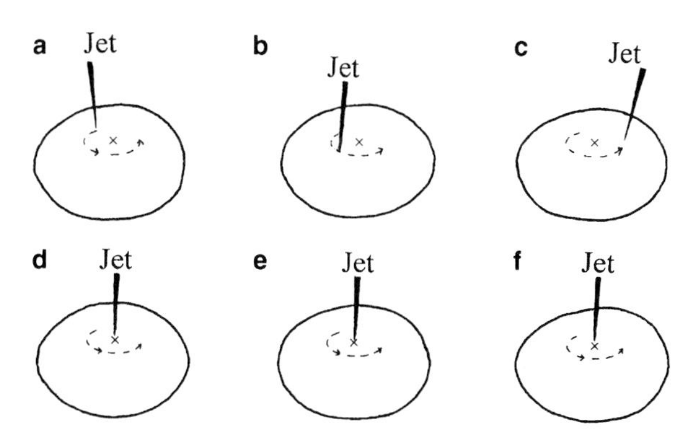

Fig. 2.27. The jet is not on the rotational pole in (**a**–**c**). Note that its orientation changes as the nucleus rotates. In (**a**), the jet tilts to the *left*, and in frame (**c**) it tilts to the *right*. If a jet is centered on the rotation pole, as is shown in (**d**–**f**), its orientation does not change as the nucleus rotates (credit: Richard W. Schmude, Jr.).

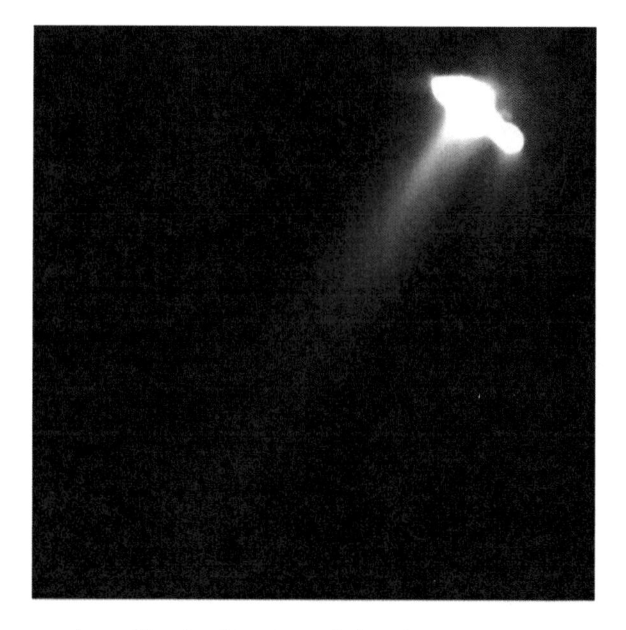

Fig. 2.28. An overexposed image of the nucleus of Comet 19P/Borrelly along with some jets. The alpha jet is the bright one. It is centered on the north pole of the nucleus and its position angle does not change as the nucleus rotates (credit: courtesy NASA/JPL-Caltech.).

One group of astronomers believes that the Alpha Jet originates from a subsurface cavity in the nucleus. See Fig. 2.29. Essentially, gas pressure builds up in the cavity and gas escapes through a crack or small hole at the surface of the nucleus. Since the pressure above the nucleus is so low, the gas escapes as a narrow fountain. Dust is entrained also in this Jet. The gas in the Jet is believed to escape at supersonic speeds. The speed of sound in the coma is probably around 400 m/s, or 900 miles/h, if the coma temperature is near room temperature. Therefore, the gas may escape at speeds above ~900 miles/h.

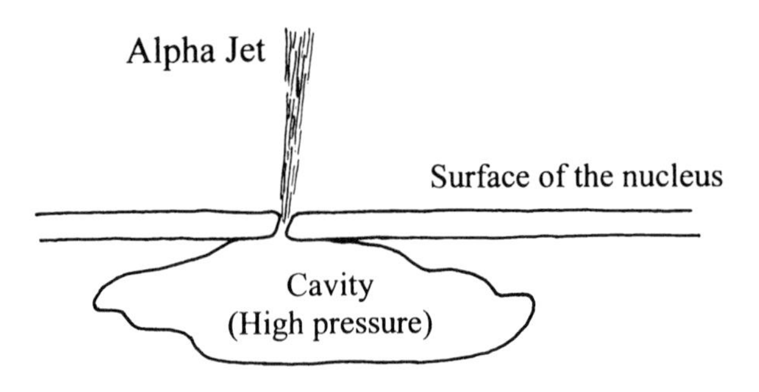

Fig. 2.29. The Alpha Jet under pressure shoots out of a small crack or hole at the surface of the nucleus. Gases in the Alpha Jet move probably at high speeds (credit: Richard W. Schmude, Jr.).

In addition to gas and dust, the coma contains charged particles. The peak density of ions in the coma exceeds 1,600 ions/cm^3. This is much lower than the number of molecules in the coma. Where do the ions come from? The most likely source is molecules on the nucleus or those in the coma which are ionized by ultraviolet light from the Sun.

Tail and Plasma Environment

Comet 19P/Borrelly does not have an impressive tail. During the early twentieth century, astronomers reported that this comet had a tail less than 1° long. During the apparition of 1994–1995, astronomers submitted over 100 estimates of the tail length to the *International Comet Quarterly*. The tail reached a maximum confirmed length of about 0.5° near perihelion in late 1994. The tail grew smaller in early 1995. It is not clear whether these lengths are for the dust or gas tail. Three excellent images of Comet 19P/Borrelly that were on the website http://www.aerith.net/comet/catalog/index-update-pictures.html show narrow tails with angular lengths of between 0.05 and 0.08°. Erik Bryssinck and Bernhard Hausler took these images between January and April of 2009. If the narrow tails in these images were gas tails, they would be around 0.009 au long. This is about a factor 30 shorter than the peak length of the tail of 1P/Halley. This result should not be surprising since 1P/Halley produces about 30 times more gas than 19P/Borrelly near perihelion.

Projections of Comet 19P/Borrelly in 2015–2026

Are you up to a challenge? If so, in 2015 Comet 19P/Borrelly may be a good object for consideration since it will be difficult to study during that year. This comet is expected to reach perihelion in 2015; however, it will also be close to the Sun at that time. The best time to image it will be before perihelion. Table 2.14 lists its predicted right ascension and declination values, along with other characteristics.

Table 2.14. Predicted positions, angular diameter values of the coma and brightness values for Comet 19P/Borrelly during early-2015. The right ascension and declination values are from the Horizons On-Line Ephemeris System. The Author computed the angular diameter of the coma and the brightness values from predicted distances and standard equations

Date (2015)	Right ascension	Declination	Angular diameter of the coma (arc-min)	Brightness (stellar magnitudes)
Jan. 15	22 h 42 min 28 s	−32° 34′ 02″	0.6	14.9
Jan. 30	23 h 15 min 22 s	−27° 57′ 41″	0.6	14.3
Feb. 14	23 h 49 min 28 s	−22° 50′ 15″	0.6	13.8
Mar. 1	00 h 24 min 43 s	−17° 12′ 48″	0.6	13.2
Mar. 16	01 h 01 min 09 s	−11° 08′ 50″	0.7	12.6

In May 2009, and in spite of the poor prospects, people were able to image its tail and coma when it was fainter than magnitude 15. The camera technology should improve between 2009 and 2015. People with charged coupled device (CCD) cameras and a moderate-sized telescope are encouraged to carry out total magnitude measurements of 19P/Borrelly in 2015. One can attain an accuracy of 1% or 0.01 stellar magnitudes with these measurements. With this accuracy, one can look for small outbursts, or monitor the brightness of the Alpha Jet.

As mentioned earlier, only one side of the nucleus faces the Sun at perihelion. Because of this, the other half remains permanently cold. Gases in the coma may condense on the cold surface, causing a change in the albedo and color of the nucleus. If there is little change in albedo, the nucleus, at aphelion, will be as bright as a 23rd magnitude star. If, however, the nucleus has an albedo of 0.5 at aphelion, it would be as bright as a 20th magnitude star. Therefore, one could measure the albedo of the southern hemisphere from brightness measurements. In addition, brightness measurements at specific times may yield information on the opposition surge.

The best chance of measuring an opposition surge for the nucleus will be in June 2012 when the nucleus will be near aphelion. According to the Horizons On-Line Ephemeris System, on June 6, 2012, at about 6:00 UT, the solar phase angle will drop to about 0.04°. At that time, the nucleus will be at a right ascension of 16 h 58 min, and at a declination of −22° 55′. It will be about 30° from a nearly full Moon. Due to the poor Moon position on June 6, a better date to measure the opposition surge would be on June 8. The nucleus will be in almost the same position as on June 6, but the Moon will be over 50° away. The solar phase angle on June 8 will be around 0.4°. The opposition surge may be measured at other times in June, 2019, and in June 2026. The solar phase angle will drop to a minimum of 0.4° and 0.7° for these times, respectively.

Part D: Comet 81P/Wild 2

Introduction

Paul Wild (Wild is pronounced Vilt) discovered a new comet in January of 1978 from photographic plates. At that time the comet was about as bright as the dwarf planet Pluto. A few days after discovery, Brian Marsden computed an orbit for this comet.

He predicted an orbital period of 6.15 years. Several people studied this comet during 1978 and reported both its brightness and position on several dates. It returned in 1983 and was given the name 81P/Wild 2. Later, calculations revealed that it made a pass of less than 0.01 au from the planet Jupiter in 1974. This changed its orbit. The perihelion distance dropped from 5 to 1.5 au. As a result, the comet developed a coma and became bright enough to be discovered with conventional photography. It has returned faithfully in 1984, 1990, 1997 and 2002. Some of its apparitions were better than others and astronomers estimated its brightness during these apparitions.

In January, 2004, The United States built Stardust spacecraft flew past its nucleus at a distance of about 240 km. It recorded several dozen images of the nucleus. It collected also thousands of microscopic dust particles. The capture of this dust was a technological feat. Since the main problem in doing so is that dust particles will vaporize upon striking a fast moving spacecraft. To circumvent this problem, scientists not only developed a trajectory which caused the probe to move at a relatively slow speed through the coma, but developed a material that captured the dust without causing it to vaporize. The Stardust spacecraft accomplished this part of the mission with its relatively slow speed of 6.1 km/s and its aerogel collectors. Aerogel is highly porous silica foam which has a density similar to that of Styrofoam. The aerogel was an ingenious development because it was able to capture the dust at a high speed without allowing it to disintegrate. In 2006, a capsule containing the dust which Stardust collected fell safely in the Utah desert.

In this Part, I describe first visual observations which include the brightness history and the light curve behavior of this comet. This is followed by a description of its nucleus, coma, dust particles and tail. Finally, I present a brief description of important events related to this comet in the second decade of the twenty-first century.

Visual Observations

Figure 2.30 shows a graph of the brightness of Comet 81P/Wild 2 against varying dates in 1996–1997. The comet reaches an average brightness of about 8th magnitude when it both closest to the Sun and Earth at nearly the same time. During early 1997, it reached a peak brightness of about 9th magnitude. At the present time, it reaches a perihelion distance of about 1.6 au.

Figure 2.31 shows a graph of the normalized magnitude, H_{10}', values of 81P/Wild 2 plotted against various dates during the five apparitions between 1978 and 2002. I used (2.1) to compute the H_{10}' values. The H_{10}' values fall generally in the range of seventh to eighth magnitude except during the apparition of 1990–1991. During that apparition, the normalized magnitudes were several tenths of a stellar magnitude fainter.

The H_{10} values of 81P/Wild 2 for 1978, 1984, 1990–1991, 1996–1997 and 2002 are 7.3, 8.1, 8.4, 7.4 and 7.9, respectively. The weighted average H_0 value (weight = the number of data points) is 7.7. These values are based on visual brightness estimates which were corrected to a standard aperture of 6.8 cm.

Figure 2.32 shows graphs plotted against time of the coma radius, DC (degree of condensation) value and the normalized magnitude, H_{10}' value. The DC value

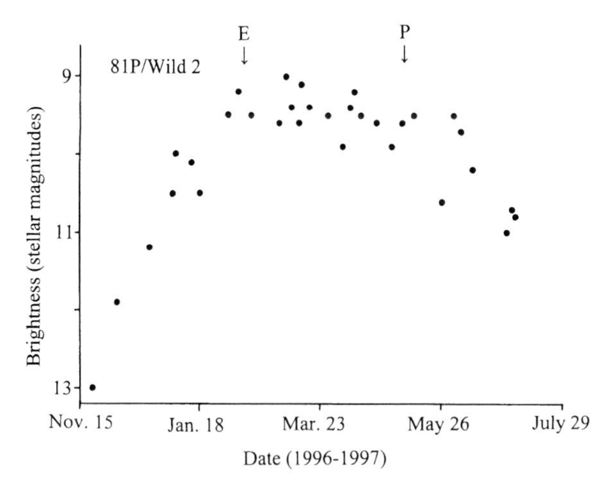

Fig. 2.30. This is a graph of the brightness of Comet 81P/Wild 2 plotted against varying dates in 1996–1997. I corrected this data to a standard aperture of 6.8 cm; the raw magnitude measurements are from the International Astronomical Union circulars. The time when the comet was closest to the Earth is shown as an E. Similarly, the time when the comet was at perihelion is shown as a P (credit: Richard W. Schmude, Jr.).

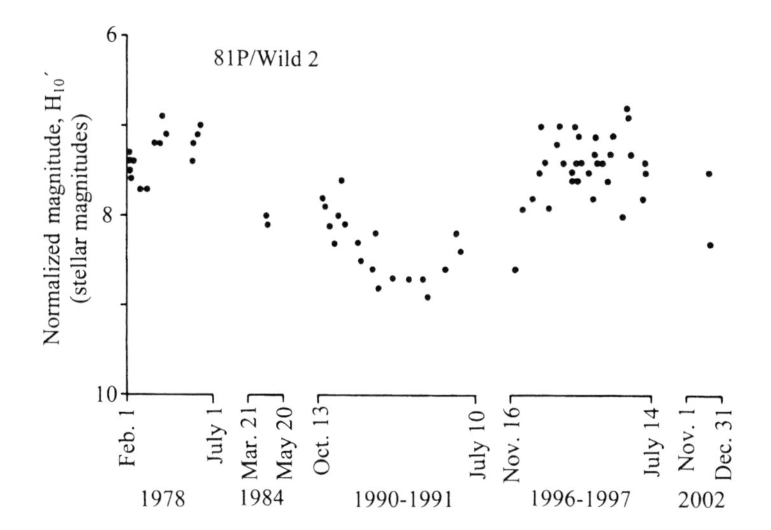

Fig. 2.31. A graph of the normalized magnitudes H_{10}' of Comet 81P/Wild 2 against varying dates for five time intervals in 1978, 1984, 1990–1991, 1996–1997 and 2002. I corrected this data to a standard aperture of 6.8 cm; the raw magnitude measurements are from the International Astronomical Union circulars (credit: Richard W. Schmude, Jr.).

describes how diffuse or condensed the coma appears, and is described further in Chap. 3. In all cases, average values for the apparition of 1990–1991 were computed for 30-day time intervals.

The biggest trend in Fig. 2.32 is the sudden drop in the size of the coma after perihelion. After perihelion, the normalized magnitude value was raised a little, which meant that the comet reflected less light during that time.

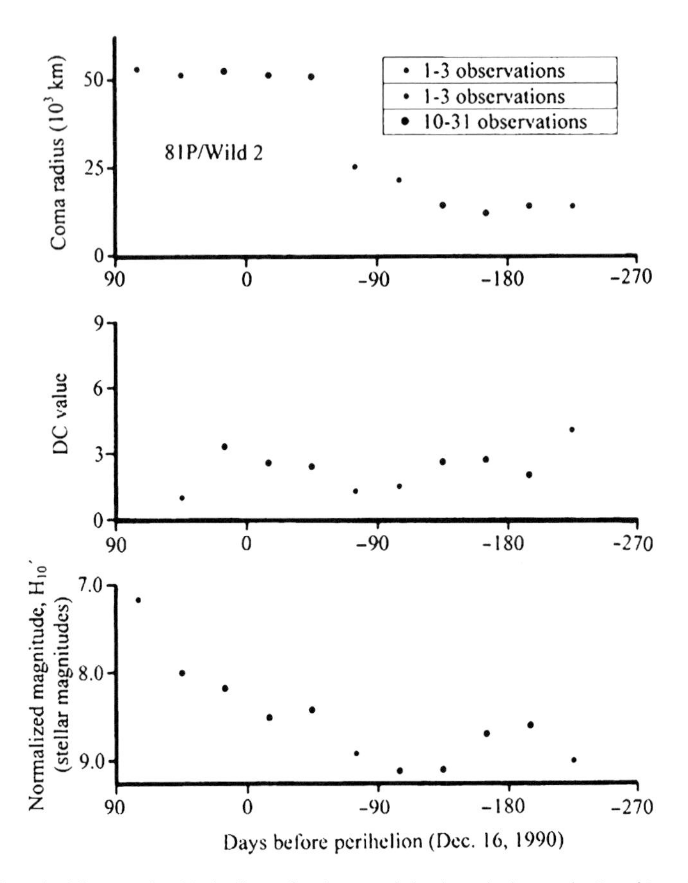

Fig. 2.32. Graphs of the coma radius, DC value (degree of condensation value) and normalized magnitudes, H_{10}', of Comet 81P/Wild 2 for the 1990–1991 apparition. Data are from the International Astronomical Union Circulars and from the *International Comet Quarterly*. The normalized magnitudes corrected to a standard aperture of 6.8 cm (credit: Richard W. Schmude, Jr.).

It was determined also that the values of the normalized magnitude H_0 and the pre-exponential factor, 2.5n, for this comet. This analysis is based on (1.6) and begins with Fig. 2.33. This is a graph of the $M_c - 5\log[\Delta]$ value against the log[r] value for the 1996–1997 apparition. Data before and after perihelion (May 6, 1997) are consistent with one another. The selected normalized magnitude, H_0, and pre-exponential factor values for this apparition are 6.5 and 13.91, respectively. The values of M_c, Δ and r are the same as in (2.1). The respective weighted average values for the normalized magnitude, H_0, and pre-exponential factor values of 81P/Wild 2, based on those determined in the 1978, 1990–1991 and 1996–1997 apparitions are 6.9 ± 0.3 and 13.5 ± 1.4. The weight for each additional value equals the change in the log[r] value. The equation that best describes the brightness, M_c, of this comet is:

$$M_c = 6.9 + 5\log[\Delta] + 13.5\log[r] \quad 1978, 1990-1991\ \textit{and}\ 1996-1997\ \textit{apparitions.} \tag{2.15}$$

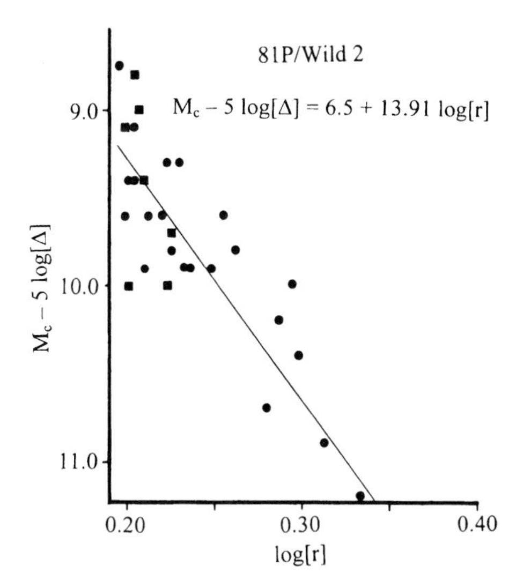

Fig. 2.33. A graph of $M_c - 5\log[\Delta]$ against log[r] for Comet 81P/Wild 2. The *filled circles* are based on brightness estimates made before this comet reached perihelion on May 6, 1997. The data was corrected to a standard aperture of 6.8 cm; the raw magnitude measurements are from the International Astronomical Union Circulars. The *squares* are based on brightness estimates made after perihelion (credit: Richard W. Schmude, Jr.).

The selected H_0 value of 81P/Wild 2 (6.9 ± 0.3) is lower than the corresponding average value of 127 numbered short-period comets (H_0 = 9.70). This means that Comet 81P/Wild 2 reflects about 13 times as much light as a typical short-period comet. The pre-exponential factor for 81P/Wild 2 (2.5n = 13.5) is close to the corresponding average value for 127 numbered short-period comets (2.5n = 13.9).

The enhancement factor for 81P/Wild 2 is 7.7 ± 0.4 stellar magnitudes. This means that the coma gets up to 1,200 times brighter than the bare nucleus near perihelion. This is nearly identical to that of 19P/Borrelly, but is less than that of 1P/Halley. This should not be surprising since 1P/Halley gets much closer to the Sun at perihelion than 81P/Wild 2.

Nucleus

Table 2.15 lists physical and photometric constants of the nucleus of Comet 81P/Wild 2. Some of these data are based on results from the Stardust spacecraft, but most of it is the result of tedious work performed at observatories around the world. The nucleus is a little smaller than that of the other three comets described in this Chapter. The mean value of its radius is 1.9 km. The nucleus may have a shape similar to that of a hamburger. Like the nuclei of the other three comets described above, the nucleus of 81P/Wild 2 reflects very little light. Consequently, it is very dark.

Figure 2.34 shows a close-up image of the nucleus. The Stardust spacecraft took this image when the nucleus was at nearly full phase. It has the shape of an ellipsoid. Figure 2.35 shows a second image taken when the nucleus was near a half

Table 2.15. Physical and photometric constants of the nucleus of Comet 81P/Wild 2

Characteristic	Value	Source
Radii	$2.6 \times 2.0 \times 1.35$ km	Howington-Kraus et al. (2005) cited in Davidsson and Gutiérrez (2006)
Radius (mean value)	1.9 km	The Author computed this value assuming a triaxial ellipsoid geometry
Density	≤ 0.8 g/cm^3	Davidsson and Gutiérrez (2006)
Surface area	46 km^2	Computed by the Author assuming a spherical shape with a radius of 1.91 km
Volume	~ 30 km^3	The Author computed the volume from the radii and assuming a tri-axial geometry
Estimated mass	1.5×10^{13} kg	Author's estimate assuming a density of 0.5 g/cm^3
Average acceleration due to gravity	~ 0.0003 m/s^2	The Author computed this value from the mean radius and the assumed mass
Rotation rate	12 or 24 h	Sekanina et al. (2004) and Davidsson and Gutiérrez (2006)
Escape speed	~ 1 m/s	The Author computed this value from the mean radius and the estimated mass
Geometric albedo (V filter)	0.04	The Author computed this value from the V(1, 0) value using (5.12) in *Uranus, Neptune and Pluto, and How to Observe Them* ©2008 by R. Schmude, Jr.
Geometric albedo (R filter)	0.08	The Author computed this value from the R(1, 0) value using (5.12) in *Uranus, Neptune and Pluto, and How to Observe Them* ©2008 by R. Schmude, Jr.
V(1, 0)	16.2	Tancredi et al. (2006)
R(1, 0)	14.9	The Author computed this value from R = 16.56 on Feb. 14, 1997, and assuming a solar phase angle coefficient of 0.046 mag./degree
Obliquity	$\sim 60°$ or $\sim 120°$	Davidsson and Gutiérrez (2006) and several others cited in this paper

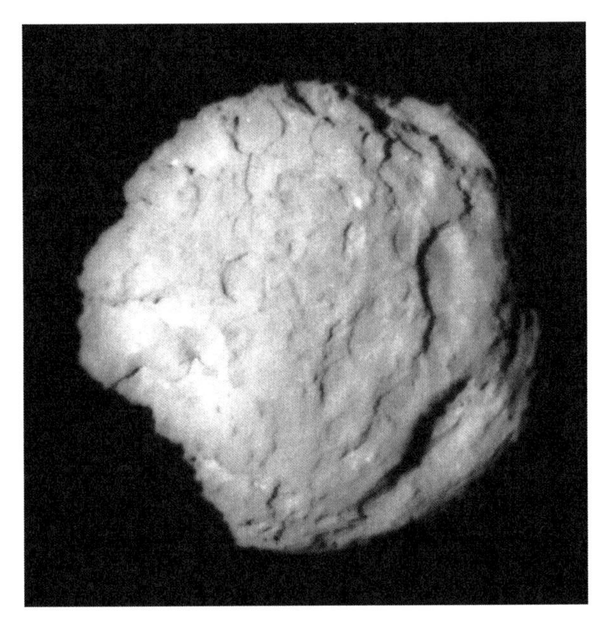

Fig. 2.34. A close-up image of the nucleus of Comet 81P/Wild 2 made by the Stardust spacecraft in 2004. The nucleus is near a full phase. Note its *roundish shape* (credit: courtesy NASA/JPL-Caltech.).

phase. In this phase, the nucleus has the shape of a potato, or of a tri-axial ellipsoid. The average radius of the nucleus is just less than 2 km.

The nature of the rotation of the nucleus is somewhat uncertain. It appears to rotate around a single axis. The general consensus is that the tilt of the axis is

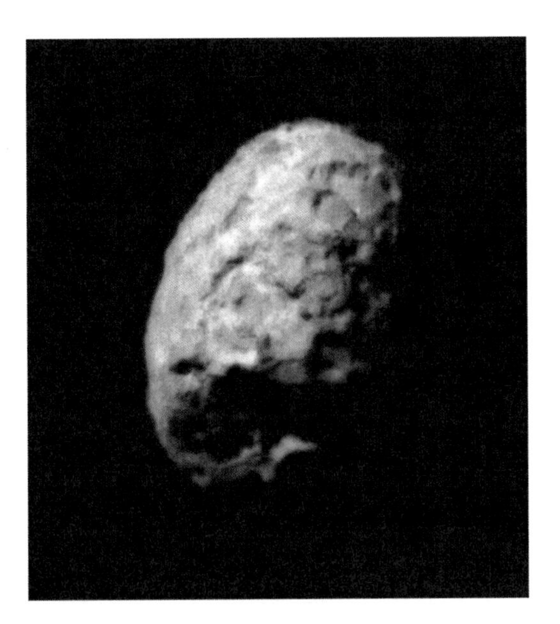

Fig. 2.35. A close-up image of the nucleus of Comet 81P/Wild 2 made by the Stardust spacecraft in 2004. The nucleus is near a half phase in this image (credit: courtesy NASA/JPL-Caltech.).

either 60° or 120° depending on the direction of rotation. The rotation period is probably between 12 and 24 h long. Future brightness measurements may reveal the rotation rate and whether the rotation axis is fixed or not.

The nucleus has several depressions in it. Some of these have a non-circular shape, like Right foot and Left foot. See Fig. 2.36. Many others have a circular shape but lack substantial raised rims. Mayo appears to be part of a large depression. See Figs. 2.34 and 2.35. The release of volatile substances is probably the mechanism that formed many of these features. See Fig. 2.36.

Coma

Table 2.16 lists photometric values and other constants of the coma. The coma has an average radius of 34,000 km which is smaller than that of both 9P/Tempel 1 and 19P/Borrelly. Comet 81P/Wild 2 also reflects less light than these comets. The respective average normalized magnitude, H_0, for these three comets, in stellar magnitudes, is 81P/Wild 2 (6.9 ± 0.3) before and after perihelion, 19P/Borrelly (6.1 ± 0.3 before perihelion and 7.2 ± 0.3 after perihelion) and 9P/Tempel 1 (5.5 ± 0.3) before and after perihelion. There may be some correlation between the coma size and the H_0 value for these comets. The average degree of condensation value for 81P/Wild 2 is 2.4, which is lower than that of 9P/Tempel 1, 19P/Borrelly and 1P/Halley. This means that its light is less concentrated than for these three comets.

Between January 18 and May 5, 1997, the solar phase angle of 81P/Wild 2 rose from 2 to 40°, but the H_{10} value barely changed. The maximum brightness drop was

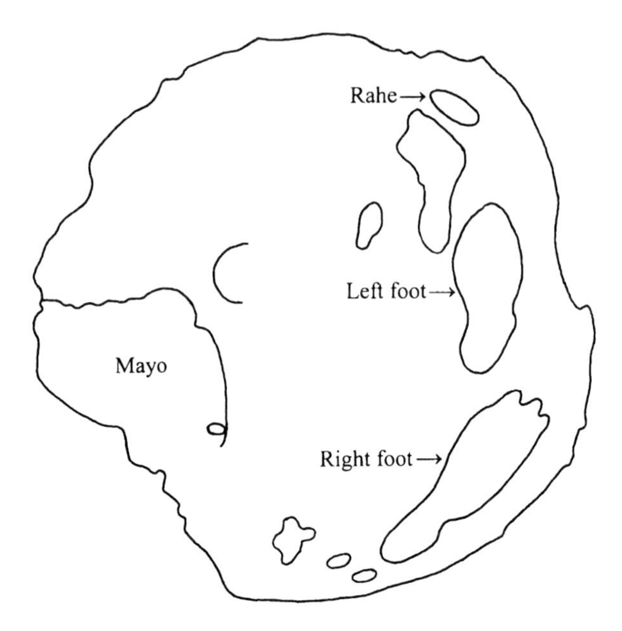

Fig. 2.36. A map of the nucleus of Comet 81P/Wild 2 made from the image in Fig. 2.34. The two depressions *left foot* and *right foot* look like two shoe prints. A third shoe print seems to lie to the *upper left* of *left foot* (credit: Richard W. Schmude, Jr.).

Table 2.16. Photometric values of the coma of Comet 81P/Wild 2

Characteristic	Value	Method
Normalized magnitude, H_{10}	7.7 ± 0.2[a]	This is based on the five apparitions between 1978 and 2003
Normalized magnitude, H_0 (before and after perihelion)	6.9 ± 0.3[a]	This is based on the 1978, 1991 and 1997 apparitions
Pre-exponential factor, 2.5n (before and after perihelion)	13.5 ± 1.4[a]	This is based on the 1978, 1991 and 1997 apparitions
Solar phase angle coefficient	≤0.01 mag./degree[a]	This is based on data from Jan. 18, 1997 to May 5, 1997
Coma radius	34,000 km	[b]
Degree of condensation or DC value	2.4	[c]
Enhancement Factor at perihelion (stellar magnitudes)	7.7 ± 0.4	[d]

[a]The Author computed all of these values from an analysis of brightness measurements published in the International Astronomical Union Circulars for the apparitions indicated in the far right column

[b]The Author computed this value from the average coma radii reported in the 1990–1991 and 1997 apparitions in *The International Comet Quarterly*

[c]The Author computed this value from data published in the *International Comet Quarterly*, based on the 1990–1991 apparition

[d]The Author computed this value from photometric data of the nucleus and of the comet. Photometric data of the nucleus is from Ferrín (2007). Photometric data of the comet is from the Author's calculations from data published in the International Astronomical Union Circulars and/or several issues of the *International Comet Quarterly*. All data used is within 15 days of perihelion for the 1978, 1990 and 1997 apparitions

0.4 stellar magnitudes for this time interval. The upper limit of the solar phase angle coefficient would be 0.4 magnitudes/(40° – 2°) = 0.01 magnitudes/degree.

The species H_2O, OH, NH, NH_2, CN, C_2 and C_3 are in the coma. The peak production rate of water is about 2×10^{28} molecules/s or 600 kg/s near perihelion. One group of astronomers reported that the production rates of NH_2 and CN were around 0.003 in comparison to that of water. This same group, however, reported a production rate of less than 0.001 for C_2 compared to that of water.

This comet has several active jets near perihelion. One jet is believed to start near a rotational pole because it does not shift much as the nucleus rotates. Jets are the main source of dust in the coma. In fact, one group of astronomers reported that dust in the jets and coma have a similar color. This is evidence that the mean particle size of the dust is similar in jets and in the coma.

Thanks to the success of the Stardust spacecraft, we have a sample of its coma dust here on Earth. As of mid-2009, scientists are continuing to analyze this dust. They have also published some results, and their findings are summarized below.

In 2004, the Stardust spacecraft collected thousands of dust particles from the coma of 81P/Wild 2. One group of astronomers estimated that the total mass of dust collected was about 3×10^{-4} g, which is the mass of about seven grains of granulated sugar. Most of the dust particles are imbedded in aerogel, but some are attached to the aluminum casing which surrounds the aerogel. Scientists have used a variety of instruments to study this dust. Some of these include electron microscopes, mass spectrometers, lasers, infrared spectrometers, Raman spectrometers and equipment that incorporates chromatography.

What are the physical characteristics of the dust? The dust particles are between 0.1 and about 300 μm across. For a comparison, a human red blood cell is about 7.5 μm across and a grain of granulated sugar is about 350 μm across. The dust particles have masses that range from less than 10^{-15} g up to about 10^{-5} g. A human red blood cell has a mass of about 10^{-10} g. Many of the dust particles are fluffy, while others are dense and cohesive. According to one study, the size distribution for particles having masses between 10^{-15} and 10^{-6} g is proportional to $D_p^{-1.72}$, where D_p is the diameter of the dust grain. Essentially, this means that for every 16-μm particle there are about 11 four-micrometer particles and about 120 one-micrometer particles; and small dust grains outnumber larger ones.

The dust grains have an interesting chemical make-up. Many of them formed in a high-temperature environment. This is surprising because comet 81P/Wild 2 is believed to have formed in the cold, outer Solar System. One group of astronomers suggested that 10% of the comet material may have formed in the inner solar nebula. The dust grains are mixtures of different solid compounds. The dust grains lack volatile substances, like water, but have plenty of silicon and carbon compounds. Olivine (a mixture of $MgSiO_4$ and $FeSiO_4$), pyroxene (a mixture of iron and magnesium silicates), troilite (FeS), forsterite (Mg_2SiO_4) and enstatite ($MgSiO_3$) are present in the grains. Some carbon compounds which are in the dust include napthalene ($C_{10}H_8$), phenanthrene ($C_{14}H_{10}$) and pyrene ($C_{16}H_{10}$). Other carbon compounds in the dust include other aromatic hydrocarbons, aliphatic hydrocarbons and compounds containing the carbonyl group and amide functional groups. The structures of naphthalene ($C_{10}H_8$) and pyrene ($C_{16}H_{10}$) are shown in Fig. 2.37. The alcohol or ether functional group is also present. An aromatic compound is

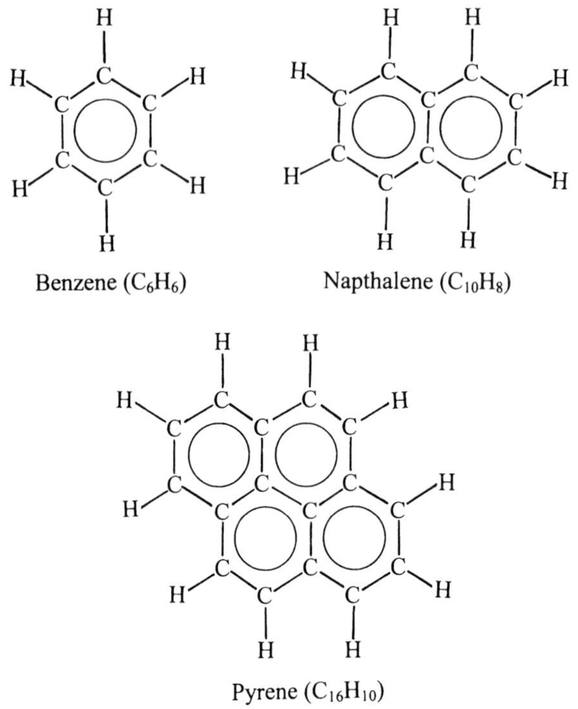

Benzene (C_6H_6)

Napthalene ($C_{10}H_8$)

Pyrene ($C_{16}H_{10}$)

Fig. 2.37. A drawing of the chemical structures of benzene, napthalene and pyrene. Napthalene and pyrene exist in the dust of 81P/Wild 2. Benzene contains six carbon and six hydrogen atoms bonded together. The *circle* in the middle of the hexagonal carbon ring represents three pairs of electrons which serve as extra bonding between the carbon atoms. Aromatic compounds have the benzene ring as part of their structure, whereas aliphatic compounds lack the benzene ring in their structure (credit: Richard W. Schmude, Jr.).

one that contains the benzene ring. Benzene (C_6H_6) is shown also in Fig. 2.37. An aliphatic compound contains carbon and hydrogen but does not contain the benzene ring.

The elements that are present in the dust are hydrogen, carbon, nitrogen, oxygen, magnesium, silicon, calcium, chromium, manganese, iron, nickel, copper, zinc, gallium, germanium and selenium. The elemental composition is similar to that of CI meteorites. With further analysis, scientists may detect additional elements.

Tail

Like Comet 19P/Borrelly, Comet 81P/Wild 2 has a faint tail. The Czech astronomers M. Tichy and Z. Moravec imaged the coma and tail on Jan. 15, 1997. This image appears on Gary Kronk's excellent website (http://cometography.com/pcomets/081P.html). A narrow tail, which I assume is the gas tail, extends about 0.9 arc-min from the coma. Using the procedure described in Chap. 4, I determined that this tail is 0.005 au long. This is similar to the length of the tail of 19P/Borrelly, but is much shorter than that of 1P/Halley.

Important Events Projected in 2012–2017

Comet 81P/Wild 2 will reach perihelion in July 2016. The best time to study this comet will be in the months of May and June 2016, when it is at least 40° from the Sun. Table 2.17 lists predicted right ascension and declination values, along with the predicted coma size and brightness values. The coma will appear to get smaller after March 1 because it will be moving farther from the Earth.

The best time to measure the opposition surge of the bare nucleus is when the solar phase angle drops to nearly 0°. This will happen on September 7, 2012, and July 21, 2017. On these dates, the solar phase angle of the nucleus will drop to respective values of 0.4 and 0.5°. On both dates, the nucleus will attain a brightness of about 22 to 23 magnitude. The Moon will have a waning crescent phase on September 7, 2012, and a waxing crescent phase on July 21, 2017.

Since the orbital inclination of 81P/Wild 2 is only 3.2°, we are always near its orbital plane. Therefore, one should search their images for a faint antitail. We will cross that comet's orbital plane on August 8, 2016.

Table 2.17. Predicted positions, coma sizes and brightness values for Comet 81P/Wild 2. The right ascension and declination values are from the Horizons On-Line Ephemeris System. The Author computed the coma diameter and brightness values from predicted distances and standard equations

Date (2016)	Right ascension	Declination	Angular diameter of the coma (arc-min)	Brightness (stellar magnitudes)
Mar. 1	5 h 12 min 07 s	21° 02′ 04″	1.1	12.3
Mar. 16	5 h 28 min 31 s	21° 47′ 02″	1.0	12.2
Mar. 31	5 h 51 min 18 s	22° 24′ 55″	1.0	12.1
Apr.15	6 h 19 min 28 s	22° 46′ 08″	0.9	11.9
Apr. 30	6 h 52 min 08 s	22° 41′ 29″	0.9	11.8
May 15	7 h 28 min 24 s	22° 02′ 51″	0.9	11.7
May 30	8 h 07 min 23 s	20° 44′ 17″	0.9	11.5
June 14	8 h 48 min 10 s	18° 42′ 47″	0.9	11.4
June 29	9 h 29 min 56 s	15° 59′ 09″	0.8	11.4
July 14	10 h 12 min 02 s	12° 37′ 58″	0.8	11.4
July 29	10 h 53 min 58 s	08° 47′ 26″	0.8	11.5
Aug. 13	11 h 35 min 27 s	04° 38′ 02″	0.8	11.6

Chapter 3

Observing Comets with the Unaided Eye and Binoculars

Introduction

The most spectacular comets are those which are visible to the unaided eye. The purpose of this Chapter is to describe the types of comet observations which one can do with the unaided eye or with binoculars. First, a description of the human eye and its capabilities is presented. This is followed by an overview of binoculars. Finally, I discuss techniques for observing comets and recording information, such as coma brightness, the degree of condensation (DC) and estimating atmospheric transparency.

The Human Eye

The human eye is a remarkable light detector. It can pick out detail on a full Moon and, at the same time, detect a nearby star that is over a million times dimmer. Photographs are unable to duplicate this feat. They will show either the Moon over-exposed or the star under-exposed. Figure 3.1 shows a diagram of the eye and some of its major parts. A discussion of the lens, pupil, retina and fovea centralis follows.

Light must travel through both the cornea and lens. Both of these help focus light onto the fovea centralis, which is the most sensitive area on the retina. Without the fovea centralis, one would not be able to read or focus on an object. The eye focuses automatically by changing the shape of the lens. This is different from a telescope where one changes manually the distance between lenses to attain a good focus.

The iris is the colored part of the eye and it controls the size of the pupil. In low and moderate light levels, the pupil appears black. The size of the pupil determines the amount of light striking the retina. When there is lots of light, the pupil gets small automatically (perhaps only 2 mm in diameter). The pupil grows in

R. Schmude, *Comets and How to Observe Them*, Astronomers' Observing Guides,
DOI 10.1007/978-1-4419-5790-0_3, © Springer Science+Business Media, LLC 2010

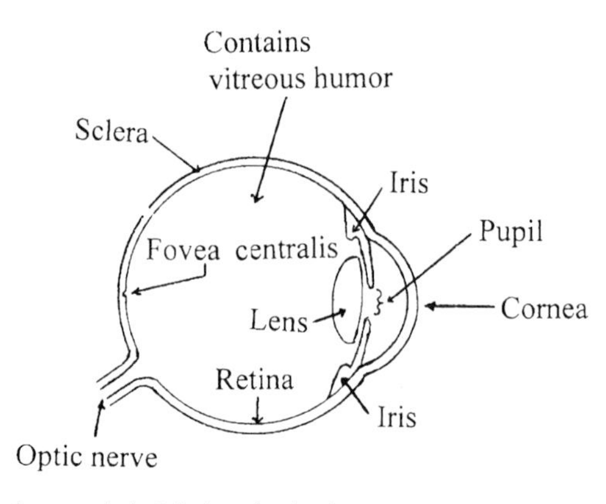

Fig. 3.1. A diagram of the human eye (credit: Richard W. Schmude, Jr.).

size, however, when the light level drops. When there is darkness for at least 10 min, the pupil reaches its maximum size. I will call this maximum size the "dark-adapted pupil size". As a general rule, the dark-adapted pupil size gets smaller with age. Based on a study of over 1,200 people, I. E. Loewenfeld found that the average dark-adapted pupil size for different ages (in parentheses) is 7.0 mm (13), 6.0 mm (35), 5.0 mm (57) and 4.0 mm (80). For a given age, though, there is a range of values. For example, people in their mid-40s have dark-adapted pupil sizes as small as 3.0 mm and as large as 7.0 mm.

The retina is the light-sensitive part of the eye. It contains two different light sensors which are called rods and cones. The retina includes the fovea centralis which is a small spot on the retina which has a concentration of cones. Rods are sensitive to low levels of light, but are not sensitive to color. Cones, on the other hand, require larger amounts of light to function. Unlike rods, they are sensitive to color. There are three types of cones namely, S, M and L, with peak sensitivities at wavelengths of 430, 530 and 560 nm. The S cones are sensitive to violet, blue and green light. The M cones are sensitive to blue, green, yellow and orange light, while the L cones are sensitive to green, yellow, orange and red light. As indicated, there is some overlap in color sensitivity. The signals from the S, M and L cones are used by the brain to distinguish color. For example, if the S, M and L cones observe light at 0, 25 and 75% of their maximum response, the brain may register a yellow color, whereas if the S, M and L cones observe light at 0, 20 and 80% the brain may register a yellow-orange color. As a result, we can recognize numerous different colors or shades of colors.

When focusing light on the fovea centralis, you obtain high-resolution color vision if there is enough light. In many cases, an object, like the gas tail of a comet, will be too dim to produce enough light for color vision. Rods are abundant on the retina and are more sensitive to low levels of light. When using inverted vision, you are focusing on one spot, but is examining light from another area. In this case, light is falling on the rods. You are able to see fainter objects with inverted vision, than when focusing on them.

Color

There are three dimensions of color, namely, hue, saturation and intensity. All of these dimensions play a role in the perceived color of an object. A discussion of these dimensions follows.

Hue refers to the dominant, or perceived, wavelength of light. Table 3.1 lists the wavelength range of different colors as seen by a calibrated spectroscope. If, for example, an object reflects light with wavelengths between 520 and 530 nm, it will appear green. In most cases, an object like a comet gives off many wavelengths of visible light. As a result, the perceived hue will be different than of the pure colors listed in Table 3.1.

In many cases, an object will reflect more of one color of light than the other and as a result, one may note only the dominant color. An example of this is green grass. These plants reflect all colors of light, but reflect more green than other colors. As a result, our eyes show shades of green when we look at grass. Light from comets behave in a similar way. The dust tail, for example, reflects all colors of visible light. It will often appear to have a yellow cast because this is the perceived color of the Sun.

In some cases though, an object will appear to have one color, but will give off other colors of light. Essentially the eye perceives a color which is the result of the color wavelengths given off. This happens every time we watch a color television picture. The yellows that we see are really some combination of blue, green and red. This can happen for the gas tail of a comet.

The second dimension of color is saturation. This is the degree to which color departs from white and approaches a pure color. Colors with a low saturation are nearly white, whereas those with a high saturation are not white. Red laser light is an example of light with a high saturation. This is because red laser light contains only red light and, hence, it departs from white light. A Fluorescent lamp, on the other hand, has a low saturation. This is because the perceived color, yellow-white, is close to white and, hence, deviates far from the true colors (red, green and violet) that it emits.

The third dimension of color, intensity, describes the amount of light reaching the eye. Light with a low intensity, regardless of the hue and saturation, will have no perceived color to our eyes. This is because light must reach a threshold before

Table 3.1. Light wavelengths for different colors based on estimates made by the Author with a calibrated spectroscope

Color	Wavelength range (nm)
Red	640–710
Orange	595–640
Yellow	580–595
Green	505–580
Blue	450–505
Violet	410–450

the color-sensitive cones in our eyes detect it. On the other hand, if too much light hits the retina, the light will appear white regardless of the hue or saturation. For example, the full Moon near zenith and the planet Saturn both have about the same hue and saturation, but when we look at them, Saturn will have a yellowish color whereas the full Moon will appear white. The full Moon appears white because it is so bright relative to Saturn, and thus too much light from it hits the retina. Most comets are dim and, as a result, our eyes may not detect color. If a comet reaches a brightness equivalent to a magnitude 1.0 star, one may be able to detect colors.

Binoculars

What instrument can be used to study a bright comet? Binoculars! I have used binoculars to study variable stars, Uranus, Neptune and comets for many years. The three advantages of binoculars are that they are portable, they allow the observer to use both of his/her eyes, and they allow one to find objects quickly. I almost always use binoculars to find a comet or other faint target before attempting to find it with a telescope. In some cases, binoculars will give a better view than the telescope. On Nov. 15, 2007, I was able to see Comet 17P/Holmes much better with my 15 × 70 binoculars than through my eight-inch telescope at 50×.

There are four common types of binoculars namely, opera glasses, porro prism, roof prism and image stabilized. Opera glasses have low magnifications (usually 2× to 5×) and have small objective lens sizes. One popular size is 3 × 25. They lack prisms and are primarily used for daytime viewing.

Porro prism and roof prism binoculars both contain prisms. The prisms invert the image upright and they make the binoculars more compact. Porro prism binoculars are generally less expensive than a pair of similar quality roof prism binoculars; however, they weigh more than a comparable pair of roof prism binoculars. Roof prism binoculars are shown in Fig. 3.2, and porro prism binoculars are shown in Fig. 3.3. The eyepieces are directly in front of the objective lenses in roof prism binoculars. This is not the case for porro prism binoculars.

Image stabilized binoculars contain both prisms and a mechanism that compensates for small movements. The image stabilized binoculars in mid-2008 weighed between 1 and 3 lb. This weight is comparable to that of my 10 × 50 binoculars. Almost all of the image stabilized binoculars require batteries. I have used this type of binoculars at night, and the view was comparable to porro prism binoculars with lenses of almost twice the diameter.

There are several binocular characteristics which one should be aware of before making a purchase. These include magnification, objective lens size, field-of-view, eye-relief, exit pupil, visibility factor, maximum size of hand-held binoculars, coatings and vignetting. Each of these characteristics is described below.

Fig. 3.2. Roof prism binoculars with a magnification of 12 and objective lenses with diameters of 25 mm (credit: Richard W. Schmude, Jr.).

Fig. 3.3. Porro prism binoculars. Note that the eyepieces are not directly behind the larger objective lenses (credit: Richard W. Schmude, Jr.).

Magnification and Objective Lens Size

The magnification and objective lens size are almost always stated as two numbers separated by a ×. The binoculars in Fig. 3.2 have 12×25 written on them (pronounced 12×25), which means that they magnify 12 times, and that the objective lenses are each 25 mm in diameter. Essentially, these binoculars make objects appear 12 times closer. Larger objective lenses allow more light to pass through which means that they make faint objects appear brighter. On the down side, larger lenses usually mean more weight. Thus there is a trade-off between brightness and weight.

Field-of-View

The field-of-view (FOV) is the size of the area that the observer sees through binoculars. Unaided eyes have an FOV of around 170°. Binoculars on the other hand, usually have FOV values of between 3° and 8°. The FOV depends on the magnification and eyepiece design. Generally, the higher the magnification, the smaller will be the FOV. Those who are starting out should get binoculars with an FOV of at least 5°. The more seasoned observer will do better with a smaller FOV and a higher magnification.

The FOV is often expressed in either degrees, feet per thousand yards or meters per thousand meters. One can use the equations below to convert to degrees.

$$\text{Feet / 1000 yards : number of feet / 52.3} = \text{FOV in degrees} \tag{3.1}$$

$$\text{Meters / 1000 meters : number of meters / 17.4} = \text{FOV in degrees} \tag{3.2}$$

For example, the binoculars in Fig. 3.2 have an FOV of 240 ft per 1,000 yards. To convert this to degrees, one would apply (3.1): $240 \div 52.3 = 4.6°$.

How accurate are the manufacturer's stated FOV values? I have tested four of my binoculars and came up with the following results: 11 × 80 (stated FOV = 4.5°, measured FOV = 5.0°±0.4°); 15 × 70 (stated FOV = 4.4°, measured FOV = 3.8 ± 0.3°); 10 × 70 (stated FOV = 5°, measured FOV = 4.5° ± 0.4°); and 8 × 21 (stated FOV = 7.1°, measured FOV = 7.3°±0.5°). Therefore, the manufacturer's stated FOV values are fairly accurate. Nevertheless, one should check the FOV of their binoculars before using them for comet studies.

How can one check the FOV of their binoculars? The best way to do this is to focus them on an object that is about 15 m or 50 ft away. Have a friend stand next to the area that you have focused on, and instruct this person to measure the diameter of the area bounded by the FOV. After this, measure the distance between the binoculars and the area on which you had focused, and divide the distance by the length of the FOV. Use Fig. 3.4 to read off the FOV. For example, I focused a pair of 10 × 70 binoculars on a railing that was 16.5 m away. The binocular FOV showed 1.307 m of the railing. The quotient of 16.5 m and 1.307 m is 12.6. This is consistent with a FOV of 4.5°.

With the use of trigonometry one may use also (3.3) to compute the FOV in degrees:

$$\text{FOV} = \text{inverse sin}[L / D]. \tag{3.3}$$

In this equation, L is the length of the FOV, D is the distance and inverse sin is the inverse sine function. The inverse sine function is often written as $\sin^{-1}$ on calculators. In the previous example, L = 1.307 m, D = 16.5 m and the FOV is 4.54°.

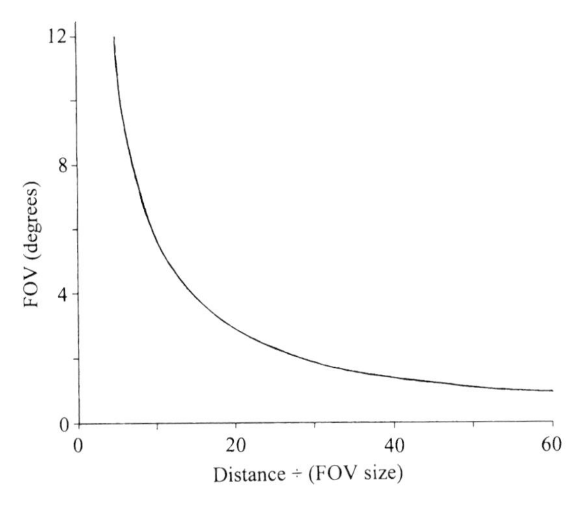

Fig. 3.4. Relationship between field-of-view (FOV) in degrees and the distance divided by the length of the FOV. For example, I focused on a railing that was 16.5 m away, and measured the length of the railing which was just inside of my FOV as 1.307 m. This distance divided by the FOV was 12.6, and this yielded an FOV of 4.5° (credit: Richard W. Schmude, Jr.).

A third way of checking the binocular FOV is to use known star separations. Figure 3.5 shows stars in three different constellations. Table 3.2 lists the distances between star pairs in Fig. 3.5. The procedure for determining the FOV of your binoculars is illustrated in the flow chart below.

Step 1: Find the widest star pair that *fits* in your binocular FOV. Use Table 3.2

↓

Step 2: Find the closest star pair that does *not* fit in your binocular FOV. Use Table 3.2

↓

Step 3: Compute the initial binocular FOV as the *average* of the separation of the star pairs in steps 1 and 2

Step 4: Compute the *uncertainty* which is ± half of the difference in the separations in the star pairs in Steps 1 and 2. The result is the binocular FOV

One should confirm the accuracy of the stated FOV of their binoculars before estimating the length of cometary features in terms of their binocular FOV. In this way, there will be more consistency between different observers.

Eye-Relief

The eye-relief is the distance between the eye and the binocular eyepiece at which one can see the entire binocular FOV. The longer the eye-relief, the farther away one can hold his/her eyes from the binocular eyepiece and still see the entire FOV. There are two advantages of having an eye-relief of at least 20 mm (0.8 in.). The first is that people with glasses can see the entire binocular FOV. I do not have to

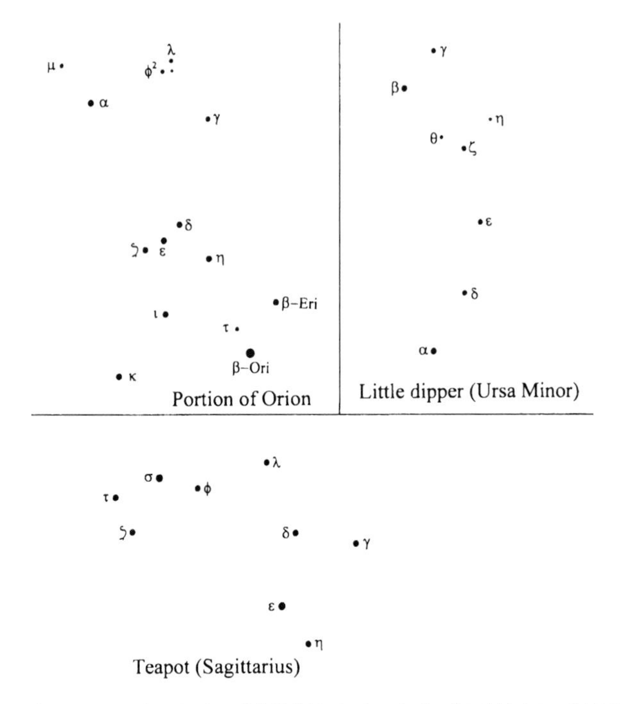

Fig. 3.5. One uses these star patterns in conjunction with Table 3.2 in the determination of the field-of-view of their binoculars. Please note that the Little dipper is located in Ursa Minor and that the teapot is located in the constellation of Sagittarius (credit: Richard W. Schmude, Jr.).

Table 3.2. Distances between different stars in three different constellations; the Author computed all distances listed in this table

Portion of Orion		Little dipper (Ursa Minor)		Teapot (Sagittarius)	
Star combination	Separation (°)	Star combination	Separation (°)	Star combination	Separation (°)
δ and ζ	2.74	β and γ	3.20	σ and φ	2.26
μ and α	2.86	α and δ	3.97	τ and ζ	2.41
η and ε	3.16	δ and ε	4.67	σ and τ	2.94
η and ζ	4.09	ε and ζ	4.77	δ and γ	3.34
ι and τ	4.53	ζ and β	5.30	σ and ζ	3.93
ι and ε	4.71	η and γ	5.55	φ and λ	4.26
φ^2 and α	4.89	ε and θ	5.70	δ and ε	4.60
ι and κ	5.29	β and η	5.82	τ and φ	4.78
ι and β-Ori	5.67	γ and ζ	6.36	γ and ε	5.54
η and β-Ori	6.31	ε and α	8.59	φ and δ	6.12
ι and β-Eri	6.92	ε and β	9.64	δ and η	6.97
τ and ε	7.30			σ and δ	8.34
δ and β-Eri	7.70				
ε and β-Eri	8.07				
ζ and β-Eri	8.80				

remove my eyeglasses when I use my 15×70 binoculars because of its long eye-relief. I must, however, remove my glasses to look through my small roof prism binoculars which have a short eye-relief. A second advantage of long eye-relief is that one is less likely to have their eyelashes come in contact with the eyepieces. Eyelashes can deposit material on the eyepiece coatings which may cause some imperfection or damage.

Exit Pupil

The exit pupil is the area at the eyepiece in which all of the light passes through. If one wants all of the light from the binoculars to fall on the retina, their pupil must be at least as large as the exit pupil. If this is not the case, some of the light from the binoculars will not enter the pupil. This is fine if one is observing during the daytime or is observing a bright object at night. If, on the other hand, one is observing a faint object at night and wants as much light as possible, the dark-adapted pupil size should be at least as large as the exit pupil. As mentioned earlier, a person's dark-adapted pupil size generally gets smaller with age. One calculates the exit pupil of binoculars by dividing the objective lens diameter by the magnification. For example, the exit pupil for a pair of 7×35 binoculars is 35 mm ÷ 7 = 5 mm. One can look at the exit pupil of binoculars by holing them 0.6 m (2 ft away) and aiming them at the daylight sky (but not towards or at the Sun). The exit pupils are the bright discs in the eyepieces.

Visibility Factor

Many people, including myself, believe that the visibility factor is a good way of rating binoculars. The visibility factor is the product of the magnification and the objective lens diameter in millimeters. For example, the visibility factor for 10×50 binoculars is 500. As a general rule, the higher the visibility factor, the better the binocular performance. Keep in mind that this factor does not consider optical coatings, the weight of the binoculars or their overall quality.

Maximum Size of Hand-Held Binoculars

What is the maximum size of binoculars that can be hand-held? This depends on several factors including binocular weight, personal strength and the frequency of short breaks between observing sessions. Heavier binoculars are harder to hold than lighter ones. In 1995, I purchased a pair of 11×80 binoculars. These binoculars weighed almost 5 lb. Later, I purchased a pair of 15×70 binoculars which only weighed 3 lb. Manufacturers are aware of the problem of weight and have made strides in lowering the weight of large binoculars.

One way to reduce arm strain is to hold the binoculars similar to what is shown in Fig. 3.6. In this case, the binoculars are held near the center of mass and the elbows are rested on a portion of a car. I have used this posture to make hundreds

Fig. 3.6. One way of holding large binoculars is shown here. The observer is holding the binoculars near their center of mass, and has his elbows resting on a portion of a car (credit: Timothy E. Abbott and Richard W. Schmude, Jr.).

of comet observations and to make tens of thousands of variable star brightness estimates.

A second factor for consideration is a person's strength. There is not much more to say other than to test out a pair. How do your arms hold out?

A final consideration should be the frequency of short breaks between observing sessions. If binoculars are to be held for 20 min straight without a short break, one will develop muscle strain with even a medium-sized pair. If, however, frequent breaks are taken, most people should be able to handle the three-pound 15 × 70 binoculars.

Coatings

Most modern binoculars have thin antireflective coatings on the optical surfaces. The purpose of the coatings is to increase the amount of light traveling through the binoculars and hence make objects brighter. An uncoated lens may reflect away 4% of the light falling on it whereas a lens with an optical coating may reflect away only 1% of the light falling on it. In other words, the coated lens allows more light to pass through. This is illustrated in Figs. 3.7 and 3.8. Note that the uncoated lens reflects more light (Fig. 3.7) than the coated lens (Fig. 3.8). Hence, optical coatings allow one to see fainter objects in the night sky.

There are at least four terms that one should be familiar with, namely, coated, fully coated, multi-coated and fully multi-coated. Binoculars with *coated* optics have at least one optical surface that is coated. All lens surfaces have a layer of antireflective coating in *fully-coated* binoculars. Binoculars that are *multicoated*

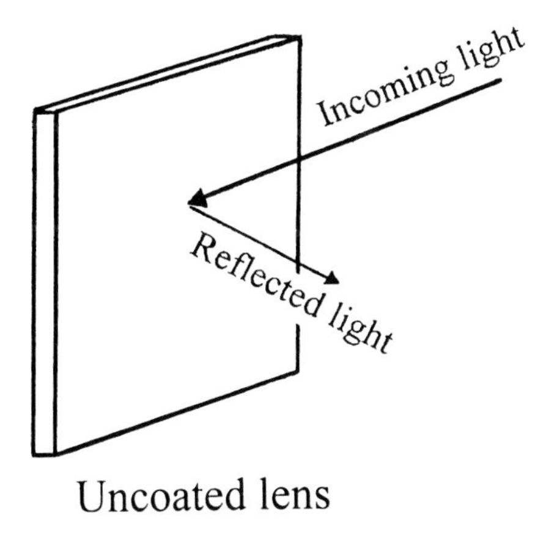

Fig. 3.7. An uncoated lens will reflect some light away and, as a result, less light travels through the lens. This, in turn, will cause objects to appear dimmer than through coated lenses (credit: Richard W. Schmude, Jr.).

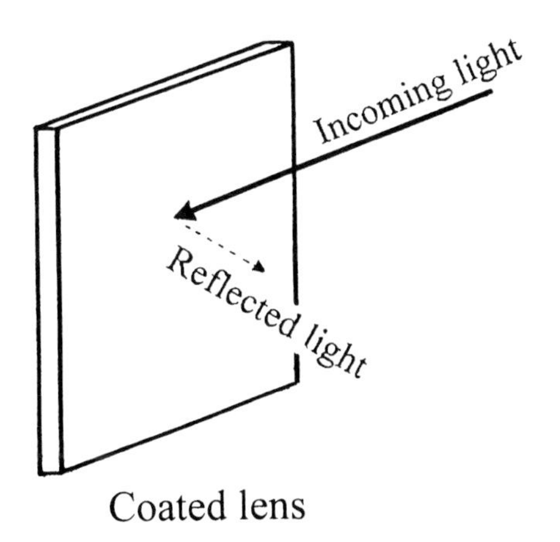

Fig. 3.8. A coated lens reflects less light away than an uncoated lens, resulting in more light traveling through it. The net effect is that dim objects appear brighter through coated lenses than through uncoated lenses (credit: Richard W. Schmude, Jr.).

have at least two different coatings on at least one lens surface. The best situation is *fully multi-coated* – where all lens surfaces have at least two different layers of antireflective coatings.

Vignetting

One limitation of binoculars is vignetting. An object near the edge of the FOV will appear dimmer through binoculars with vignetting. One can test for this by focusing on a star and then moving the binoculars so that the star is at the center and then at the edge of the FOV. In many cases, the star will appear brighter when it is centered. The difference in one of my pairs is very noticeable. One may also use the star pairs in Table 3.3 to test for vignetting. Each pair has stars of nearly the same brightness. One focuses the binoculars so that one star is in the center of the FOV and the other one is near the edge. See Fig. 3.9. The star near the edge may appear dimmer than the one in the center. If vignetting is significant, one should be consistent in placing objects in the center of the FOV when making brightness and tail-length estimates of comets. I believe that vignetting is not a serious problem as long as one centers objects in the FOV.

Table 3.3. Star pairs to test for vignetting in binoculars

Star pair and constellation	Hemisphere	Separation (°)	Approximate right ascension	Approximate declination
ε-Ori and ζ-Ori	N or S	1.4	5 h 38 min	2°S
φ-Psc and υ-Psc	N	3.0	1 h 17 min	26°N
κ-Cyg and ι-Cyg	N	2.5	19 h 24 min	53°N
HD216446 and HD217382	N	1.2	22 h 51 min	84°N
ι-Oct and κ-Oct	S	1.1	13 h 18 min	85°S
HD74772 and HD73634	S	1.3	8 h 41 min	43°S
υ-Lib and τ-Lib	S	1.7	15 h 38 min	29°S

Each pair contains stars of nearly the same brightness. The Author computed all star separation values in this table. The right ascension and declination values are from *Sky Catalogue 2000.0*, Volume 1, 2nd edition ©1991 by Alan Hirshfeld, Roger W. Sinnott and François Ochsenbein

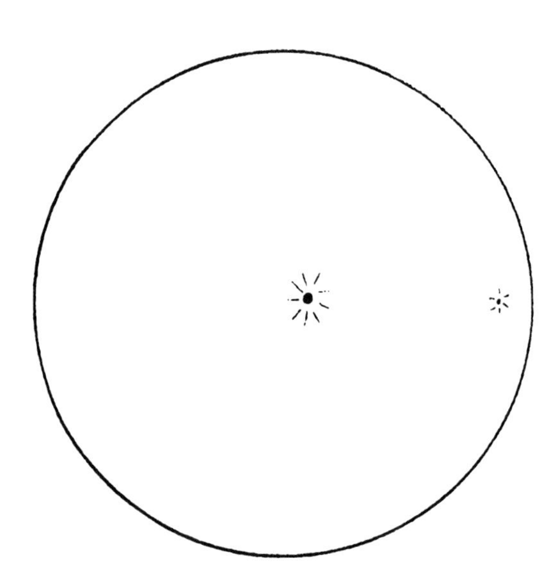

Fig. 3.9. One way to test for vignetting is to focus binoculars on one of the star pairs in Table 3.3. Arrange the binocular FOV so that one of the stars is near the FOV center and the other is near the edge as is shown. If the central star is much brighter than the star at the edge, there is a lot of vignetting in the binoculars (credit: Richard W. Schmude, Jr.).

Observing Comets

In this section, I will discuss what the observer should look for and the types of measurements that one can do to help in the better understanding of comets. The most important observation that people can do with binoculars is to estimate the brightness of a comet, and, to this end a discussion of comet brightness estimates and how one can construct a model (or equation) describing comet brightness is presented. The sources of error in brightness estimates will be discussed, followed by a section on drawing comets. The chapter ends with a discussion on how to evaluate the atmospheric transparency and how to correct brightness estimates for atmospheric extinction.

Coma Brightness

Over half of the comets that I have observed have lacked a visible tail. Even in those cases where tails were seen, I felt that most of the light from the comet comes from the coma. Keep in mind that photographs of comets usually show an overexposed coma and as a result, give the misleading impression that most of the light that a comet reflects is from its tail. I believe that in most cases, over 80% of a comet's light comes from its coma and, as a result, one can estimate coma brightness and report this as the comet brightness. If, however, the tail reflects more light than the coma, one must be careful. In this case, one must state whether the brightness estimate corresponds to just the coma or the entire comet.

Why should one estimate the brightness of the coma? These estimates may be used to determine how quickly the coma brightens as it approaches the Sun. This, in turn, can serve as a baseline for outbursts. Brightness estimates may yield also information on the age of a comet. In some cases, brightness estimates may yield information on the size and rotation rate of the comet's nucleus.

Before one makes a brightness estimate you must have a reliable source of magnitudes of comparison objects. The International Comet Quarterly (ICQ) staff has a list of recommended sources of such magnitudes and a few of these are listed in Table 3.4. Other recommended sources are at (http://www.cfa.harvard.edu/icq/ICQRec.html). One should never use sources for which the available star magnitudes are all brighter than the comet. Instead, one comparison object should be a little dimmer than the coma and the other should be a little brighter than the coma.

There are several ways of estimating coma brightness. These are summarized in Table 3.5, a few of which need additional explaining. For years, people have used the Reverse-binocular Method to estimate the brightness of the Moon during a total lunar eclipse. Essentially, one views a very bright object with binoculars by looking through one of the objective lenses instead of the eyepieces. (One can look through both objective lenses in opera glasses or roof prism binoculars.) When an object is viewed through binoculars in reverse, it appears much smaller and dimmer. Therefore, there is no need to de-focus stars or the comet. One then compares the brightness of the object with objects seen with the unaided eye.

Table 3.4. International Comet Quarterly (ICQ) recommended sources of magnitude ranges of comparison objects

Source	Magnitude range
Planet magnitudes in the Astronomical Almanac	−4 to +8
Arizona Tonantzintla Catalog (found in Sky & Telesc. Vol. 30, No. 1, p. 21)	−1 to +5
Star magnitudes in the Astronomical Almanac (UBVRI standard stars) *listed for 2002 or earlier almanacs*	0 to +7
Star magnitudes in the Astronomical Almanac (u, v, b, y and Hβ standard stars)	+1 to +7
Star magnitudes in the Astronomical Almanac (UBVRI standard stars) *listed for 2003 and later almanacs*	+9 to +16
Yale Bright Star Catalog	−1 to +6
Johnson V photometry by Brian Skiff at Lowell Observatory; see http://www.kusastro.kyoto-u.ac.jp/vsnet/catalogs/skiffchart.html (Bright BVRI)	+6 to +8
Johnson V photometry by Brian Skiff at Lowell Observatory; see http://www.kusastro.kyoto-u.ac.jp/vsnet/catalogs/skiffchart.html (Stars near variables; over 50 different locations listed)	+4 to +17 mostly +8 to +13

Additional sources are listed at: http://www.cfa.harvard.edu/icq/ICQRec.html

Table 3.5. Methods of estimating the brightness of the coma of a comet

Method	Symbol	Description
Sidgwick	S	Compare the in-focus coma to out-of-focus comparison stars. The out-of-focus stars must be about the same size as the coma
Bobrovnikoff	B	Compare the out-of-focus comet to out-of-focus comparison stars. All out-of-focus objects should appear to have the same size
Morris	M	Compare the slightly out-of-focus comet to out-of-focus comparison stars
Beyer	–	De-focus comet and comparison stars and note the order of disappearance
In-focus	–	Compare the in-focus comet to in-focus comparison stars
Direct comparison	–	Compare the comet with Messier objects (galaxies, nebulae, open star clusters or globular star clusters) or other similar objects
Reverse-binocular	–	For bright comets, view them through binoculars in reverse and compare to in-focus comparison stars. Subtract MF from the comet magnitude. Compute MF from: MF = 5 log[P] + 0.31, where P = magnification[a]

[a]I carried out 24 measurements of the Moon in different phases using the Reverse binocular technique with both binoculars and telescopes. The average discrepancy between the measured and predicted brightness of the Moon was 0.10 magnitudes with a standard deviation of 0.68 magnitudes. Although this result is consistent with this equation, the observer using the Reverse-binocular method should make several measurements and take an average. In this study, the predicted V filter brightness values of the Moon are from Schmude (2001) and are corrected to m_v values using (3.12)

Once a match is found, one subtracts MF, defined in Table 3.5, from the magnitude value of the unaided-eye object(s) to get the true brightness of the bright object. For example, if one finds that the coma through 7×35 binoculars in reverse is equal in brightness to a magnitude 1.5 star with the unaided eye, the MF is computed as:

$$\begin{aligned} MF &= 5 \log [7] + 0.31 \\ &= 5 \times 0.845 + 0.31 \\ &= 4.23 \times 0.31 \\ &= 4.54 \end{aligned}$$

and the coma magnitude = 1.5 − 4.54 = −3.04 or −3.0. The reverse-binocular technique should only be used for very bright comets.

Perhaps the most accurate way of estimating coma brightness is to use fuzzy objects like globular star clusters as comparison objects – the second-to-last method in Table 3.5 (Direct Comparison). In this method, one defocuses either the comet or the extended object(s) until it (they) are the same size as the comet. It is important to use comparison objects with known B and V filter magnitudes. Table 3.6 lists a few extended objects with known brightness values. In all cases, I have used (3.12) to convert V filter magnitudes to brightness values that our eyes see. A more complete list can be found in *Sky Atlas 2000.0* Vol. 2©1985 by Alan Hirshfeld et al. One problem with this method is that there are not many extended objects brighter than 7th magnitude.

In the first five methods in Table 3.5, one may estimate coma brightness by using comparison stars. The advantage of this method is that there are thousands of stars brighter than 7th magnitude. Let me explain. By way of explanation, let's say that one observes a comet and notes that the coma is dimmer than star A (magnitude = 6.01) but is brighter than star B (magnitude 6.91). Furthermore, the

Table 3.6. Total V filter brightness values in stellar magnitudes of selected galaxies, open star clusters and globular star clusters

Object	Right ascension (2000.0)	Declination (2000.0)	Type	Brightness, m_v (stellar magnitudes)
M31	0 h 42.7 min	41° 16 min	Galaxy	3.7
M33	1 h 33.9 min	30° 39 min	Galaxy	5.8
M34	2 h 42.0 min	42° 47 min	Open star cluster	5.3[a]
M79	5 h 24.5 min	−24° 33 min	Globular star cluster	8.1
M36	5 h 36.1 min	34° 08 min	Open star cluster	5.3[a]
M81	9 h 55.6 min	69° 04 min	Galaxy	7.1
M68	12 h 39.5 min	−26° 45 min	Globular star cluster	8.3
M53	13 h 12.9 min	18° 10 min	Globular star cluster	7.9
M5	15 h 18.6 min	2° 05 min	Globular star cluster	5.9
M13	16 h 41.7 min	36° 28 min	Globular star cluster	6.0
M22	18 h 36.4 min	−23° 54 min	Globular star cluster	5.3
M2	21 h 33.5 min	−0° 49 min	Globular star cluster	6.6

Coordinates are from *Sky Catalogue 2000.0* Volume 2©1985 by Alan Hirshfeld and Roger W. Sinnott. The brightness values, m_v are what our eyes would see. They are computed from (3.12) and B and V magnitudes in *Sky Catalog 2000.0* Volume 2©1985 by Alan Hirshfeld, Roger W. Sinnott and Francois Ochsenbein

[a]Computed using an assumed B-V value of 0.7

difference between the coma and star A is twice that of the coma and star B. The magnitude is computed as:

Star A (magnitude = 6.01)

↓ One step

↓ One step

Coma

↓ One step

Star B (magnitude = 6.91)

$$\begin{aligned}\text{Coma brightness} &= (6.91-6.01)/3\,\text{steps} = 0.30\,\text{magnitudes/step}\\ &= 6.01+2\,\text{steps}\times 0.30\,\text{magnitudes/step}\\ &= 6.61 \text{ or } 6.6\end{aligned}$$

Visual magnitudes are reported to one decimal place. Keep in mind that the lower the object's magnitude value, the brighter is the object.

Some binoculars have an adjustable magnification. If you are using such a pair, use the lowest magnification to estimate the comet's brightness. Lower magnifications reduce the size of the comet, thus making it easier to estimate its brightness.

How accurate are visual brightness estimates of comets? I carried out two experiments related to this question. In the first experiment, I estimated the brightness of several Messier objects using either the Sidgwick or the Bobrovnikoff methods. (The Messier objects include galaxies, open star clusters, nebulae and globular star clusters.) I then compared my estimated magnitudes to V filter photoelectric magnitude measurements. The difference between the average estimated brightness and the V filter result for 30 different Messier objects was 0.26 stellar magnitudes for the Sidgwick method. The corresponding difference for 33 different Messier objects for the Bobrovnikoff method was 0.10 stellar magnitudes. In other words, the Sidgwick method yielded an average result which was 0.26 stellar magnitudes fainter than the photoelectric brightness value. The Bobrovnikoff method yielded an average result which was 0.10 stellar magnitudes fainter than the photoelectric brightness value.

In a second experiment, I compared unaided-eye visual brightness estimates corrected to a standard aperture of 6.8 cm with photoelectric V filter measurements for comets 17P/Holmes and Hale-Bopp (C/1995 O1). As it turned out, the V filter brightness measurements were 0.2 magnitudes brighter than the unaided-eye estimates. These differences were smaller than the estimated uncertainty of 0.5 stellar magnitudes for visual brightness estimates.

One problem with visual brightness estimates is that they may have inconsistencies due to the size and type of instrument used. Astronomers are aware that people using large instruments tend to under-estimate the brightness of a comet compared to those using smaller ones. There are also inconsistencies of measurements made with refractors and reflectors. Morris (1973) carried out a study of brightness estimates made through different pieces of equipment. He selected a standard aperture of 6.78 cm (hereafter rounded off to 6.8 cm). He then derived equations that yielded correction factors for different apertures and telescope types. These equations are:

$$M_c = m_v - 0.066/\text{cm} \times (A - 6.8\,\text{cm}) \quad \text{Refractors} \tag{3.4}$$

$$M_c = m_v - 0.019/\text{cm} \times (A - 6.8\,\text{cm}) \quad \text{Reflectors} \tag{3.5}$$

In both equations, M_c is the brightness corrected to an aperture of 6.8 cm, m_v is the visual brightness estimated with an aperture of A centimeters. My analysis of Comets 1P/Halley, 9P/Tempel 1, 19P/Borrelly and 81P/Wild 2 confirm (3.4) and (3.5) for telescope diameters of up to 0.6 m.

I conducted a similar study for unaided-eye brightness estimates. The goal of this study was to develop an equation that converts unaided-eye brightness estimates to a standard aperture of 6.8 cm. Unaided-eye estimates were compared to those made with binoculars and telescopes to a standard aperture of 6.8 cm. Data from ten different comets were included in the evaluation. The resultant equation is:

$$M_c = m_v + 0.24 \text{ stellar magnitudes} \tag{3.6}$$

In this equation, M_c is the brightness corrected to an aperture of 6.8 cm, and m_v is the unaided-eye brightness estimate.

Two common equations that describe the brightness of a comet are:

$$M_c = H_{10} + 5\log[\Delta] + 10\log[r] \tag{3.7}$$

$$M_c = H_0 + 5\log[\Delta] + 2.5n\log[r]. \tag{3.8}$$

These are the same (1.5) and (1.6) in Chap. 1.

Equation (3.8) is a more flexible model for brightness and, hence, is the preferred one to use provided that sufficient data are available. If, on the other hand, only a few brightness measurements are available, of if the data cover a range of log[r] below 0.1, (3.7) would be the preferred one to use.

There are other equations which may be used to model comet brightness. They require lots of data and are used for a few comets. Until those equations become more widely used, there is little need to discuss them further.

Equations (3.7) and (3.8) work well as a staring point for the understanding and prediction of comet brightness. An example of how to come up with photometric constants of a comet based on (3.8) is presented below.

Example of Analysis

My brightness estimates of Comet Hyakutake (C/1996 B2) are listed in Table 3.7. The decimal dates in Universal Time (UT) are listed in the first column and my brightness estimates are listed in the second column. The decimal date is computed by dividing the number of minutes past 0:00 UT by 1,440, the number of

Table 3.7. The Author's brightness estimates of Comet Hyakutake (C/1996 B2)

Date (1996)	m_v	M_c	Δ (au)	r (au)	$M_c - 5\log[\Delta]$	log[r]
April 3.05	2.8	3.1	0.331	0.848	5.5	−0.072
April 4.05	2.7	3.0	0.366	0.830	5.2	−0.081
April 6.05	2.9	3.2	0.419	0.784	5.1	−0.106
April 7.13	2.7	3.0	0.443	0.750	4.8	−0.125
April 8.04	3.1	3.4	0.500	0.720	4.9	−0.143
April 10.05	3.2	3.5	0.561	0.682	4.8	−0.166
April 11.05	3.3	3.6	0.588	0.644	4.8	−0.191
April 13.04	3.4	3.7	0.647	0.621	4.6	−0.207
April 17.04	2.8	3.1	0.739	0.523	3.8	−0.281

minutes in a day, and adding this quotient to the date. For example, if a measurement is made at 1:15 UT on April 3 the decimal date would be April 3.0 + (75 min after 0:00 UT ÷ 1,440 min/day) = April 3.05. Since I used 2.1 cm binoculars which are similar to refractors, I used (3.4) to correct to the standard aperture of 6.8 cm. For example, the first brightness estimate in Table 3.7 corrected to the standard aperture is:

$$\begin{aligned} M_c &= 2.8 - 0.066\,/\,\text{cm} \times (2.1\text{ cm} - 6.8\text{ cm}) \\ &= 2.8 - 0.066\,/\,\text{cm} \times (-4.7\text{ cm}) \quad \text{(note the units cm cancel)} \\ &= 2.8 - -0.3102 \\ &= 3.1102 \text{ or } 3.1. \end{aligned}$$

Values of M_c for all estimates are listed in the third column of Table 3.7.

Equation (3.8) can be rearranged as:

$$M_c - 5\log[\Delta] = 2.5n\log[r] + H_0 \tag{3.9}$$

Equation (3.9) is in the form of $y = mx + b$ which is the equation for a straight line. In this case, y is $M_c - 5\log[\Delta]$; the slope, m, is 2.5n, x is log[r]; and the y-intercept is H_0. The brightness data are fitted then to (3.9). Values of Δ and r are listed in the fourth and fifth columns in Table 3.7. One plots $M_c - 5\log[\Delta]$ versus log[r] to compute the slope (2.5n) and the y-intercept (H_0). Accordingly, I have computed values of $M_c - 5\log[\Delta]$ and log[r], and these are listed in the sixth and seventh columns in Table 3.7. Values of $M_c - 5\log[\Delta]$ versus log[r] are plotted in Fig. 3.10. Since the data appear to fall along a straight line, I used a linear least squares routine to find the best-fit linear equation. My result is:

$$M_c - 5\log[\Delta] = 6.68\log[r] + 5.8. \tag{3.10}$$

The slope is $2.5n = 6.68$ and the y-intercept is $H_0 = 5.8$. I prefer to report H_0 to the nearest tenth of a magnitude if the data are based on visual brightness estimates, as was the case in Table 3.7. If, on the other hand, the brightness data are accurate to 0.01 magnitude, then I believe that H_0 should be reported to the nearest 0.01 magnitude.

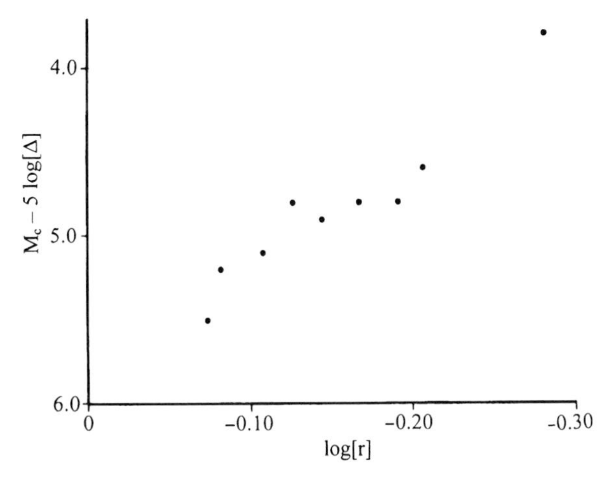

Fig. 3.10. A plot of $M_c - 5\log[\Delta]$ versus log[r] for my nine brightness estimates of Comet Hyakutake (C/1996 B2) (credit: Richard W. Schmude, Jr.).

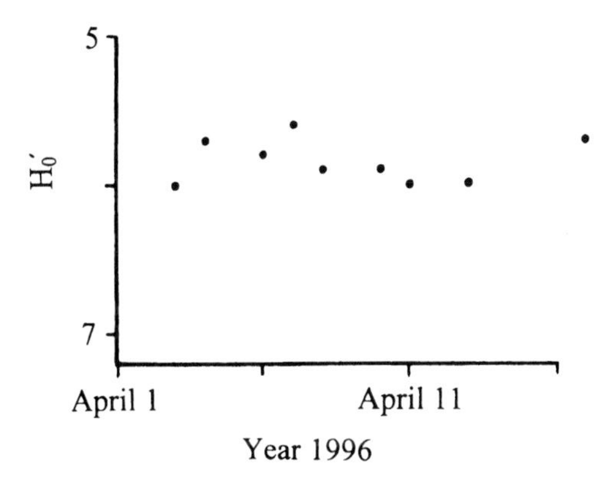

Fig. 3.11. A plot of H_0' values (normalized magnitudes) for Comet Hyakutake (C/1996 B2) on different dates in April 1996. I used (3.11) along with the values in Table 3.7 and 2.5n = 6.68 to compute the H_0' values (credit: Richard W. Schmude, Jr.).

One may rewrite (3.8) as:

$$H_0' = M_c - 5\log[\Delta] - 2.5n\log[r]. \tag{3.11}$$

In this equation, $H_0' = H_0$. One can insert $2.5n = 6.68$ into (3.11) and calculate H_0' based on the M_c values in Table 3.7. The purpose of doing this is to see whether there is any increasing or decreasing trend in H_0' values with time. Accordingly, I did this, and the H_0' values for the different dates are plotted in Fig. 3.11. Since the H_0' values fall along a nearly straight line, it shows that (3.10) is a good model for the brightness of Comet Hyakutake between April 3 and April 17, 1996.

One must realize that comets may brighten at a rate that does not depend on a strict distance relationship. As a result, not all comets will brighten according to (3.8). In these cases, H_0' will trend upwards or downwards with time. A few comets have had complicated lightcurves which did not follow (3.8).

Sources of Error in Estimates of Visual Brightness of a Comet

There are several sources of error associated with the estimates of the visual brightness of a coma. These are: extinction, magnification, color difference, changing background light level and changing values of the solar phase angle.

If the comparison star is at a different altitude than the coma, extinction error will creep into the brightness estimates. In cases where the comet and comparison stars are at least 45° above the horizon, extinction errors will be minimal. If the comet is less than 45° above the horizon, one should select comparison stars which are at nearly the same altitude as the comet. If the comparison stars are at a different altitude than the comet, an extinction correction should be made. At the end of this chapter, I describe how to correct for this error in the "Extinction Correction" section.

The magnification may affect a comet's perceived brightness. High magnifications spread out light resulting in lower surface brightness. In many cases, dim portions of a comet will be missed at high magnification. In one case, the reported brightness values of Comet 19P/Borrelly ranged from magnitude 9 to magnitude 11. In *Comets: A Descriptive Catalog* ©1984 by Gary W. Kronk, the writer (Kronk) notes that the brighter estimates usually came from lower-magnifications, and the fainter estimates came from higher magnifications. Because of this, I believe that people should state the magnification used in all comet brightness estimates. I am not aware of any studies of how magnification affects perceived comet brightness.

A third source of error is due to differences in color between the comparison stars and the comet being studied. The human eye is more sensitive to green light than to red light. As a result, the m_v value (which is the symbol for visual magnitude) instead of the V filter magnitude should be used. According to a large study carried out by Stanton (1999), the m_v and V filter magnitudes for over 20 people are related to each other through:

$$m_v = V + 0.21(B - V) \tag{3.12}$$

where m_v is the visual magnitude, V is the V filter brightness in stellar magnitudes, B is the B filter brightness in stellar magnitudes and B – V is the color index. The B and V are brightness values made with filters transformed to the Johnson B and V system, respectively.

A changing background light level can affect the brightness estimate, especially for faint comets. The Moon, a bright planet or star, the Milky Way Galaxy, the Zodiacal Light, twilight and artificial light may affect the sky background. If the comet is in a brighter part of the sky than the comparison star(s), the comet's magnitude will be fainter than what it is in actuality. The best way of reducing error due to different sky background is to select comparison star(s) that are as close to the comet as possible.

A fifth source of error is changing values of the solar phase angle. The solar phase angle coefficient expresses how fast an object dims with increasing solar phase angle. For example, Uranus' moon Titania dims at a rate of 0.02 magnitudes/degree.

Therefore, when the solar phase angle increased from 20 to 60°, Voyager 2 imaged this moon to dim by 0.8 magnitudes. Ferrín (2005) lists solar phase angle coefficients for the nuclei of ten different comets with a mean value of 0.046 magnitudes/degree. Once a comet develops a coma and tail, the solar phase angle coefficient may drop to less than 0.01 magnitudes/degree. I have estimated an upper limit of 0.005 magnitudes/degree for the solar phase angle coefficient of the coma of Comet 1P/Halley. See Chapter 2. Similarly, I have estimated an upper limit of 0.008 magnitude per degree for coma of Comet Hale-Bopp. If a comet has a solar phase angle coefficient of 0.005 magnitudes/degree, and if the solar phase angle changes by 50°, this would cause a 0.25 magnitude change.

Drawing Comets

Why should anybody make a comet drawing when there are so many people taking such wonderful digital images of comets? The most obvious reason is that expensive equipment is not needed. A second advantage is that there is no need to do image processing. Another consideration is that our eyes can perceive a greater brightness range than images. Faint jets may not appear in an image which shows faint tail detail, and a close-up image of the coma may not show faint detail in the tail. When drawing a comet, you can examine both the tail and the coma. A final reason for making a comet drawing is for historical consistency. Drawings are the only pictures that we have of comets which appeared before the late nineteenth century.

When making a drawing, I recommend that one select a location free from artificial light and make a photocopy of a page from a star atlas of where the comet is expected to be at the time of the drawing. A dark location is needed to reduce unwanted artificial light and to yield a nearly consistent level of scattered light. In some cases, natural light may be a good thing. Stephen O'Meara, for example, reports that sometimes he is able to see comet detail in twilight that he would not have seen in a dark sky. One should avoid making tail length measurements when the Moon is very bright. A photocopied page from a star atlas is an excellent way of estimating the position angle of the visible tail(s).

Some equipment that I would recommend for making drawings include a clipboard, a soft lead pencil (HB lead is fine), an eraser, a red flashlight, a clock, toilet paper and binoculars. On many occasions, I have used binoculars to locate a fuzzy comet. Many comets have faint shadings, and the toilet paper may be the best way to reproduce them. Essentially one can use toilet paper to smear the lead to make faint smudges. One can gently erase a smudge to create features in a drawing. The clipboard keeps the drawing in place and serves as a hard surface on which to make the drawing. The red flashlight helps preserve night vision. The clock is used to record the time of the drawing.

The easiest way to draw a comet is to make a negative drawing. This is where a white background is used for the dark sky and the brightest portion of the comet is drawn darkest. See Fig. 3.12. I use the toilet paper to smear out faint shadings like what is found in the tail. Of course, one may construct a positive drawing later

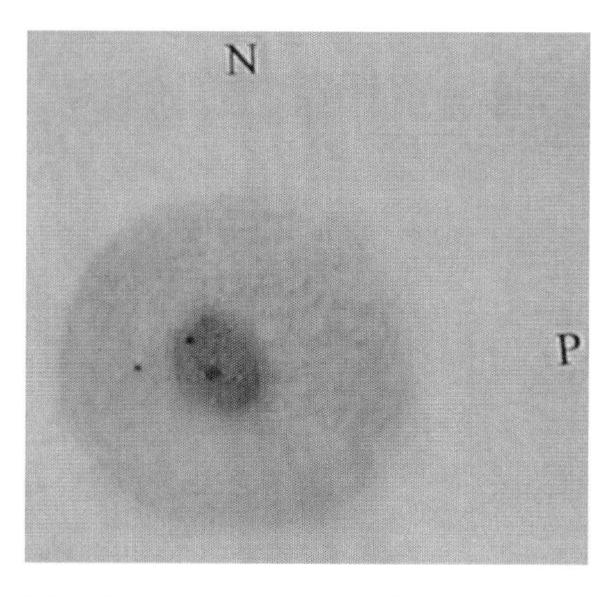

Fig. 3.12. A negative drawing of Comet 17P/Holmes I created on October 29, 2007. This drawing is called a negative because the *dark sky* background is *white* and the brightest part of the comet is drawn darkest. The N and P are the north and proceeding directions, respectively. The approximate dimensions of the drawing are 12 by 13 arc-min (credit: Richard W. Schmude, Jr.).

from the negative drawing. Figure 3.13 shows a positive drawing of comet 17P/ Holmes.

One may also elect to make an intensity drawing. In this case, one draws an outline of the comet and all of the albedo features in it, and then assigns intensity values. The European intensity scale ranges from 0 (white) to 10 (black) while the ALPO intensity scale is 0 (black) to 10 (white). Be sure to state which scale that you are using. I also recommend describing the contrast with the symbols "lc" (low contrast), "mc" (moderate contrast) and "hc" (high contrast). An example of an intensity drawing is shown in Fig. 3.14.

When making a drawing, one should focus on the four parts of a comet, namely, the central condensation, coma, dust tail and gas tail. I will discuss what to draw and look for in each of these parts.

The central condensation is usually the brightest part of the coma. This region is often called the "nucleus" or the "nuclear condensation". These terms can be misleading. It is doubtful that anybody has ever seen the nucleus of a comet. In almost all cases, the central condensation contains material close to the nucleus. If the central condensation is bright enough, look for color. A bluish color is consistent with lots of gas, whereas a yellow or orange color is consistent with lots of dust. One should also note the shape of the nuclear condensation and its location within the coma. When the nucleus splits, the central condensation may appear bar-shaped. After a few days, two or more bright areas will be present. This was observed in 1976 when Comet West (C1975 V1) broke apart. One needs to be careful of background stars that might be confused for the central condensation. One should also look for jets. Jets can be used to determine the rotation period of the

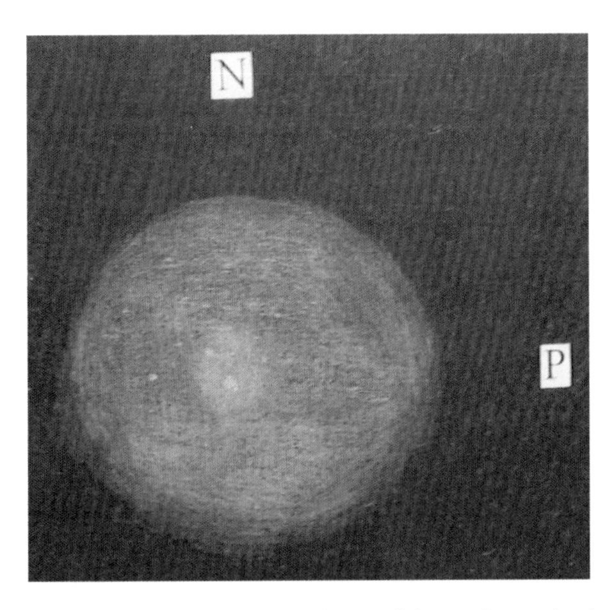

Fig. 3.13. A positive drawing of Comet 17P/Holmes I created. This drawing is called a positive because it shows things as they are; that is, the *dark sky* background is *dark* and the brightest part of the comet is drawn *bright*. The approximate dimensions of the drawing are 11 by 12 arcmin (credit: Richard W. Schmude, Jr.).

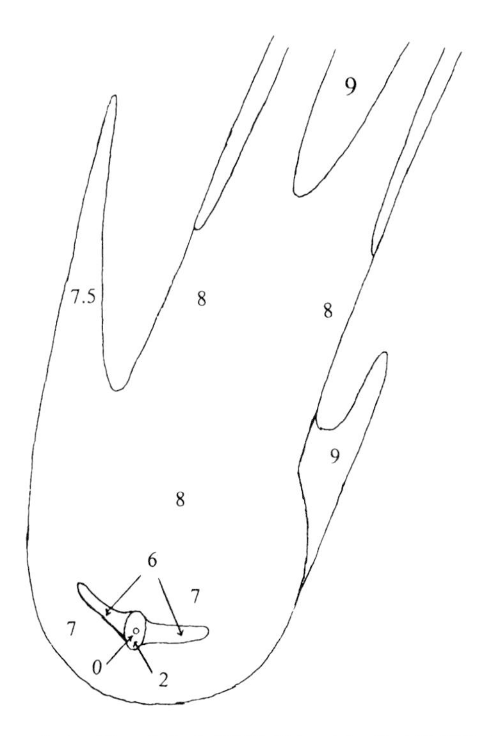

Fig. 3.14. This is an intensity drawing of Comet Hyakutake (C/1996 B2) I created on March 25, 1996, with 11 × 80 binoculars. In this drawing, the intensity scale is 0 = *white* and 10 = *black*. The frame is approximately 3° by 4° (credit: Richard W. Schmude, Jr.).

nucleus. In the case of 19P/Borrelly, one jet was even used in establishing the orientation of the rotational axis of the nucleus. This was discussed in Chap. 2, Part C. In some cases, a jet can change within a few hours; hence, if a jet is present, one should measure its position angle and size every hour. The position angle is measured from the north direction and can range from 0 up to 359°. Recall from Chap. 2, Part A, that the position angles are: north = 0°, east = 90°, south = 180°, west = 270°. For example, if a jet points halfway between south and west then its position angle would be 225°.

One should also look for faint arcs around the central condensation. Comet Hale-Bopp displayed several of these arcs. I used the arc separation in that comet to measure how fast the arcs were moving from the nucleus. This is discussed in Chap. 4. One should attempt also to estimate the diameter of the central condensation. This may be accomplished with either an eyepiece having a reticule, or by estimation in terms of stars with a known separation. In the second method, one notes the size of the central condensation in terms of the separation between two nearby stars. Later, one measures the separation of the stars from a star atlas and then computes the angular size of the central condensation. The *Millennium Star Atlas* ©1997 by Roger W. Sinnott and Michael A. C. Perryman is good for star positions.

One may also estimate the brightness of the central condensation. This is accomplished by comparing its brightness to that of nearby stars. The same procedure and rules outlined in the "Coma Brightness" section should be followed here.

The coma or head is usually the brightest part of the comet. There are at least five characteristics that one should note when observing the coma. These are the color, shape, size, degree of condensation and brightness. A description of the first four of these characteristics follows. The brightness characteristic was discussed earlier in this chapter.

The color of the coma can yield information on its gas-to-dust ratio. A bluish color is consistent with a high gas concentration, whereas a yellow or orange color is consistent with a high dust concentration. Kermit Rhea reported a bluish color for the coma of Comet Bradfield (C/1987 P1) and John Bortle reported a similar color for Comet Austin (C/1984 N1). When making color estimates, the comet should be bright enough to show color. A good rule-of-thumb is that the comet should be at least seven stellar magnitudes brighter than the limiting magnitude of the instrument in order for it to show color.

In addition to color, one should also note the shape of the coma. Some comets have a circular coma, while others have a parabolic, elliptical or pear-shaped coma. Three different shapes are illustrated in Fig. 3.15 along with at least two comets with the reported shape. The shape is probably determined by a combination of the solar wind, the comet's orbital speed and the composition of the coma.

The size of a coma may yield information on the nucleus. In order to compute the coma dimension, one simply needs to estimate its angular size. The easiest way to do this is to compare its size to the distance between two stars. One then measures the angular separation of two stars in a star atlas and determines the angular size of the coma. One computes the size by using the angular size in arc-minutes and the coma distance, through the equation:

$$\text{Size} = 0.000291/\text{arc-minutes} \times (\text{angular size in arc-minutes}) \times \text{distance}. \quad (3.13)$$

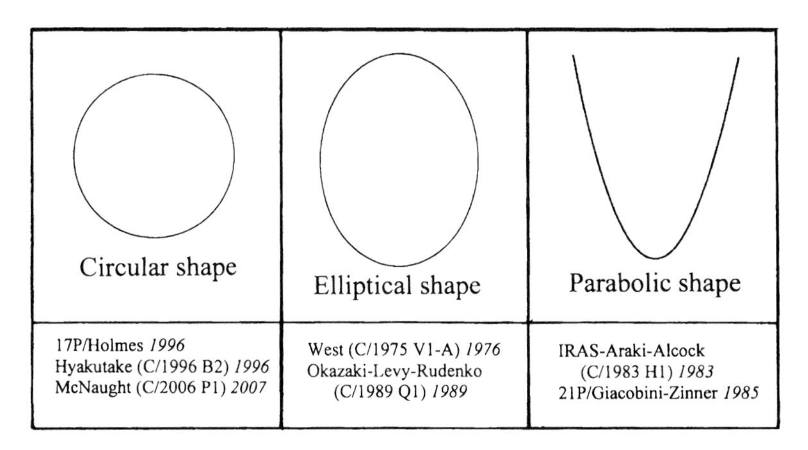

Fig. 3.15. Different shapes of the coma for different comets. The comets listed below each shape had comas with that shape for at least part of the year in parentheses. Observations are from Bortle (1983, 1985), Machholz (1996), Set Vol. 113, No. 4, P.89, Wallentinsen (1981) and the Authors personal records. The year of the coma observation is in italics (credit: Richard W. Schmude, Jr.).

The 0.000291 represents the number of radians in 1.0 arc-min. This equation is valid for angular sizes less than 600 arc-min or 10°. For example, if the coma has an angular diameter of 9.0 arc-min and it is 120,000,000 km from the Earth, its size would be:

$$\begin{aligned} \text{Size} &= 0.000291/\text{arc-minute} \times (9.0\ \text{arc-minutes}) \times 120{,}000{,}000\ \text{km} \\ &= 314{,}280\ \text{km} (\text{or rounded off to } 310{,}000\,\text{km; note arc-minutes cancel}) \end{aligned}$$

Before describing the fourth coma characteristic, degree of condensation, I would like to present an example of the expanding coma of Comet 17P/Holmes:

During late 2007, I was able to measure the expansion rate of the coma of Comet Holmes with just a pair of binoculars, a see-through ruler and the *Millennium Star Atlas* ©1997 by Roger W. Sinnott and Michael A. C. Perryman. Let me explain how I did this. On 15 different dates between October 29 and November 15, 2007, I estimated the coma diameter in terms of the distance between nearby stars. After making each estimate, I went inside and measured the exact separation between the two reference stars using my see-through ruler and the *Millennium Star Atlas*. With this measurement, I computed the angular separation between the two stars and then determined the coma diameter. My coma diameter measurements are summarized in Table 3.8. I then computed the actual size of the coma from (3.13). I used the Horizons On-Line Ephemeris System to determine the comet distances. For example, the coma diameter on October 29.1, 2007, was computed as:

$$\begin{aligned} \text{Size} &= 0.000291/\text{arc-minute} \times (8.0\ \text{arc-minutes}) \times 244 \times 10^6\ \text{km} \\ &= 0.568032 \times 10^6\ \text{km or} \\ &= 0.57 \times 10^6\,\text{km}. \end{aligned}$$

Table 3.8. Estimates of the coma size of Comet 17P/Holmes made by the Author in late 2007

Date (2007)	Coma diameter (arc-min)	Coma diameter (10^6 km)	Comet-Earth distance (10^6 km)	Coma radius (10^6 km)
Oct. 29.1	8.0	0.57	244	0.28
Oct. 30.18	9.0	0.64	243	0.32
Nov. 1.12	13.2	0.933	243	0.467
Nov. 2.15	13.5	0.955	243	0.477
Nov. 3.16	14.6	1.03	243	0.516
Nov. 4.09	15.9	1.12	243	0.562
Nov. 6.14	18.5	1.31	243	0.654
Nov. 7.23	20.8	1.47	243	0.735
Nov. 8.11	20.8	1.47	243	0.735
Nov. 9.04	22.8	1.61	243	0.806
Nov. 10.08	24.8	1.75	243	0.877
Nov. 11.01	25.6	1.81	243	0.905
Nov. 12.11	26.7	1.89	243	0.944
Nov. 13.09	27.4	1.94	243	0.969
Nov. 15.01	32.4	2.30	244	1.15

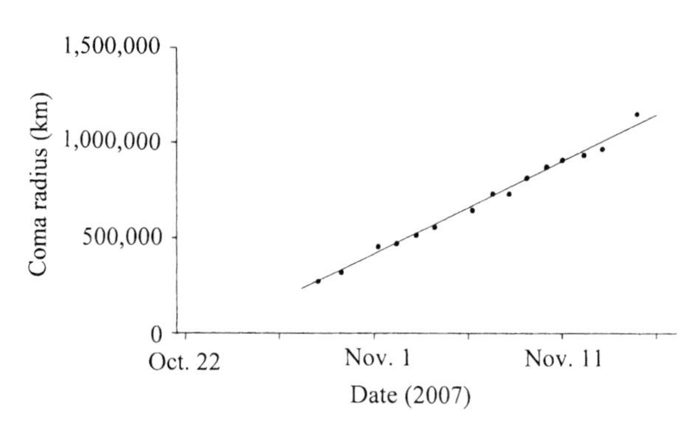

Fig. 3.16. Coma radius values on different dates for Comet 17P/Holmes. I used 15 × 70 binoculars to make all coma size estimates and converted them to coma radii in kilometers (credit: Richard W. Schmude, Jr.).

Once I computed all of the coma diameters, I converted them into radii. Finally I plotted the coma radii versus the date, and computed the best-fit straight line. See Fig. 3.16. The resulting equation is:

$$\text{Coma radius} = -16{,}800\,\text{km} + 48{,}500\,\text{km}(\text{t} - -\text{Oct.}\,23.0, 2007\,in\,days) \quad (3.14)$$

Where t is the date and (t – Oct. 23.0, 2007) is the number of days after Oct. 23.0, 2007. This equation shows that the radius increased by 48,500 km/day. This is equivalent to 0.56 km/s or almost 1,300 miles/h.

A fourth coma characteristic that one should note in observing the coma is the degree of condensation (DC). This is rated on a scale from 0 to 9. If the coma has an equal brightness from its center to its edge, DC = 0, whereas, if the central condensation is point-like with little or no coma surrounding it, DC = 8 or 9. See Fig. 3.17. The DC value often rises as a comet approaches the Sun. Figure 3.18

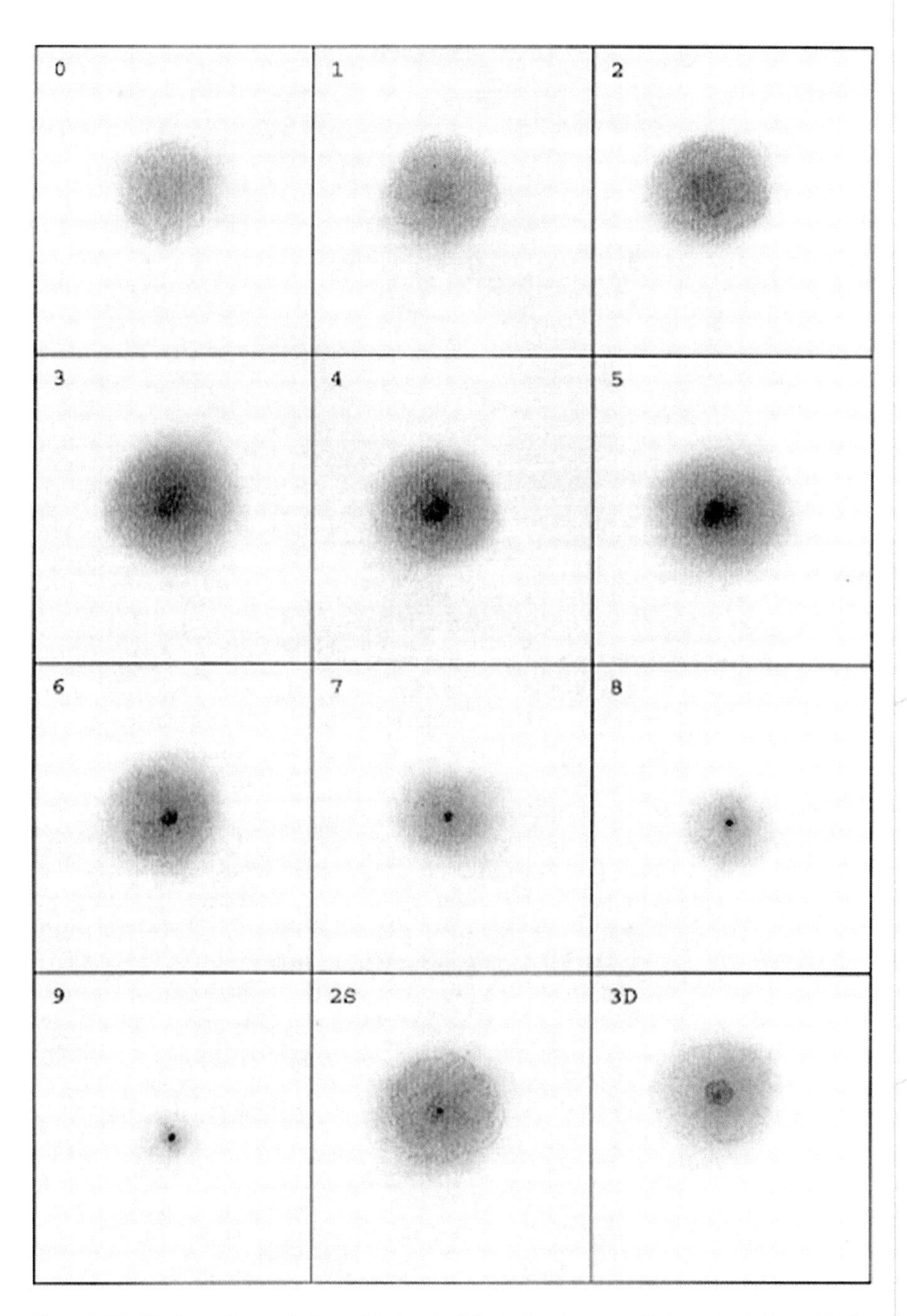

Fig. 3.17. This diagram illustrates the Degree of Condensation (DC) value for various comas. This is reproduced with permission from *Observing Guide to Comets* © 2002 by Jonathan Shanklin (credit: Jonathan Shanklin).

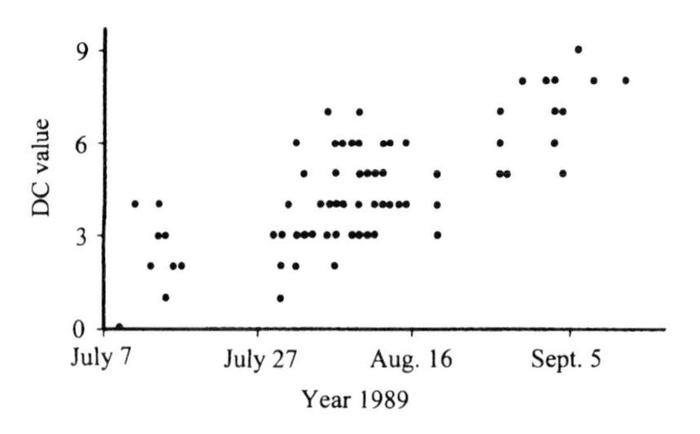

Fig. 3.18. Estimates of the DC value of Comet 23P/Brorsen-Metcalf made by members of the Association of Lunar and Planetary Observers (credit: Richard W. Schmude, Jr.).

shows DC estimates of Comet P/Brorsen-Metcalf made by members of the Association of Lunar and Planetary Observers (ALPO). This comet reached its closest distance to the Sun on September 11, 1989. Figure 3.18 shows that the DC values increased as Comet 23P/Brorsen-Metcalf approached the Sun.

After analyzing the coma, one should study the tail. Many comets may not have a visible tail. Others will display a gas tail, a dust tail or both types of tails. The gas tail is often straight and narrow. The dust tail will often be broad and curved except when the Earth lies near the comet's orbital plane. In this case, it will appear narrow and straight, and both the gas and dust tails will appear as a single feature. When observing a comet, look for the tail and, if possible, try to distinguish the gas tail from the dust tail. The tail should be drawn on the photocopied page of the star chart. Use the background stars as guides in making the drawing. Later, be sure to measure the position angle of the tail. The best place to measure the position angle is where the tail meets the coma. One may measure the tail length from the drawing by using star separations, or may estimate the tail length in terms of the FOV of the binoculars. If several measurements of the tail are planned try to use the same piece of equipment and magnification for each measurement. Since the surface brightness drops with increasing magnification, you may introduce inconsistencies into the data if using different magnifications. When I studied Comet Hale-Bopp (C/1995 O1), I used the same pair of binoculars for all of my tail-length estimates. If possible, try to note the color of the tail. In many cases, the tail will show no color because its surface brightness is below the color-detection threshold of the human eye.

The length of a comet's tail depends on its actual length and its orientation. If our line of sight is perpendicular to a linear tail, we will see its true length. In most cases though, the tail will not be perpendicular to our line of sight. In these cases, we would view the tail at an angle and, hence, it would look shorter than what it is in actuality. The process of using the angular size of a tail to compute its actual length depends on the tail's distance, shape and orientation. This is discussed further in Chap. 4.

If the gas tail is visible, one should inspect it for any irregularities. The gas tail can change shape or length within minutes. On Feb. 6, 1980, the gas tail of Comet

Table 3.9. A few comets that had antitails

Comet	Date seen/imaged	Antitail length (°)	Source
Arend-Roland (C/1956 R1)	April 1957	15	Sky and Telescope 73, No. 4, p. 456–457
Austin (C/1984 N1)	Sept. 1984	~0.4	Sky and Telescope 68, No. 5, p. 482
Bradfield (C/1987 P1)	Dec. 1987	~0.5	Sky and Telescope 75, No. 3, p. 334–335
19P/Borrelly	Dec. 1994	~0.01	Sky and Telescope 90, No. 2, p. 108
Hale-Bopp (C/1995 O1)	Jan. 1997	2	Sky and Telescope 93, No. 4, p. 28
Skiff (C/1999 J2)	April and May 2000	0.1	IAU Circular #7415
LINEAR (C/2000 WM_1)	Nov. 2001	~0.01	Icarus 197, p. 183–202
LINEAR (C/2003 K4)	Nov. 2004	~0.2	http://www.yp-connect.net/~mmatti/
Lulin (C/2007 N3)	Jan. 2009	~0.2	http://www.spaceweather.com/comets/lulin/o8jan09/Paul-Mortfield1.jpg

61P/Shajn-Schaldach changed from being straight to being bent in just 30 min. Comet 1P/Halley underwent over a dozen similar events in 1985–1986. Changes in the gas tail probably are more likely to occur during solar maximum. Solar maximum is expected to occur around 2012 and 2023.

When the Earth passes within a few degrees of the comet's orbital plane, one should look for an antitail. This is a feature that points towards the Sun. Material in the comet's orbit is responsible for this feature. Table 3.9 lists a few comets that had antitails. One should also note the relative intensity of this feature. Some comets have bright antitails while others have faint ones, or lack them completely.

Once the drawing is finished, one should note the date (in Universal Time), the size and magnification of their binoculars, his or her location and the sky transparency. A brief discussion of sky transparency and how to estimate it follows.

Sky Transparency

Sky transparency is a measure of the amount of light that is getting through our atmosphere. While a thin haze may help one get a better view of a bright planet like Mars, it may be a problem for comets because of their lower surface brightness. More importantly, the atmospheric transparency can affect the perceived brightness of a diffuse comet. As a result, it is important for the observer to determine his or her transparency. There are two ways of estimating the transparency.

One method is to determine the magnitude of the faintest star which is at the same altitude as the target. This star should be at the limit of visibility in direct vision. The problem with this method, however, is that one may not find a star meeting such criteria.

A second method of estimating the limiting magnitude is to determine the faintest star in a constellation or a grouping that *is* visible and the "brightest" star in it that is *not* visible. While this may seem confusing, one can estimate the limiting magnitude by studying the star patterns in Figs. 3.19 and 3.20 and following the calculations described in this example. Using Fig. 3.19, let's say that one is able to see the stars β, α, γ, δ and 7-Triangulum, but is unable to see the others. In

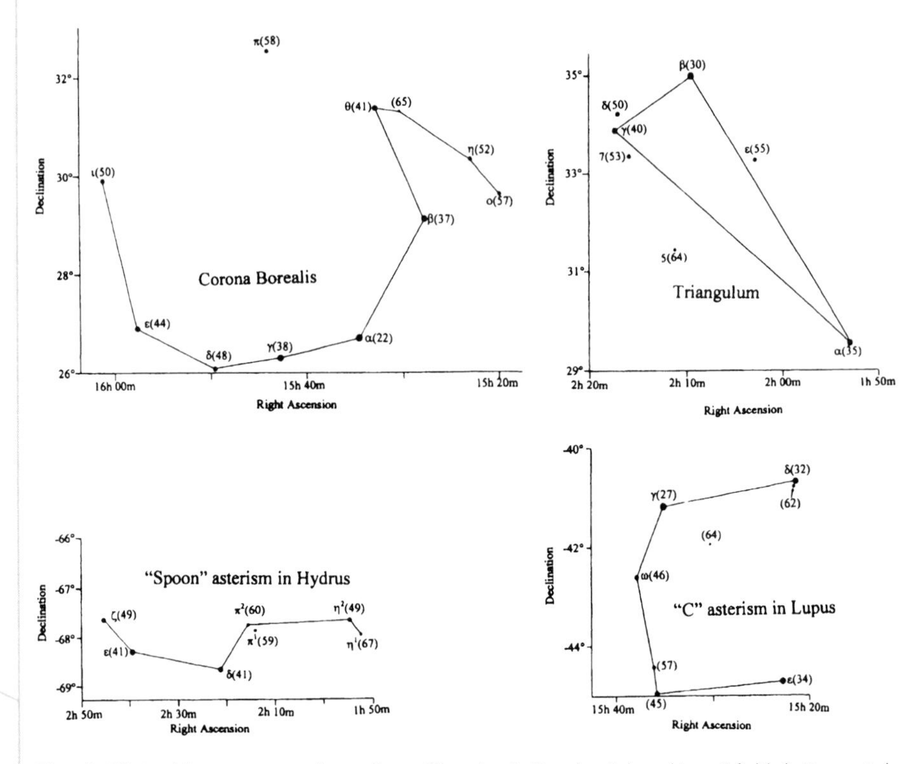

Fig. 3.19. Four different star patterns in the constellations of Corona Borealis, Triangulum, Hydrus and Lupus. To find the limiting magnitude, one determines the faintest star that can be seen and the "brightest" star that cannot be seen. The limiting magnitude is the average of the stellar magnitudes of these two stars (credit: Richard W. Schmude, Jr.).

this example, the faintest star seen is 7-Triangulum at a stellar magnitude of 5.3. The "brightest" star that is not visible is ε-Trianguli at a stellar magnitude of 5.5. The limiting magnitude is computed as the average of these two stellar magnitudes (faintest star seen and "brightest" star not seen) which is magnitude 5.4. Note that each star in Figs. 3.19 and 3.20 has its Greek letter designation along with its brightness in stellar magnitudes, with the decimal point eliminated to avoid confusion. All brightness values are reported to the nearest tenth of a magnitude. In all cases, I have used (3.12) and V filter magnitude measurements to compute the stellar magnitudes.

One can also determine the limiting magnitude by counting the number of stars in a portion of the sky. Figure 3.21 shows four portions of the sky bounded by bright stars in four different constellations. One determines the limiting magnitude by examining one of the four areas and counting the number of stars that are visible. Stars that define the area are included in the count. One then uses Table 3.10 to determine the faintest star visible and the "brightest" star that is not visible and computes the limiting magnitude in the same way as is described in the previous paragraph. For example, let's say that one examines the Cetus polygon and sees ten stars. One then examines the fifth and sixth

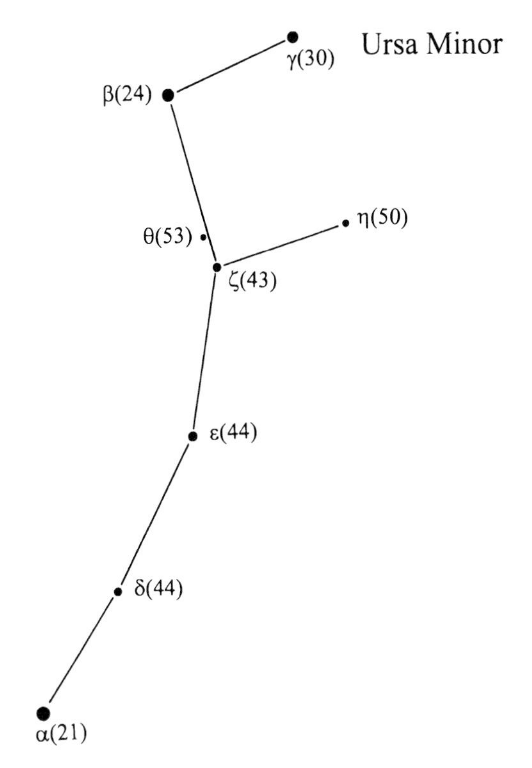

Fig. 3.20. Stars in the Little dipper (located in the constellation of Ursa Minor) along with corresponding brightness values in stellar magnitudes. To find the limiting magnitude, one determines the faintest star that can be seen and the "brightest" star that cannot be seen. The limiting magnitude is the average of the stellar magnitudes of these two stars (credit: Richard W. Schmude, Jr.).

columns of Table 3.10 and notes that the faintest star seen is at magnitude 5.7 and the "brightest" star not seen is at magnitude 6.4. The limiting magnitude is computed as the average of these two values. It comes out to be 6.05 which is reported as 6.1.

When estimating the limiting magnitude one should keep two things in mind. The limiting magnitude should be for the same altitude as the comet of interest, and one should use direct vision when determining their limiting magnitude. In this regard, the star patterns in Figs. 3.19–3.21 will usually not be at the same latitude as the comet of interest. In this case, the limiting magnitude must be considered "uncorrected" until an extinction correction factor is added. Figure 3.22 shows a chart that one can use to determine this correction factor. Let's say that one observes a comet at an altitude of 20°, and determines an uncorrected limiting magnitude of 5.3 from stars at an altitude of 30°. Figure 3.22 shows that the correction factor is -0.2 magnitudes and, hence, the limiting magnitude is 5.1. Essentially one adds the value in Fig. 3.22 to their uncorrected value to get the corrected limiting magnitude.

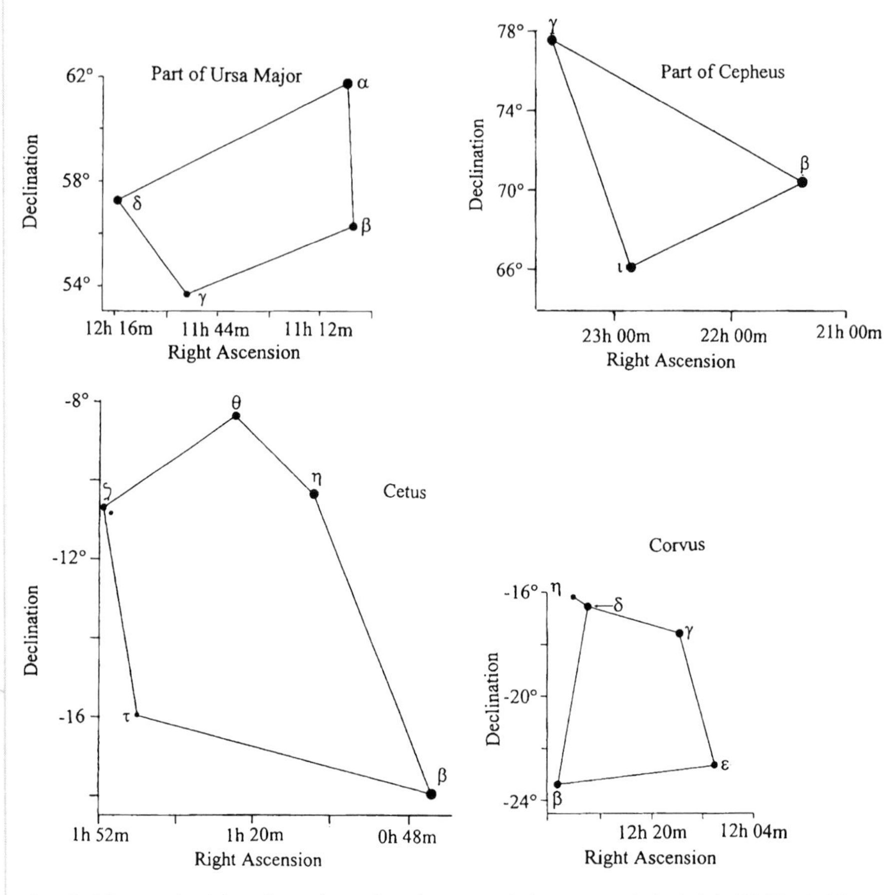

Fig. 3.21. Figures described in Table 3.10 that may be used in estimating the limiting magnitudes (credit: Richard W. Schmude, Jr.).

Extinction Correction

One can also use Fig. 3.22 for brightness estimates of the coma. For example, let's say that the estimated brightness of the coma is equivalent to a star of magnitude 1.5, and that the coma (or comet) is 6° above the horizon and the comparison stars are 20° above the horizon. One matches up the 6° and 20° and obtains –1.7 from Fig. 3.22. The corrected brightness of the coma is 1.5 + –1.7 = –0.2 stellar magnitudes.

Table 3.10. Limiting magnitudes in four different constellations

Part of Ursa Major		Part of Cepheus		Cetus		Corvus	
Number of stars visible	Magnitude	Number of Stars visible	Magnitude	Number of Stars visible	Magnitude	Number of Stars visible	Magnitude
1	2.0	1	3.1	1	2.3	1	2.6
2	2.4	2	3.4	2,3	3.7	2	2.8
3	2.4	3	3.7	4	3.8	3	2.9
4	3.3	4	4.6	5	4.0	4	3.3
5	5.5	5	4.8	6	4.6	5	4.4
6	5.8	6	5.0	7	5.1	6	5.2
7	5.9	7	5.1	8	5.6	7, 8	6.1
8	6.1	8	5.2	9, 10	5.7	9	6.5
9 or 10	6.3	9	5.6	11–13	6.4	10	6.7
11	6.5	10, 11	5.7	14	6.7	11	7.1
12	6.8	12	5.8	15	7.0	12	7.5
13	7.1	13, 14	6.1				
		15–17	6.3				
		18, 19	6.4				
		20	6.8				

Refer to Fig. 3.21. I computed all star magnitude values in this table from (3.12) (published in Stanton, 1999) and B and V magnitudes published in *Sky Catalogue 2000.0*, Volume 1, 2nd edition ©1991 by Alan Hirshfeld, Roger W. Sinnott and François Ochsenbein

Comparison star(s) altitude (degrees)

Comet altitude (degrees)	4	6	8	10	13	16	20	25	30	35	40	45	50	55	60	70	80	90
4	0	-0.9	-1.5	-1.8	-2.2	-2.4	-2.6	-2.7	-2.8	-2.9	-2.9	-2.9	-3.0	-3.0	-3.0	-3.0	-3.0	-3.0
6	0.9	0	-0.6	-1.0	-1.3	-1.5	-1.7	-1.8	-1.9	-2.0	-2.0	-2.0	-2.1	-2.1	-2.1	-2.1	-2.1	-2.1
8	1.5	0.6	0	-0.4	-0.7	-0.9	-1.1	-1.2	-1.3	-1.4	-1.4	-1.4	-1.5	-1.5	-1.5	-1.5	-1.5	-1.5
10	1.8	1.0	0.4	0	-0.3	-0.5	-0.7	-0.8	-0.9	-1.0	-1.1	-1.1	-1.1	-1.1	-1.2	-1.2	-1.2	-1.2
13	2.2	1.3	0.7	0.3	0	-0.2	-0.4	-0.5	-0.6	-0.7	-0.7	-0.8	-0.8	-0.8	-0.8	-0.8	-0.9	-0.9
16	2.4	1.5	0.9	0.5	0.2	0	-0.2	-0.3	-0.4	-0.5	-0.5	-0.6	-0.6	-0.6	-0.6	-0.6	-0.7	-0.7
20	2.6	1.7	1.1	0.7	0.4	0.2	0	-0.1	-0.2	-0.3	-0.3	-0.4	-0.4	-0.4	-0.4	-0.5	-0.5	-0.5
25	2.7	1.8	1.2	0.8	0.5	0.3	0.1	0	-0.1	-0.2	-0.2	-0.2	-0.3	-0.3	-0.3	-0.3	-0.3	-0.3
30	2.8	1.9	1.3	0.9	0.6	0.4	0.2	0.1	0	-0.1	-0.1	-0.1	-0.2	-0.2	-0.2	-0.2	-0.2	-0.3
35	2.9	2.0	1.4	1.0	0.7	0.5	0.3	0.2	0.1	0	0	-0.1	-0.1	-0.1	-0.1	-0.2	-0.2	-0.2
40	2.9	2.0	1.4	1.1	0.7	0.5	0.3	0.2	0.1	0	0	0	-0.1	-0.1	-0.1	-0.1	-0.1	-0.1
45	2.9	2.0	1.4	1.1	0.8	0.6	0.4	0.2	0.1	0.1	0	0	0	0	-0.1	-0.1	-0.1	-0.1
50	3.0	2.1	1.5	1.1	0.8	0.6	0.4	0.3	0.2	0.1	0.1	0	0	0	0	-0.1	-0.1	-0.1
55	3.0	2.1	1.5	1.1	0.8	0.6	0.4	0.3	0.2	0.1	0.1	0	0	0	0	0	0	-0.1
60	3.0	2.1	1.5	1.2	0.8	0.6	0.4	0.3	0.2	0.1	0.1	0.1	0	0	0	0	0	0
70	3.0	2.1	1.5	1.2	0.8	0.6	0.5	0.3	0.2	0.2	0.1	0.1	0.1	0	0	0	0	0
80	3.0	2.1	1.5	1.2	0.9	0.7	0.5	0.3	0.2	0.2	0.1	0.1	0.1	0	0	0	0	0
90	3.0	2.1	1.5	1.2	0.9	0.7	0.5	0.3	0.3	0.2	0.1	0.1	0.1	0.1	0	0	0	0

Fig. 3.22. This chart allows one to select the proper correction factor for his or her uncorrected limiting magnitude and to correct for extinction for a visual brightness estimate. The conversion factors here must be added to the uncorrected limiting magnitude value. In all cases, correction factors are in stellar magnitudes. A correction factor must be added if the comet and star(s) used in estimating the limiting magnitude are at different altitudes. The values of 8° and above in this chart are based on an extinction coefficient of 0.25 magnitudes/air mass. Values of 6°, 4° and 2° are based on the same extinction coefficient except that −0.3, −0.2 and −0.1 magnitudes were added to them to account for the curvature of the Earth and its atmosphere. In all cases, I computed the values in this chart. The −0.3, −0.2 and −0.1 magnitude values are based on my measurements of Venus as it set on January 15, 2009 (credit: Richard W. Schmude, Jr.).

Chapter 4

Observing with Small Telescopes

Introduction

With a small telescope (lens diameter of up to 12 in. or 0.30 m) one should be able to carry out many useful comet projects. While some features of a comet cannot be measured because of a lack of scale, this should not dissuade one from conducting studies of comet features. In the text which follows, various comet features which can be measured are presented, together with my beliefs of what needs to be done in order to maximize the scientific potential of drawings and images. In conducting such studies, for example, one should be able to measure the rotation period of the nucleus and the growth rate of any jet(s). With the proper filters, measurements could be made of the gas-to-dust ratio in the coma, and with a CCD camera and an electronic calculator, the orientation, shape and length of the gas and dust tails as the comet moves around the Sun could be monitored. Other areas of study are presented below.

First, the equipment which one may use to study comets and how to compute the image scale are discussed.

Equipment

In the next few pages, I will discuss telescopes, mounts, eyepieces, filters, finders, bino viewers, giant binoculars and the atmospheric dispersion corrector. My goal is to help the reader optimize his/her system for the study of comets. Several tables in this section list references to product reviews. These reviews are written to inform of the strengths and weaknesses of various pieces of commercially available equipment.

R. Schmude, *Comets and How to Observe Them*, Astronomers' Observing Guides,
DOI 10.1007/978-1-4419-5790-0_4,

Telescopes

The refractor, Newtonian and Schmidt-Cassegrain Telescopes are described below. I will also describe the three functions of an astronomical telescope – light gathering power, resolution and magnification.

Refractors, Newtonians and Schmidt-Cassegrains

Modern refractors have either achromatic or apochromatic objectives. Achromatic refractor objectives contain two lenses. One is made of crown glass and the other is made of flint glass. Crown and flint glass bend light differently and, when used together, they reduce the amount of false color. Apochromatic refractors contain objectives made of two or more lenses. One of the lenses is made of either extra-low dispersion glass or fluorite. Apochromatic refractors have less false color than achromatic refractors but are more expensive. Light enters the front of both types of refractor telescopes and the observer focuses it into the eyepiece. See Fig. 4.1.

A second type of telescope is the Newtonian Reflector. It contains a large (or primary) mirror at the back end of the tube and a second smaller one (secondary mirror) near the front of the tube. Light moves through the front of the tube, hits the primary mirror and is bent towards the secondary mirror which reflects the light into the eyepiece. See Fig. 4.1.

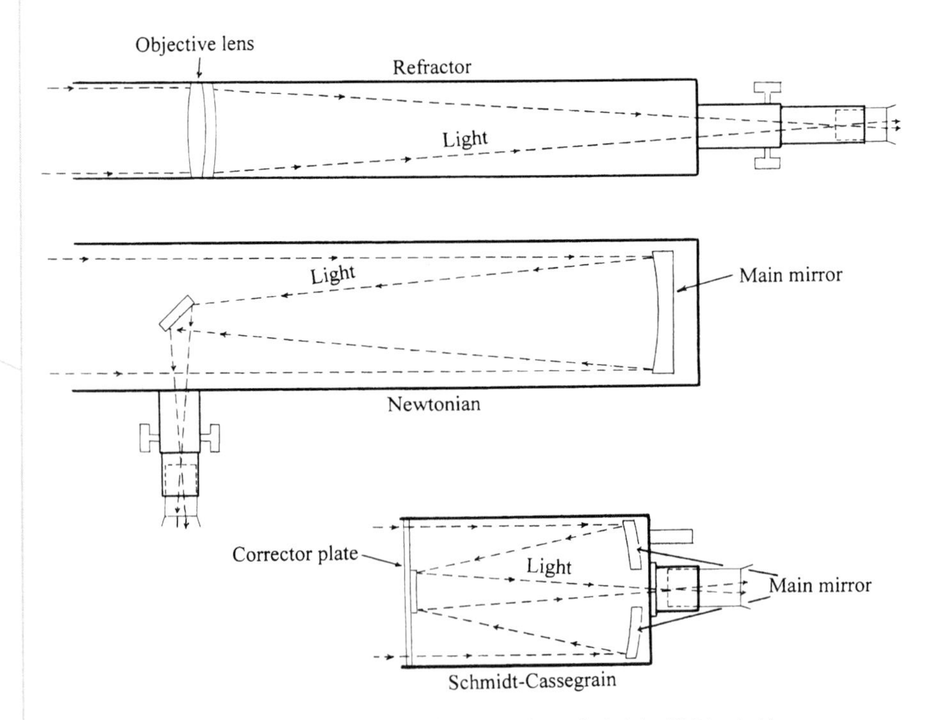

Fig. 4.1. Light paths for the refractor, Newtonian and Schmidt-Cassegrain telescopes (credit: Richard W. Schmude, Jr.).

Fig. 4.2. Newtonian (*left*) and Schmidt-Cassegrain (*right*) telescopes. Note how much shorter the Schmidt-Cassegrain telescope is compared to the Newtonian. The Newtonian is on a Dobsonian mount and the Schmidt-Cassegrain is on an equatorial mount (credit: Truman Boyle).

A third type of telescope is the Schmidt-Cassegrain. It contains two primary mirrors. Since this telescope has a folded optical path, it is short in size. See Fig. 4.1. Figure 4.2 shows a 0.2 m (8 in.) Schmidt-Cassegrain and a 0.2 m (8 in.) Newtonian Reflector. Notice the smaller size of the Schmidt-Cassegrain. The optical path of a Schmidt-Cassegrain is shown in Fig. 4.1. Dew will collect often on its corrector plate and, hence, one should insert a dew shield on the front of the telescope. A homemade dew shield can be made out of either poster board or cardboard.

I have used all three types of telescopes and believe that any of them can be used effectively in the study of comets. Table 4.1 lists advantages and disadvantages of each type. Table 4.2 gives lists of product reviews of several types of telescopes published in previous issues of the magazines *Sky and Telescope* and *Astronomy*. Product reviews in *Sky News*, a Canadian astronomy magazine, can be found online at http://www.SkyNews.ca (click on news and then click on reviews).

Light Gathering Power, Resolution and Magnification

The three functions of an astronomical telescope are to gather light, resolve fine detail and magnify an image. Of course, its most important function is to gather light. How does a telescope do this? All light entering one's eye must travel through the pupil. Light not striking the pupil will not reach the retina. See Frame a of

Table 4.1. Strengths and weaknesses of the Refractor, Newtonian, Schmidt-Cassegrain and Maksutov Telescopes.

Type	Strengths	Weaknesses
Refractor	Closed tube, low maintenance, and convenient location of eyepiece	Lens absorbs ultraviolet light Diameters larger than seven inches are not portable Expensive per inch of aperture
Newtonian	Low-cost per inch of aperture Can be used for ultraviolet photometry	Periodic maintenance required Difficult to collimate for low f-numbers (focal ratios) Eyepiece can be in an inconvenient location
Schmidt-Cassegrain or Maksutov	Low-cost per inch of aperture Closed tube Long focal length Portable Eyepiece is in a convenient location	May require more time to reach ambient temperature than Newtonians

Fig. 4.3. The situation is different for a telescope. Essentially almost all of the light striking the objective lens or primary mirror is focused into a small area which then enters the pupil. A 0.2 m (8 in.) telescope can gather 1,600 times as much light as a person's eye alone. See Frame b of Fig. 4.3. This is why one is able to see fainter objects through the telescope than with just his or her eyes.

Resolution is the smallest angular separation that can be seen. It affects the focus. All telescopes show point sources of light as a "blur". In addition, thin lines of light are blurred a little through the telescope. Poorly constructed (or poorly aligned) telescopes will show a larger blur than well constructed ones. Generally, large telescopes show a smaller blur than small ones. Therefore, telescopes showing the smallest blur have the best resolution and, hence, the sharpest focus.

There are several ways of defining resolution. One common way is in terms of how well a telescope splits double stars in visible light. The Dawes Limit is based on this. It expresses the telescope resolution value, R, as:

$$R = \frac{4.57\,\text{arc - seconds} \times \text{inches}}{D}. \tag{4.1}$$

In this equation D is the telescope diameter. Since we must cancel units, the unit for D must be inches. For example, the Dawes limit of a 0.20 m (8 in.) telescope is 0.57 arc-sec. For comparison, the resolution of the unaided eye (20/20 vision) is 60 arc-sec. Therefore, the 0.20 m telescope can resolve detail over 100 times smaller than the unaided eye. Equation (4.1) is valid only for well-aligned telescopes. The atmosphere will often limit the resolution to 1 or 2 arc-sec.

One can also describe a telescope's resolving capacity in terms of its line resolution. This describes the thinnest line of light that can be seen. Cassini's Division in Saturn's rings is an example. I have seen this dark strip almost to the edge of Saturn's disc (in good seeing) on several occasions with my 0.1 m (4 in.) refractor. Cassini's Division is usually less than 0.3 arc-sec wide near Saturn's disk. This is well below the Dawes Limit of my telescope. The line resolution value of a telescope is much smaller than the Dawes Limit. Therefore, the smallest feature that a telescope can resolve depends partly on the nature of the feature. For the purposes of studying diffuse

Table 4.2. (A) Product reviews of different kinds of telescopes in *Sky and Telescope (S & T) Magazine*; diameters are approximate. (B) Product reviews of different kinds of telescopes in *Astronomy (Ast.) Magazine*.

Refractors (apochromatic unless stated otherwise)			Newtonians			Schmidt-Cassegrains (SC), Maksutovs (M) and others		
Vendor	Diameter	Reference	Vendor	Diameter	Reference	Vendor-type[a]	Diameter	Reference
Several	10 cm (4 in.)	May 02, p. 44	Meade	15–25 cm (6–10 in.)	Dec. 02, p. 48	Celestron-SC	28 cm (11 in.)	Feb. 02, p. 49
Takahashi	10 cm (4 in.)	Jun. 02 p. 48	Orion	11 cm (4.5 in.)	Jun. 03, p. 46	Meade, Orion-M	11–14 cm (4–5 in.)	Mar. 02, p. 44
TMB Optical								
TALscopes	10 cm (4 in.)	Apr. 03, p. 56	Discovery	32 cm (12.5 in.)	Nov. 03, p. 54	Questar-M	9 cm (3.5 in.)	Nov. 02, p. 49
Stellarvue	8 cm (3 in.)	Sept. 03, p. 50	Hardin Optical	25 cm (10 in.)	May 04, p. 96	Meade-SC	20 cm (8 in.)	Mar. 03, p. 50
Telescope Eng. Company	14 cm (5.5 in.)	Dec. 03, p. 54	Several	25 cm (10 in.)	May 04, p. 104	Celestron-SC	36 cm (14 in.)	Mar. 04, p. 54
Orion	8 cm (3 in.)	Feb. 04, p. 60	Several	10 cm (4 in.)	Oct. 04, p. 96	Several-M	9 cm (3.5 in.)	Apr. 05, p. 98
William Optics	11 cm (4.4 in.)	Jun 04 p. 94	Orion	20 cm (8 in.)	Nov. 04, p. 86	Meade (RCX400)	30 cm (12 in.)	Feb. 06, p. 78
Several	10 cm (4 in.)	Jun. 04, p. 104	Orion	9 cm (3.5 in.)	Apr. 05, p. 88	Celestron-SC	20 cm (8 in.)	Mar. 06, p. 74
Tele Vue	6 cm (2.4 in.)	Dec. 04, p. 102	Several	7–11 cm (3–4.5 in.)	Dec. 05, p. 86	Celestron-SC	15 cm (6 in.)	Aug. 06, p. 86
Several	6–7 cm (2–3 in.)	Dec. 04, p. 112	Meade	25 cm (10 in.)	Oct. 06, p. 80	STF-M	18 cm (7 in.)	Feb. 07, p. 76
William Optics	7 cm (2.6 in.)	May 06, p. 76	Parks	20 cm (8 in.)	Aug. 07, p. 64	Celestron-SC	15 cm (6 in.)	Dec. 07, p. 34
William Optics	13 cm (5 in.)	May 07, p. 77	Orion	15 cm (6 in.)	Sept. 08, p. 36	Astro Systeme-Astrograph	20–40 cm (8–16 in.)	Jun. 08, p. 37
Tele Vue	13 cm (5 in.)	Jul. 07, p. 66	Obsession	32 cm (12.5 in.)	Feb. 09, p. 38	Meade	15 cm (6 in.)	Mar. 09, p. 38
Borg	8–10 cm (3–4 in.)	Mar. 08, p. 40	Sky-Watcher	30 cm (12 in.)	May 09, p. 34			
TMB	9.2 cm (3.6 in.)	Mar. 09, p. 36	Orion	30 cm (12 in.)	July 09, p. 34			
William Optics	9 cm (3.6 in.)	May 07, p. 74						

(continued)

Table 4.2. (continued)

Refractors (apochromatic unless stated otherwise)			Newtonians			Schmidt-Cassegrains (SC), Maksutovs (M), Maksutov-Cassegrain (MC), Schmidt-Newtonian (SN), Ritchey-Chrétien (RC), and Klevtsov-Cassegrain (KC)		
Vendor	Diameter	Reference	Vendor	Diameter	Reference	Vendor-type[a]	Diameter	Reference
Konus	10-15 cm (4-6 in.)	July 02, p. 66	Konus	11 cm (4.5 in.)	July 02, p. 66	RC Optical Systems-RC	32 cm (12.5 in.)	Dec. 02, p. 80
Tele Vue	8 cm (3 in.)	Sept. 02, p. 66	Celestron & Swift	11 cm (4.5 in.)	Nov. 02, p. 72	LOMO America-MC	7–20 cm (2.8–8 in.)	May 02, p. 62
Several[b]	8–9 cm (3–4 in.)	Oct. 02, p. 68	Orion	11–20 cm (4.5–8 in.)	May 03, p. 90	Celestron-SC	20 cm (8 in.)	Jan. 03, p. 84
Tele Vue	6 cm (2.4 in.)	Nov. 04, p. 90	DGM Optics	9 cm (3.6 in.)[e]	Oct. 03, p. 82	Meade-SC	20 cm (8 in.)	Feb. 03, p. 82
Several[c]	8 cm (3 in.)	Mar. 06, p. 86	Orion	11 cm (4.5 in.)	Jan. 04, p. 84	Celestron-SC	13 and 20 cm (5 and 8 in.)	Mar. 03, p. 90
Vixen	8 cm (3 in.)	June 06, p. 90	Orion	20 cm (8 in.)	May 04, p. 86	Meade-SN	20 cm (8 in.)	Aug. 03, p. 96
Tele Vue	6 cm (2.4 in.)	Sept. 06, p. 78	Celestron	20 cm (8 in.)	Aug. 04, p. 88	TAL instruments-KC	15 and 20 cm (6 and 8 in.)	Mar. 04, p. 90
Stellarvue	10 cm (4 in.)	Oct. 06, p. 80	Orion	25 cm (10 in.)	Jan. 05, p. 82	Konus USA-MC	13 cm (5 in.)	Apr. 04, p. 84
Orion	10 cm (4 in.)	Apr. 07, p. 70	Celestron	8 cm (3 in.)	Feb. 05, p. 92	Meade-SC	25 cm (10 in.)	July 04, p. 88
Meade	15 cm (6 in.)	Sept. 07, p. 86	Star Structure Telescopes	32 cm (12.5 in.)	May 05, p. 78	Celestron-SC	20 cm (8 in.)	Aug. 04, p. 88
William Optics	7–9 cm (3 in.)	Oct. 07, p. 70	Starmaster Portable Telescopes	20 and 28 cm (8 and 11 in.)	Apr. 06, p. 90	Meade-SC	36 cm (14 in.)	Mar. 05, p. 78
Tele Vue	10 cm (4 in.)[d]	Jan. 08, p. 74	Meade	25 cm (10 in.)	May 07, p. 72	Meade-RC	30 cm (12 in.)	Feb. 06, p. 84

Orion	10 cm (4 in.)	Oct. 08, p. 78	Telescopes of Vermont	15 cm (6 in.)	June 07, p. 74	Celestron-SC	15 cm (6 in.)	Mar. 07, p. 76
			Parks Optical	15 and 20 cm (6 and 8 in.)	Nov. 07, p. 70	Celestron-SC	28 cm (11 in.)	Aug. 07, p. 72
			Obsession Telescopes	46 cm (18 in.)	Apr. 08, p. 68	Meade-SC	20 cm (8 in.)	Dec. 08, p. 70

[a] Type = telescope type
[b] Celestron, Orion, Stellarvue and Swift
[c] Stellarvue, Vixen and William Optics
[d] Nagler-Petzval telescope
[e] Off-axis Newtonian

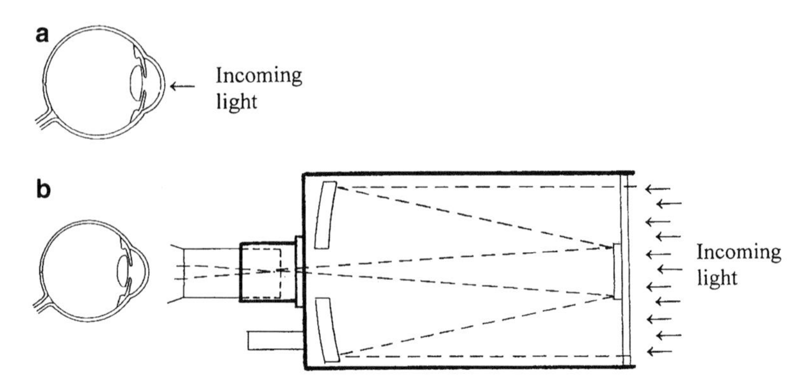

Fig. 4.3. (**a**) The only light entering our eye passes through the pupil. (**b**) Almost all of the light passing through the telescope is focused into an area smaller than the pupil and hence, passes into the eye (credit: Richard W. Schmude, Jr.).

features in comets, I believe that the Dawes Limit is appropriate. In many cases though, one will not attain even the Dawes Limit due to poor alignment, imperfections in the telescope optics, low contrast or atmospheric turbulence (seeing).

Since the late 1990s, astronomers have improved the resolution of their images by a factor of three or more. They have done this by taking many very short exposures and selecting the best ones. In some cases, they have stacked several images to make one high-resolution image. The atmospheric seeing does not blur short exposures as much as longer ones. Essentially, the seeing is nearly frozen. Astronomers have used this technique for comets with great success. For example, in 2007 Philip Good took 40 short exposures of Comet 17P/Holmes in four different filters (luminance, red, green and blue) to make the image in Fig. 4.4. This image shows a bright central condensation surrounded by the coma. Mike Salway took many 0.0006 s exposures of the International Space Station (ISS). Two of his best frames are shown in Fig. 4.5. Atmospheric seeing had little effect on the short exposures and, hence, Mike was able to "freeze" nearly the seeing. This was in spite of the rapid motion of the ISS across the sky.

The least important characteristic of a telescope is magnification. Magnification, M, is computed from:

$$M = \frac{\text{Telescope focal length}}{\text{Eyepiece focal length}} \tag{4.2}$$

For example, a telescope with a focal length of 2.0 m (or 2,000 mm) and an eyepiece of 20 mm focal length will yield a magnification of 2,000 mm ÷ 20 mm = 100×. One must use the same units for the telescope and eyepiece focal lengths to compute magnification. Useful magnifications range from about one-fifth to twice the telescope aperture in mm. The lower magnification limit for Newtonians is due to the secondary mirror obstruction creating a dark spot in the image. One may be able to go to lower magnifications with refractors provided that the goal is to study a relatively bright comet. The upper limit of magnification is due to a variety of reasons, including the limiting resolving power of telescopes.

Fig. 4.4. An image of Comet 17P/Holmes made by Philip Good on Oct. 25, 2007. This is a composite of 40 images taken in each of the four filters (luminance, *red, green* and *blue*) (credit: Philip Good).

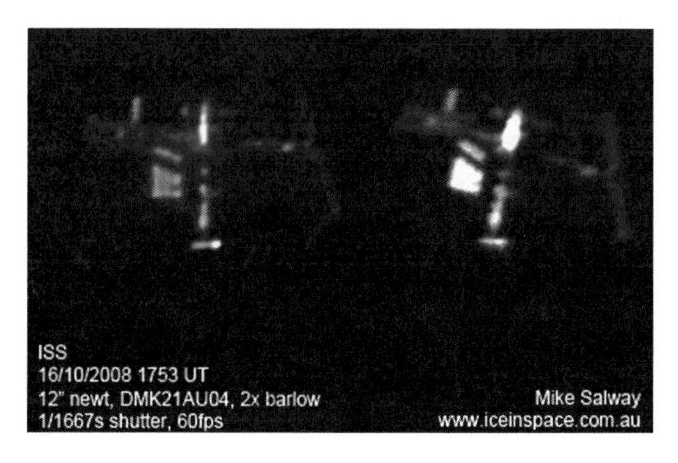

Fig. 4.5. An image of the International Space Station. These are two of the best single frames. The resolution of these images is well below 1.0 arc-sec and hence, the seeing was nearly frozen (credit: Mike Salway).

Table 4.3 lists different telescope diameters along with their light gathering power (LGP) compared to the unaided eye, the faintest comet (in stellar magnitudes) that can be studied with that LGP and the approximate useful magnification range. The limiting magnitudes for comets are based on my own experience. The LGP values are with respect to a pupil diameter of 0.5 cm or 0.2 in. The beginner may

Table 4.3. Light gathering power (LGP) of certain diameter telescopes compared to human eye pupils, approximate limiting magnitudes for comet viewing and approximate useful magnifications for different sized telescopes

Telescope diameter meters (inches)	Light gathering power compared to 0.5 cm pupils	Faintest comet (stellar magnitudes)[a]	Approximate useful magnifications
0.1 (4)	410	9–10	20× to 200×
0.15 (6)	930	10–11	30× to 300×
0.20 (8)	1,700	10.5–11.5	40× to 400×
0.25 (10)	2,600	11–12	50× to 500×
0.36 (14)	5,100	12–13	70× to 700×
0.51 (20)	10,000	12.5–13.5	100× to 1,000×
0.61 (24)	15,000	13–14	120× to 1,200×
0.76 (30)	23,000	13.5–14.5	150× to 1,500×

[a] This is my limit for studying comets. Detection limits are fainter

not reach these limits whereas the veteran comet observer probably will exceed them. People with high-quality instruments probably will exceed these limits.

A telescope's focal ratio will often be shown on the telescope. The focal ratio is the focal length of the objective lens or primary mirror divided by its aperture. For example, my 0.10 or 100 mm refractor (aperture = 100 mm) has a focal length of 1,000 mm and, hence, its focal ratio is 1,000 mm ÷ 100 mm = 10. This is written as f/10.

Mounts

Almost all telescopes require a mount. The mount should be sturdy and must be able to support the weight of the telescope, counter weights and any attachment(s) to it. In many cases, it must also move the telescope to compensate for Earth's rotation. The mount is often the most expensive part of a telescope system. A discussion of the three types of mounts and of setting circles follows.

Types of Mounts

Three common types of mounts are the ball-and-socket, altitude-azimuth and equatorial mounts. A ball-and-socket mount allows the telescope to move in any direction. This mount is usually portable and easy to set up. An example of this mount is the red Astroscan® telescope. One drawback is that there is no commercially available ball-and-socket mount to my knowledge which holds celestial objects as Earth rotates.

Altitude-azimuth mounts move the telescope in two directions – altitude and azimuth. Azimuth is movement along the horizon whereas altitude is the up-and-down movement. The Dobsonian mount is a type of altitude-azimuth mount. The Newtonian telescope in Fig. 4.2 has a Dobsonian mount.

An equatorial mount, moves the telescope in both right ascension and declination. Two examples of this kind of mount are the German equatorial and the fork equatorial mount. A telescope on an equatorial mount moves in right ascension and declination. Because of this, one of its axes must be pointed at a celestial pole. In the northern hemisphere, Polaris is very close to the north celestial pole, and, hence, one of the axes is pointed near this star.

There are many commercially available altitude-azimuth and equatorial mounts which can hold (or "track") stars. In some cases, these mounts can carry out more complex movement such as holding or tracking the Moon.

The automated mount points the telescope to a selected target. It can also move the telescope so that the target does not drift as a result of Earth's rotation. The automated mount can either be an equatorial or an altitude-azimuth mount. Keep in mind that many comets are recent discoveries and will not be in computer data bases of automated mounts. In this case, setting circles or star hopping may have to be used to find the comet.

In purchasing a mount, one should spend some time in examining product reviews. Portability, storage and set-up time should be considered also. Product reviews of telescope mounts are listed in Table 4.4. In most cases contact information for the vendors is published in the reviews.

Setting Circles and Finding Objects

Telescopes with an equatorial mount will usually have setting circles. See Fig. 4.6. One setting circle disc shows right ascension and the other shows declination. If the mount is set up so that one of the axes is pointed at a celestial pole, the declination

Table 4.4. Product reviews of telescope mounts in *Sky and Telescope* (S&T) and *Astronomy* (Ast.)

Vendor	Product name	Type	Reference
Pacific Telescope Corp.	Sky Watcher EQ6	EQ	(S&T) Oct. 02, p. 45
Paramount	GT-1100 ME	EQ	(Ast.) Apr. 03, p. 88
Orion	Sky View Pro	EQ	(S&T) Aug. 03, p. 56
Losmandy	Gemini System	EQ	(S&T) Oct. 03, p. 50
Vixon	Sphinx Go To	A	(S&T) Jul. 05, p. 84
Several[a]	See note a	A	(S&T) Jul. 05, p. 98
Celestron	CG-5	A	(S&T) Aug. 05, p. 82
Vixen	Sphinx automated mount	A	(Ast.) Nov. 05, p. 94
Vixen	Skypod mount	A	(Ast.) Dec. 07, p. 98
Astro-Physics	Mach1 GTO	A	(S&T) Dec. 07, p. 36
iOptron Corp.	Cube "Go To"	A	(S&T) Feb. 08, p. 34
Astro-Tech	Voyager mount X	AA	(S&T) Jul. 08, p. 36
AstroTrac Ltd.	AstroTrac TT320 mount	–	(S&T) Oct. 08, p. 38
iOptron Corp.	MiniTower	A	(S&T) Dec. 08, p. 48

EQ equatorial mount, *A* automated or go to mount, *AA* Altitude-Azimuth mount

[a]Celestron (CG-5; CGE), Losmandy (GM-8, G-11), Sky-Watcher (EQ6 SkyScan), Takahashi (EM-11 Temma II), Vixen (Sphinx)

Fig. 4.6. Setting *circles* on a German equatorial mount (credit: Richard W. Schmude, Jr.).

circle will give the correct declination. The right ascension is a different matter. One must set the right ascension circle for *each* observing session. This is accomplished by the following procedure: First, aim the telescope at a convenient bright object in the sky – let's say the star Sirius. Second, center Sirius in the eyepiece, focus the telescope and look up its coordinates which are: right ascension (2000) = 6 h 45 min, declination (2000) = −16° 43′. Third with these coordinates, check the declination to make sure that it is correct, and then move the right ascension dial until it is at 6 h 45 min. This must be done quickly because the Earth rotates about 0.25° per minute of time. Finally, with the right ascension and declination set properly, point the telescope so that it matches the right ascension and declination of the target, using a wide-field eyepiece. The target should be in the field.

There will occasionally be a bright daytime comet, like McNaught (C/2006 P1) in early 2007. In this case, the Sun (with a Sun filter *mandatory*!) may be used to get an approximate setting for right ascension and, if necessary, Venus or the Moon may be used for a more exact setting. Figure 4.7 shows the Sun's declination and Figs. 4.8–4.10 show the Sun's right ascension for different dates throughout the year. The Sun's coordinates repeat nearly each year and can be used to make approximate calibrations on the setting circles during the daytime. One should pre-focus the telescope. If it is not pre-focused, it will be much more difficult to find a target in the daytime.

Pre-focusing may be accomplished in the following way: First, focus the telescope on a distant object like the Moon. This can be accomplished in the day or at night. After this, make pencil marks on the drawtube where the eyepiece e.g. 25 and 9.7 mm, come into focus. During the day, move the focuser until it lines up with the 25 mm line on the drawtube. See Fig. 4.11. Finally, insert the eyepiece into the telescope.

Comets close to the Sun require extra precautions. The steps that I would take in this situation would be to pre-focus the telescope, position it so that the Sun is just

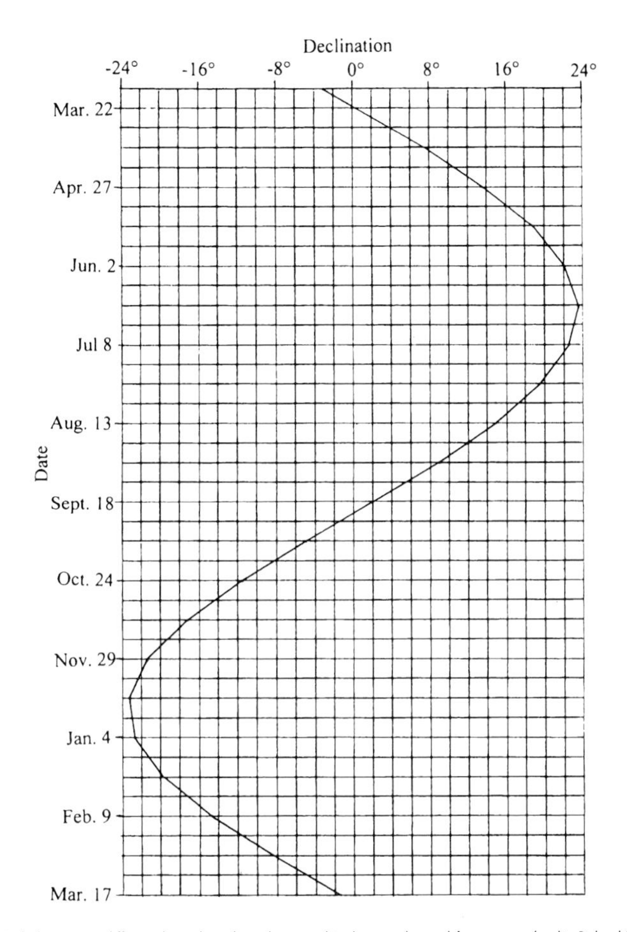

Fig. 4.7. Sun's declination on different dates throughout the year; this chart can be used for any year (credit: Richard W. Schmude, Jr. and the Astronomical Almanac).

above an obstruction like a roof, use the Sun to set right ascension and then point the telescope at the target's coordinates. For safety reasons, *never* look at an object near the Sun unless the Sun is *completely covered* by an obstruction. I have used this technique many times to find Venus when it was within 30° for the Sun. One can use this technique when the comet is north, south or east of the Sun. If it is west of the Sun, it can not be used since the Sun "moves" west.

Eyepieces

Eyepieces come in two standard sizes – 1.25 in. (32 mm) and 2 in. (51 mm). Some small, older telescopes also have 24.5 mm (0.965 in.) eyepieces. One should consider several factors when purchasing eyepieces. Eyepiece coatings and weight are two

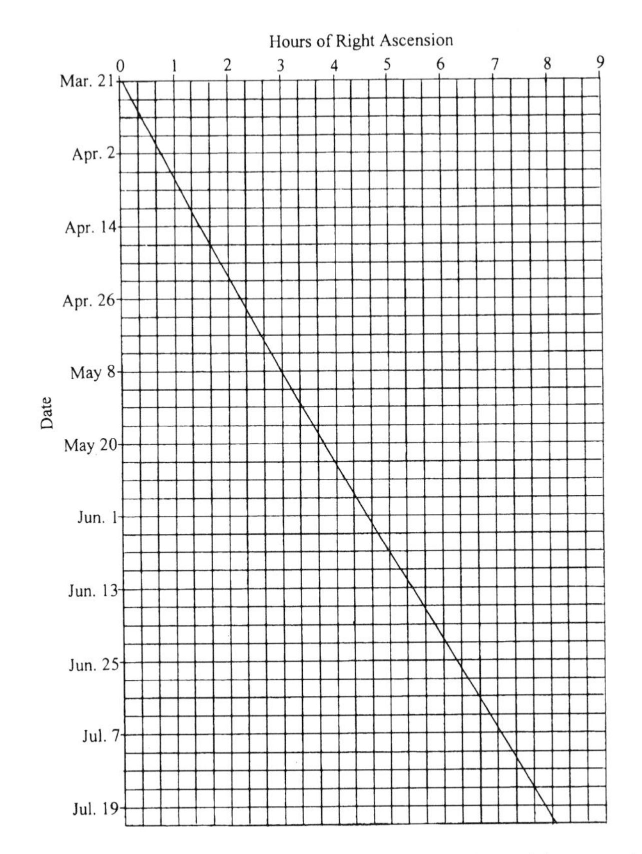

Fig. 4.8. Sun's right ascension from late-March through late-July (credit: Richard W. Schmude, Jr. and the Astronomical Almanac).

important characteristics. The coating terms for eyepieces are the same as for binoculars. See Chap. 3. Eyepieces can weigh quite a bit. The heavier ones will require substantial rebalancing. In addition to coatings and weight, one should consider also three other eyepiece characteristics when making a purchase, namely, focal length, eye-relief and apparent field. These characteristics are discussed below.

The focal length is expressed in millimeters (mm). The focal length determines both the magnification and the exit pupil. As mentioned earlier, the exit pupil is the diameter of the circular area where light passes through the eyepiece. The telescope exit pupil is computed as:

$$\text{Telescope exit pupil} = \frac{\text{Telescope objective diameter (in mm)}}{\text{Magnification}} \quad (4.3)$$

The same rules for binocular exit pupil apply to telescopes.

A small exit pupil may give an inferior view. By way of explanation, the maximum magnification is determined usually by the atmospheric seeing or the

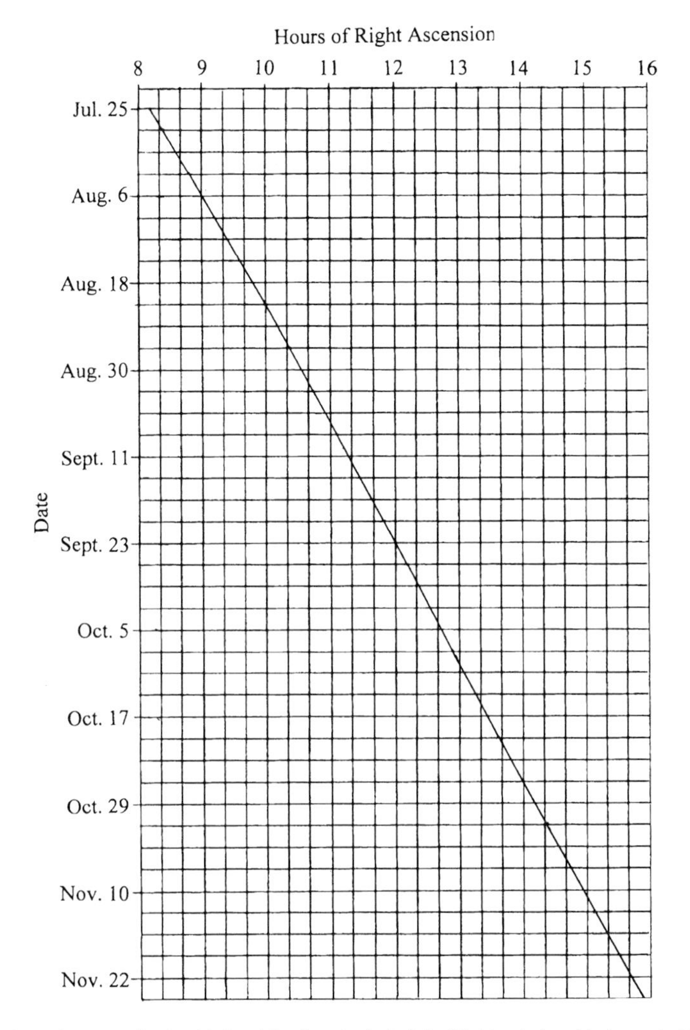

Fig. 4.9. Sun's right ascension from late-July through late-November (credit: Richard W. Schmude, Jr. and the Astronomical Almanac).

presence of floaters. Floaters are tiny pieces of material in the eye which interfere with views through small exit pupils. (I have noticed floaters when my exit pupil drops below 0.6 mm.) If a target is very bright, like Venus, a maximum magnification of 50× per inch of aperture (or 2× per mm of aperture) may be used. Comets are usually much dimmer than Venus and hence, lower magnifications are more suitable.

Veteran comet observer John Bortle recommends a magnification of 5× to 10× per inch of aperture (200× to 400× per meter of aperture) for comets. Therefore, magnifications of 40× to 80× are best for viewing a comet with a 0.2 m (8 in.) telescope.

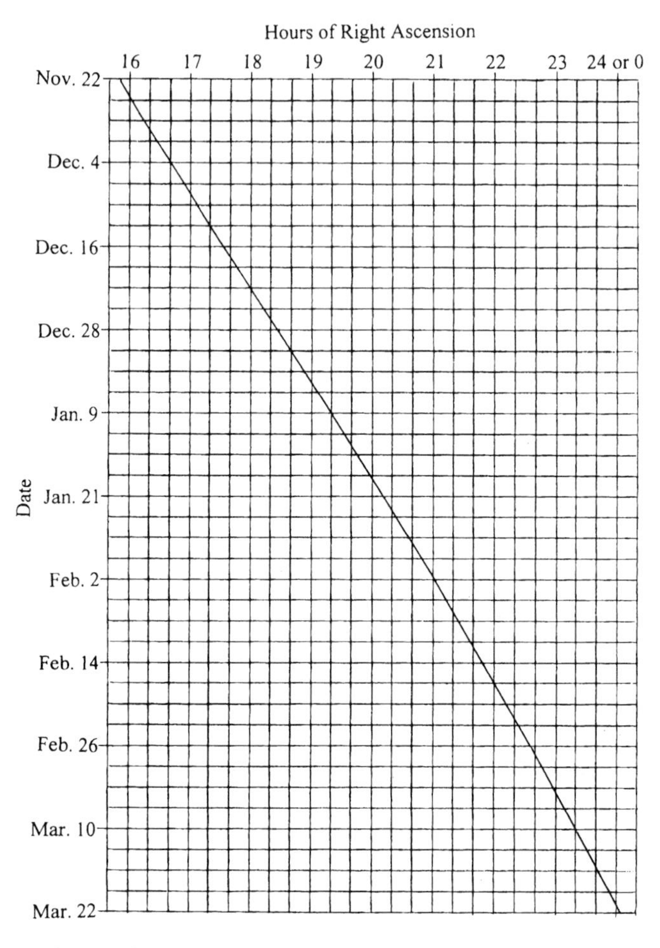

Fig. 4.10. Sun's right ascension from late-November through late-March (credit: Richard W. Schmude, Jr. and the Astronomical Almanac).

Another eyepiece characteristic is eye-relief. The eye-relief is the maximum distance between the observer's eye and the top of the eyepiece, where the entire field-of-view is visible. See Fig. 4.12. Most eyepieces have eye-relief values of between 10 and 30 mm (0.4–1.2 in.). As a general rule, the longer the focal length, the longer the eye-relief. One can use eyeglasses to look through eyepieces with an eye-relief of about 20 mm or more. If one has eyepieces with short eye-relief, he or she may be able to use them even with vision problems.

In this regard, people with corrective lenses are usually either nearsighted or farsighted. Many also have astigmatism. Corrective lenses are made in such a way as to correct these conditions. A telescope can be focused to correct for nearsightedness or farsightedness but until recently could not correct for astigmatism. One company (Tele Vue) has developed a corrector plate for that problem. This plate fits on certain eyepieces and will correct for astigmatism. Therefore, astronomers with

Fig. 4.11. An image showing marks on the drawtube where my 25 and 9.7 mm eyepiece come into focus. I move the drawtube until it lines up with the appropriate mark and the telescope is in focus (credit: Richard W. Schmude, Jr.).

astigmatism who use this corrector plate can look through a telescope without using their corrective eyeglasses. This corrector plate is especially useful for eyepieces with a short eye-relief.

The apparent field is another important eyepiece characteristic. This is the field when one looks through the eyepiece (without a telescope). The apparent field is reported in degrees. The vendor should know the apparent field. It is usually around 40 or 50°, but can be as high as 100°. The telescope FOV is approximately:

$$\text{Telescope(FOV)} = \frac{\text{Apparent field of eyepiece}}{\text{Magnification}} \tag{4.4}$$

For example, if one uses a Plössl eyepiece with an apparent field of 50° and a focal length of 8 mm with a telescope having a focal length of 1,200 mm, the magnification is 1,200 mm ÷ 8 mm = 150×. The FOV of the telescope with this eyepiece equals 50° ÷ 150 = 0.33°. One can verify the apparent field by recording the length of time it takes for a star "to move" across the telescope field. See Fig. 4.13. Essentially one aims the telescope at a star with a declination between 5°N and 5°S, and records the length of time it takes for the star "to move" across the field. This time (in seconds) is multiplied by 0.25 arc-min/sec to compute the telescope field in arc-minutes.

Filters

Filters are devices which block out some wavelengths of light but allow others to pass through. The full-width at half transmission (FWHT) defines the wavelength

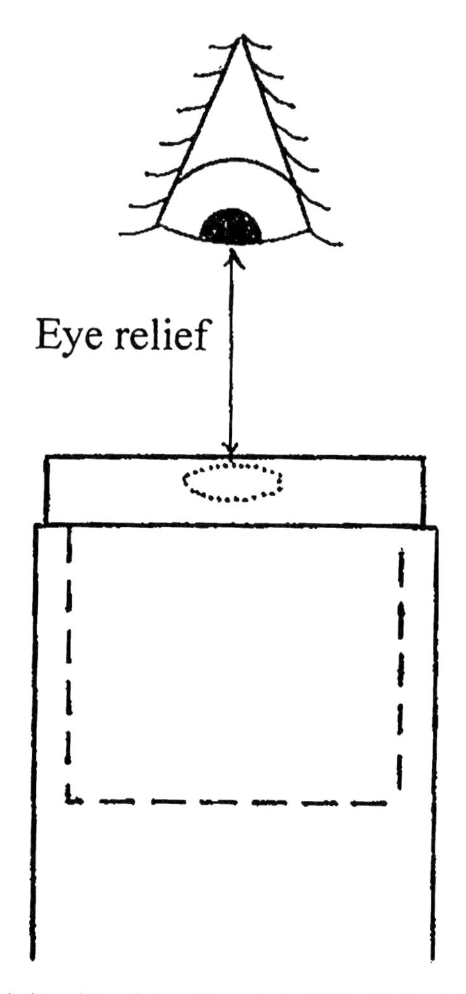

Fig. 4.12. The eye relief is the distance between the eye and the eyepiece (credit: Richard W. Schmude, Jr.).

range over which the transmission is 50% of the peak transmission. Some filters have a FWHT of 100 nm and hence, allow a wide range of wavelengths to pass through. Others have FWHT values of less than 15 nm and allow a narrow range of wavelengths to pass through. The amount of light that passes through a filter depends on both the FWHT and the peak transmission. A few commercially available filters also transmit light in two or more wavelength regions. Frame a of Fig. 4.14 illustrates a yellow filter. In this example, the FWHT is ~15 nm.

Two ways that filters enhance our view of comets are they: (1) eliminate chromatic aberration and (2) reduce the amount of unwanted scattered light.

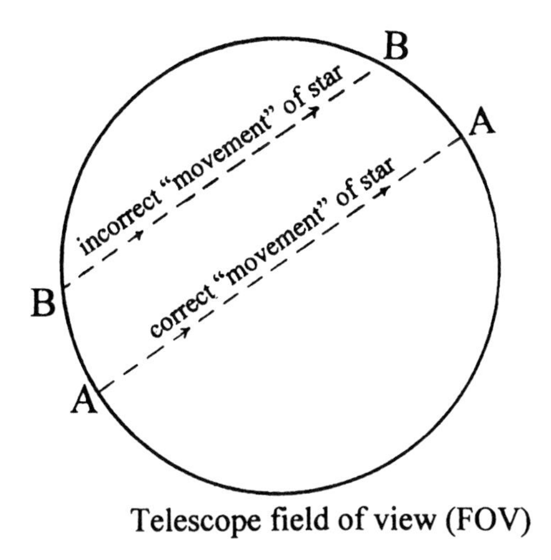

Fig. 4.13. One can determine the telescope's field of view (FOV) by measuring the time it takes a star to cross the diameter of the FOV (segment A-A). If the star does not cross the center of the FOV (such as segment B-B) then the time is invalid (credit: Richard W. Schmude, Jr.).

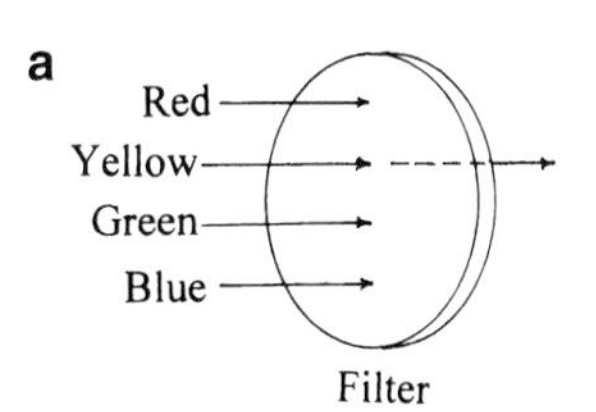

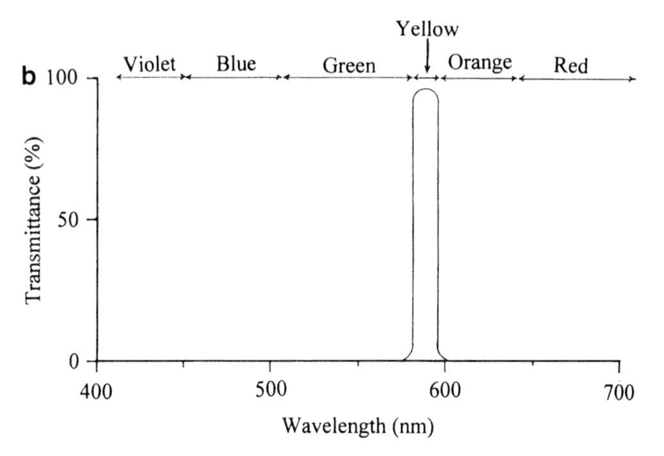

Fig. 4.14. (**a**) A *yellow filter* allows *yellow light* to pass through but blocks other colors (or wavelengths) of light. (**b**) Intensity of transmitted light versus the wavelength (or color) for a *yellow filter* (credit: Richard W. Schmude, Jr.).

The minus-violet filter blocks out wavelengths shorter than about 450 nm (violet light). When used with an achromatic refractor, this filter reduces the false violet color. In other words, it reduces chromatic aberration. In one product review, this filter is reported to enhance the contrast. There are several vendors that sell it. One product review of the minus-violet filter is in *Sky and Telescope* magazine (Jan. 2002, p. 57). One disadvantage of this filter is that it cuts out violet light making objects a bit dimmer.

Much of the light passing through the eyepiece can drown out important detail in a comet. In many cases, one can use a filter to cut out most of this interference. The objective of a light pollution filter is to reduce the amount of unwanted scattered light. In an ideal case, this filter will block out all of the unwanted scattered light without blocking out light from the target. A discussion of scattered light and of the two types of light pollution filters follows.

Three sources of scattered light are airglow, natural light and artificial light. Airglow occurs when high-energy light, or the solar wind, strikes atoms and molecules in our atmosphere. The atoms and molecules then emit visible light. Kevin Krisciunas carried out a study of the sky brightness on the slope of Mauna Kea (altitude = 2,800 m or 9,200 ft). Since there was no artificial light or moonlight, he essentially measured the airglow. He found that the airglow was a factor of two brighter near solar maximum in 1991 than in 1986 and 1995 – 2 years that were near solar minima. Therefore, airglow is probably lower near solar minimum.

Natural light also brightens the sky. This includes light reflected by the Moon, the brighter planets, stars and that which is reflected or scattered by dust in outer space. Light from dust causes both the Zodiacal light and the Gegenschein.

Artificial light is a third source of scattered light. Some artificial light, like low-pressure sodium bulbs, give off light in just a small wavelength range. See Fig. 4.15.

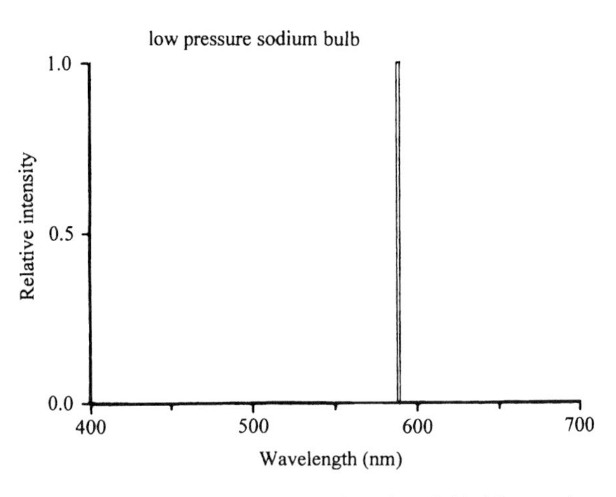

Fig. 4.15. Relative intensity versus wavelength for a low-pressure sodium lamp. Almost all of the light is emitted at a wavelength of 589 nm. Small amounts of light may be given off at other wavelengths but are not drawn in the spectrum (credit: Richard W. Schmude, Jr.).

Fluorescent tubes give off light at several wavelengths and incandescent bulbs give off light in all wavelengths. See Figs. 4.16 and 4.17. If most of the scattered artificial light is from low-pressure sodium light it can be filtered out under clear, haze-free skies. It is more difficult to filter out light from fluorescent and incandescent lamps, however. Aerosols and haze in our atmosphere tend to spread out the wavelengths of scattered light. As a result, it is easier to filter out unwanted light in nearly haze-free skies than in hazy skies.

Figure 4.18 shows a transmission versus wavelength graph for the Lumicon Deep Sky Filter. It allows in a lot of the visible light. Orion's SkyGlow Filter is similar. Most of the light given off by both low- and high-pressure sodium bulbs and mercury vapor bulbs is blocked by both filters. These filters will not perform well when there is a bright Moon in the sky since the Moon reflects all colors of light.

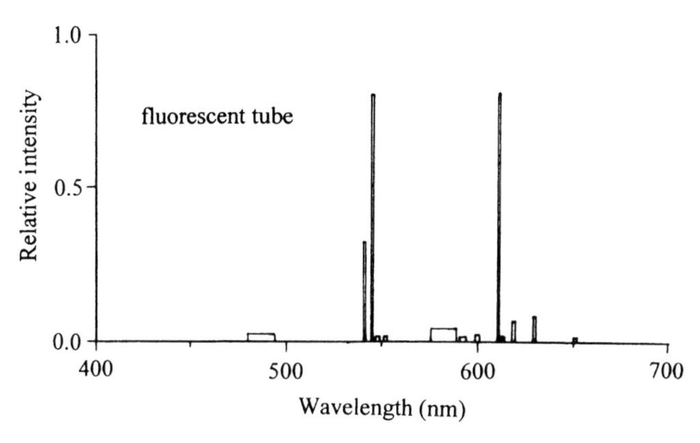

Fig. 4.16. Relative intensity versus wavelength for the fluorescent lamp in my classroom. All intensity values are estimates made by the Author. Note that more colors of light are given off than in the low pressure sodium lamp (credit: Richard W. Schmude, Jr.).

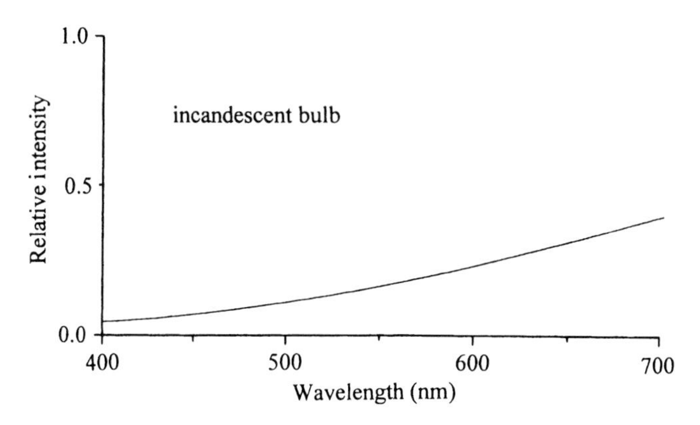

Fig. 4.17. Relative intensity versus the wavelength of an incandescent lamp. Note that all wavelengths of visible light are given off (credit: Richard W. Schmude, Jr.).

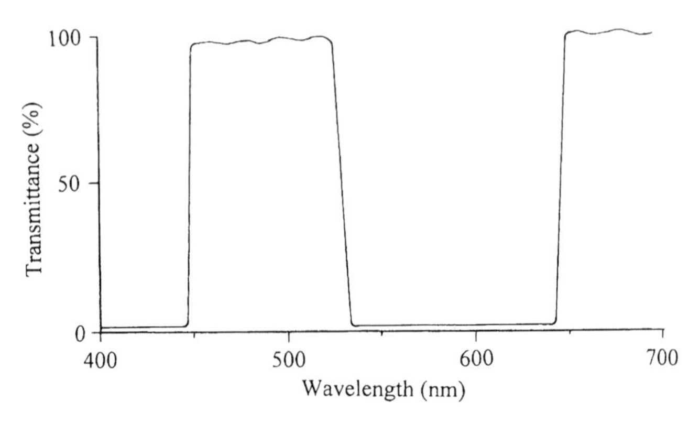

Fig. 4.18. A transmission versus wavelength graph of a Lumicon Deep Sky light pollution filter (credit: courtesy of Lumicon International).

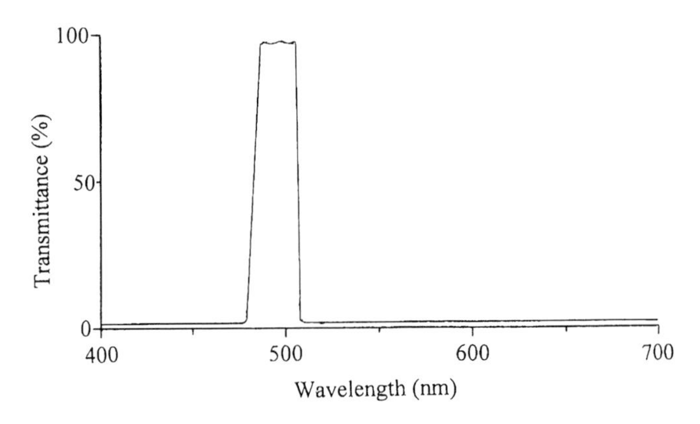

Fig. 4.19. A transmission versus wavelength graph of a Lumicon UHC light pollution filter (credit: courtesy of Lumicon International).

The Lumicon UHC Filter and Orion's UltraBlock Filter block out most light given off by artificial sources but allow specific wavelengths to pass. See Fig. 4.19 pertaining to the Lumincon UHC Filter. Light that is transmitted includes oxygen III light (496 and 501 nm), and hydrogen-beta light (486 nm). Since most of the light given off by planetary nebulae is not blocked by the UHC or UltraBlock filters, they yield excellent views of these features. One may also be able to use them to study or image comets with lots of gas. If one plans to image comets in the UHC or UltraBlock filters, he or she should use also an infrared blocking filter. This will eliminate the problem of infrared light contaminating the image.

Some features give off specific wavelengths of light and, hence, a filter which transmits just those wavelengths will reduce interfering light and enhance the contrast. The gaseous tail, for example, usually has strong C_2 emissions near 511 and 514 nm and, hence, a filter that transmits these wavelengths but blocks others will enhance the gas tail. One should realize that this enhancement is due to a reduction of scattered light around the comet. A filter will not make objects brighter, but it can improve the contrast and visibility.

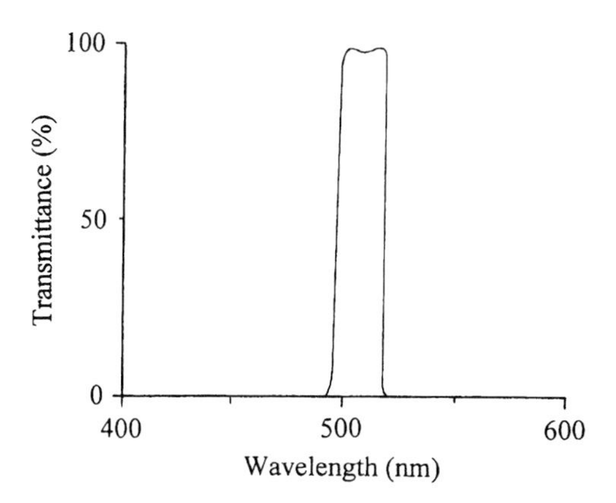

Fig. 4.20. A transmission versus wavelength graph of the Lumicon Swan band filter (credit: courtesy of Lumicon International).

One commercially available comet filter is the Lumicon Swan-Band (or comet) filter. It transmits light between ~495 nm and ~516 nm, but blocks out most other wavelengths. See Fig. 4.20. Much of the light from oxygen and molecular carbon (C_2) pass through it. If a comet has lots of gas, it will probably be more distinct in this filter. This is because it will not block out much light from the gaseous comet but will block out most of the scattered light surrounding it. The Swan-Band Filter, however, will not improve the view of a dusty comet. This is because it will block out most of the yellow, orange and green light reflected by the dust.

I have used the Swan-Band Filter to study Comet Lulin (C/2007 N3). This filter improved my view of the outer coma and central condensation of the comet on Feb. 6, 2009.

There are also several filters which transmit specific wavelengths. The Sulfur-II Filter is one example. It transmits light which sulfur emits and, hence, one could monitor the development of sulfur features with this filter.

Finders

The purpose of the finder is to help the observer find his or her target. The finder has a wide field-of-view and is attached to the main telescope. Regardless of the type, it should be in a convenient location and it should have the same orientation as the telescope. Three common types of finders are "1×" finders, optical finders and laser finders. Each of these is described.

The "1×" finder includes the Telrad and the peep-hole device. A Telrad finder shows a red bulls-eye on a transparent screen, and hence both the bulls-eye and any objects in the sky are visible. I have used this finder on several occasions and have found it useful. Three limitations of the Telrad are (1) its limiting magnitude is a little less than that of the unaided eye, (2) it is susceptible to dew and (3) it requires batteries.

An optical finder (or finderscope) is a small telescope usually with cross hairs. One advantage of it is that its limiting magnitude is greater than that of the unaided eye. The recommended size is at least one-fourth of the telescope diameter. A 50 mm (2 in.) finder works well with a 0.20 m (8 in.) telescope. A straight through finder works best for Newtonians. If one uses a star diagonal consistently, a right angle finder (a finder with a star diagonal) will work well. Figure 4.21 shows a right angle finder. Finders are prone to attracting dew. Therefore if one plans to use a finder for over 30 min, it may be worthwhile to insert a dew shield on it. See Fig. 4.22.

Fig. 4.21. A right angle finder on a telescope with a *star diagonal* (credit: Richard W. Schmude, Jr.).

Fig. 4.22. A homemade dew-shield on a finder (credit: Richard W. Schmude, Jr.).

The laser finder is a recent development. Essentially one mounts a green laser on the telescope and aligns it so that it points to the same area as the telescope. Point the laser at a target and the telescope should be positioned correctly. At least one vendor, Orion, sells mounting brackets for a green laser. Two drawbacks of this pointer are that it can be dangerous and it should not be used at a dark site where others are observing. In addition, laser finders or pointing should *never* be used near airports or when aircraft are scheduled to be flying or are perceived to be flying in the area of observation.

Bino Viewers and Giant Binoculars

Three advantages of using bino viewers (sometimes called binocular viewers) for viewing comets are (1) one uses both eyes and, hence, there is no distraction from an unused eye, (2) they give a pseudo three-dimensional look and (3) they reduce the problem of floaters.

There are several commercially available bino viewers. Product review references for a few models are summarized in upper portions of Table 4.5. (Table 4.5 also contains product review references for binocular mounts and giant binoculars.) When purchasing a bino viewer, it is important to check the adjustable intraocular distance and focusing. Since people's eyes are spaced differently, one must be able to move readily the eyepieces either closer together or farther apart. The focus must also be checked since light must travel an extra distance in a bino viewer. Two models in one review were reported to require about 100 and 125 mm (4 and 5 in.) of drawtube movement before they would come into focus. Many telescope focusers are unable to move the drawtube this distance. Therefore, before purchasing a bino viewer, make sure that it focuses in *your* telescope.

Three minor drawbacks of bino viewers are light loss, extra weight and cost. One reviewer, Alan Dyer, reported a light loss of 0.5 magnitudes in one model. This loss was due to the light being split in half for each eye and the extra optical

Table 4.5. Product reviews of bino viewers, binocular mounts and giant binoculars in *Sky & Telescope* (S&T) and *Astronomy* (Ast.)

Product	Vendor	Reference
Bino Viewer	Baader Planetarium[a], Tele Vue	(S&T) Sep. 02, p. 46
Bino Viewer	Denkmeier Optical	(S&T) Mar. 05, p. 88
Bino Viewer	Baader Planetarium, BW Optik, Celestron, Denkmeier, Lumicon, Siebert Optics, Tele Vue	(S&T) Mar. 05, p. 98
Bino Viewer	Denkmeier Optical Inc.	(Ast.) Jan. 06, p. 94
Binocular Mount	Trico Machine Products	(S&T) Jan. 02, p. 55
Binocular Mount	Bigha	(S&T) Dec. 06, p. 90
Binocular Mount	Farpoint Astronomical Research	(S&T) May 08, p. 34
Giant Binoculars	Jim's Mobile Inc. (JMI)	(Ast.) Feb. 04, p. 90
Giant Binoculars	Vixen Optics	(Ast.) Nov. 08, p. 72
Giant Binoculars	Jim's Mobile Inc. (JMI)	(S&T) Sept. 05, p. 96 & Jul. 07, p. 12
Giant Binoculars	Garrett Optical	(S&T) May 09, p. 37

[a]US dealer is Astro Physics

surfaces. A bino viewer and two eyepieces can easily weigh 4 lb (or 1.8 kg) and, hence, it brings extra stress to the focuser. A final drawback is cost. The prices in 2008 ranged from about 200–1,600 US dollars. One must also obtain two eyepieces for each desired magnification.

One can also use giant binoculars to view comets. Several models of six inch binoculars are available commercially. One company (Jim's Mobile Incorporated – JMI) manufactures giant binoculars with apertures of up to 0.40 m (16 in.). A reference to a product review for JMI's six inch binoculars is listed in Table 4.5. In addition, some people have made their own giant binoculars. See *Sky and Telescope* magazine (Nov. 1984, p. 460) and (Feb. 1993, p. 89).

Atmospheric Dispersion Corrector

Many comets can be seen only when they are at low altitudes. This is a problem because our atmosphere acts like a weak prism. Essentially, blue light is bent more than red light which in turn causes objects to have a blue tinge on one side and a red tinge on the other. More importantly, this prismatic effect degrades the view. This effect is small for objects more than 45° above the horizon, but increases for objects at lower altitudes. An atmospheric dispersion corrector can correct this problem. Essentially this device contains a small prism which can reverse the prismatic effect of the atmosphere.

For low lying comets (less than 20° above the horizon) one can use two of these devices to eliminate large atmospheric dispersion effects. A product review for one model of an atmospheric dispersion corrector is in *Sky and Telescope*, June 2005, p. 88.

Image Scale

The image scale (or plate scale) defines the scale of the image, drawing or photographic plate. As indicated in the Introduction to this Chapter, a scale is necessary for one to measure the size of comet features. Since most comet images are made with digital cameras, I will focus on these instruments and describe the prime focus technique.

Prime focus images have a large field-of-view (FOV). The entire coma and tail will often fit in a prime focus image. When making this type of image, one may use either a camera lens or a telescope. In either case, the image scale, S, in units of arc-seconds/pixel is:

$$S = \frac{206{,}265 \text{ arc - seconds} \times n}{(1000\,\mu\text{m}/\text{mm} \times f)} \tag{4.5}$$

where n is the pixel size in micrometers (μm) and f is the focal length of the telescope or lens in mm. Camera lenses are described in terms of their focal length instead of their diameter. For example, a 50 mm lens has a focal length of 50 mm, whereas a 50 mm telescope has a diameter of 50 mm. The FOV of the image, F_I is:

$$F_I = S \times N \tag{4.6}$$

where N is the number of pixels in the camera chip. Two examples using (4.5) and (4.6) are worked out as follows:

Example 1: Camera with a 100 mm telephoto lens. If one uses a camera having an 800 by 800 pixel chip to image a comet, and if each pixel is 8 μm across, the image scale would be:

$$S = \frac{206,265 \text{ arc - seconds} \times 8\,\mu\text{m / pixel}}{(1000\,\mu\text{m / mm}) \times 100\,\text{mm}}$$

The μm and mm units cancel to yield:

$$S = \frac{206,265 \text{ arc - seconds} \times 8 \text{ / pixel}}{(1000) \times 100}$$

or

$$S = \frac{1,650,120 \text{ arc - seconds / pixel}}{100,000} = 16.5 \text{ arc - seconds / pixel.}$$

The angular size of the full image, F_I, is computed as:

$$F_I = (16.5 \text{ arc - seconds / pixel}) \times 800 \text{ pixels.}$$

The pixel units cancel and then we have:

$$F_I = 16.5 \text{ arc - second} \times 800$$

or

$$F_I = 13,200 \text{ arc - seconds} \left(\text{or } 3.67 \text{ degrees}\right).$$

In this example, one would conclude that the image covers an angular length of 3.67°.

Example 2: Camera attached to a telescope with a focal length of 2,000 mm. Let's say that the same camera as in example 1 is used. The scale for a prime focus image is computed as:

$$S = \frac{206,265 \text{ arc - seconds} \times 8\backslash\mu\text{m/pixel}}{(1000 \backslash\mu\text{m/mm}) \times 2000 \text{ mm}}$$

The μm and mm units cancel to yield:

$$S = \frac{206,265 \text{ arc - seconds} \times 8 \text{ / pixel}}{(1000) \times 2000}$$

or

$$S = \frac{1,650,120 \text{ arc - seconds / pixel}}{2,000,000} = 0.825 \text{ arc - seconds / pixel}$$

The angular size of the full image, F_I, is computed as:

$$F_I = (0825\,\text{arc-seconds/pixel}) \times 800\,\text{pixels}.$$

The pixel units cancel and then we have:

$$F_I = 0.825\,\text{arc-second} \times 800$$

$$F_I = 660\,\text{arc-seconds (or 0.183 degrees)}.$$

In this example, one would conclude that the image covers an angular length of 0.183°.

A camera with a 100 mm telephoto lens yields a much larger FOV (3.67°) than that of a telescope. Therefore, if a comet has a 3° tail, one would be able to image the whole object in one image with the lens. This would not be possible with the telescope. The telescope, however, allows one to study much smaller detail since its image scale (0.825 arc-sec/pixel) is much smaller than that of the telephoto lens. Furthermore, many comets have faint tails with lengths of less than 0.1°. For these comets, a telescope would be better since it will gather more light than a camera lens.

If one desires a larger FOV, he or she should use a camera with a bigger chip, or use a shorter focal length lens or use a 35 mm film camera. A conventional 35 mm film camera and a 50 mm lens will yield a photograph with a field of 27° by 38°. These large fields are appropriate for imaging long tails. One may consider using also a medium or large format camera to image long tails. Bear in mind that these cameras are more expensive than 35 mm cameras and require special film.

If one desires to zoom in on fine detail, he or she would need to use the eyepiece projection technique. This can be accomplished with either a Barlow lens or an eyepiece. As the eyepiece focal length drops, the image scale drops, but the required exposure time increases.

Making an Observation or Measurement

Three organizations which are devoted to the study of comets are the British Astronomical Association (BAA), the Association of Lunar and Planetary Observers (ALPO) and the group that publishes the *International Comet Quarterly* (ICQ). Information that must be included with any submission of comet observation to any such organization are: Observer's name and contact information, the nature and results of the particular observation or study, date and time involved, telescope size and magnification, filters used (if any), the comet's name, sky transparency and seeing conditions. If drawings and images are submitted, the sky directions and the angular scale must be given also. This information is discussed below.

The observer should always state his or her name and contact information on *each* observation. The main reason for this is in the event somebody wants more information from the observer. Contact information should include full name, an e-mail address, postal address and a telephone number. All of these are needed in the event the observer moves or changes his/her e-mail address etc. Coordinators of the BAA and ALPO Comet Sections and the ICQ people also need the observer's name and location so that proper credit can be given in any published report.

The date and time involved in the observation(s) are needed for all comet observations. There are over 30 times zones in the world. To make matters more complicated, many of these time zones go on daylight savings time for part of the year. For these reasons and others, both the ALPO and BAA comet sections report all times in Universal Time (UT). This time standard may also shift the date. Conversion factors for computing Universal Time are given in Table 4.6 for a few time zones. The values in this table are added to the local (military calculated) time. Military time starts at midnight and goes up to 23:59. For example, if a comet observation is made on Dec. 5 at 10:00 pm (or 22:00 military time) in San Diego, CA (pacific time zone; standard time), the correct date and time would be 10:00 pm = 22:00 + 8:00 = 30:00 h − 24 h = 6:00 UT on Dec. 6. The date changes because one passes midnight when adding 8 h. One must also write the year on all comet observations.

One should report the type and size of their telescope along with the magnification used. This is especially important if comet magnitudes are reported because aperture corrections must be made. Aperture corrections are described in Chap. 3. I also believe that the magnification may affect the brightness value.

The observer should also note which filters, if any, were used. Give as much information as possible. A brand name, vendor name and filter number would be ideal. Different filters have different spectral responses and, hence, it is important to give specific information about filters whenever possible.

One should also write the name of the comet for each observation. The name should be on every drawing or data sheet since these are often separated. In many cases, observations are arranged in chronological order and, hence, drawings (even if they are stapled) may become separated. As the ALPO Jupiter Coordinator, I organized Jupiter images and drawings in this way. I broke apart stapled sets of drawings many times.

One must note the limiting magnitude. This is discussed in Chap. 3.

Table 4.6. Conversion to Universal Time. These factors are added to the local (military calculated) time

Time zone	Standard time (add this amount-hours)	Daylight Savings Time (add this amount-hours)
Atlantic	4	3
Eastern	5	4
Central	6	5
Mountain	7	6
Pacific	8	7
Hawaii	10	9
Western Europe	−1	–
India	−5.5	–

There are several ways of reporting the seeing. In the first method, seeing is rated on a scale of 0 (poor) to 10 (perfect). This scale is used by members of the ALPO. Very recently, word descriptions were given to each number; more information about this is in *Saturn and How to Observe It* ©2005 by Julius Benton. A second method – developed by Eugene Antoniadi – is similar. The seeing is rated on a scale of I (excellent) to V (poor). This scale is often used in Europe. A third method is to estimate the size of a star. For example, if a star appears to have an angular diameter of 2.0 arc-sec, we say that the seeing is 2.0 arc-sec. (One can also image a star and measure its size from the image.)

One can also estimate the seeing by estimating how much of the Cassini Division is visible when $B > 15°$. (The Cassini Division is the dark gap between Saturn's bright outer A ring and its bright inner B ring.) If the Division is completely visible, the seeing is excellent, but if it is partially visible, the seeing is average. If this Division is barely visible or not visible at all, the seeing is poor. The advantages of this scale are (1) it is based on observing a planet instead of a star, (2) it is quick to use and (3) one can compute the exact width of the Cassini Division. See Fig. 4.23.

One can also use binary stars as a way of estimating the seeing. For example, Epsilon-1-Lyrae and Epsilon-2-Lyrae had respective separations of 2.1 (2003) and 2.2 (2004) arc-sec. If these binary stars can be resolved, the seeing is better than about 2 arc-sec. One thing to keep in mind, though, is that the separation between two stars in a binary star system changes over time. Therefore, find the current separation of the binary star before using this method.

If one draws, photographs or images a comet or part of one, the sky directions must be shown. One may use nearby stars and a star atlas to report directions. Alternatively one may use the following procedure: To find the south direction, nudge the telescope north (N), and the portion of the comet that moves towards the edge of the telescope field-of-view is the south direction (S). See Fig. 4.24. To find the preceding edge (P), turn off the telescope drive and the direction of movement will be the preceding edge. See Fig. 4.25. The preceding edge is usually not due west because the Earth's axis is tilted. Others can determine the east and west directions from the direction of the preceding edge and the north direction.

I believe that the observer must determine angular scale for his/her drawings and images. Without an angular scale, it is not possible to make quantitative measurements of the comet and its features. Many of the comet drawings and images that I have seen posted online lack this information. An angular scale can

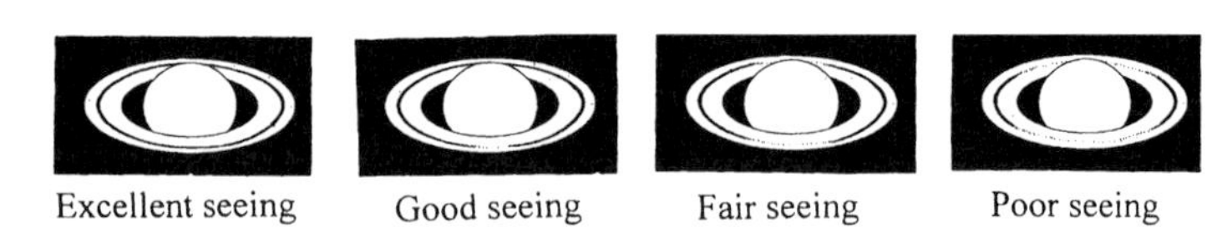

Fig. 4.23. If Saturn's rings are tilted then one can estimate the seeing by looking at Cassini's division, which is the dark band in the rings. If Cassini's division can be seen to the edge of Saturn then the seeing is excellent (credit: Richard W. Schmude, Jr.).

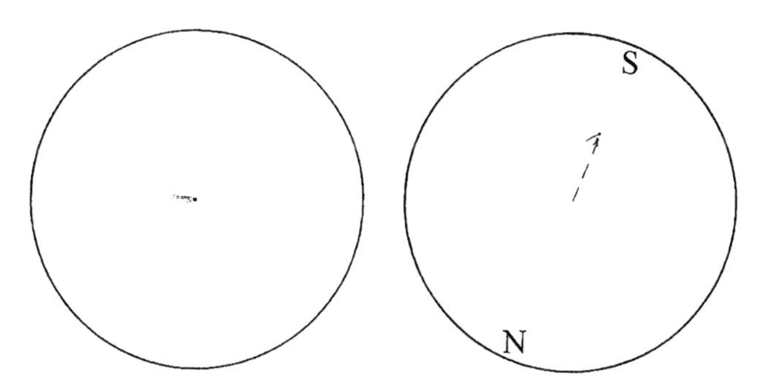

Fig. 4.24. To find the south direction of the sky, just nudge the telescope northward and the object will appear to move in a southerly direction (credit: Richard W. Schmude, Jr.).

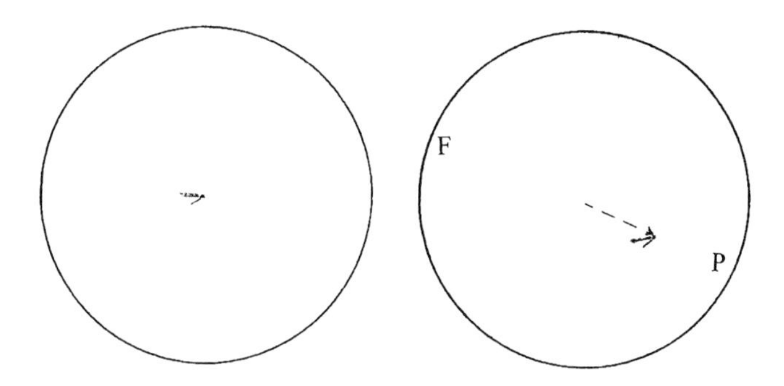

Fig. 4.25. To find the preceding-P (or nearly west) direction then turn off the telescope drive and the comet will appear to move in the preceding direction (credit: Richard W. Schmude, Jr.).

be converted to a scale in kilometers or miles and, hence, one can measure more accurately various features in a comet.

The simplest way to compute an angular scale is to compute the telescope field-of-view, FOV (4.4), and note the size of the FOV in relation to the comet. For example, if the telescope FOV is 20 arc-min across, write a line that is one-tenth the size of the field next to the drawing. This line would represent 2 arc-min.

A second way of estimating the angular scale is to make a drawing or image of the comet at low magnification and include bright background stars. Then one could measure the distance between the bright background stars, use a star atlas or the appropriate software and determine the scale.

A third method of determining the angular scale would be to compute the camera FOV using the procedure discussed earlier.

A fourth method of measuring the angular scale would be to photograph or image a target like Mars or Jupiter and then image the comet without changing the camera focus. I did this with Comet Hale-Bopp (C/1995 O1) on March 16, 1997.

A fifth way of determining the angular scale would be to cut off the clock drive, and time how long it takes for the comet to drift in 1 min. One then uses Table 4.7 to determine the angular distance. If, for example, a comet is at a declination of 20°N, it will drift an angular distance of 846 arc-sec in 1 min of time.

Once the angular scale is computed, it must be placed in a prominent location near the image or drawing or in the figure caption. A good example is Fig. 4.26. This figure shows both the inner coma of Comet Hale-Bopp (C/1995 O1) and the scale bar of 100,000 km at the bottom of the image.

Table 4.7. Drift rates for different declinations. I calculated all of these values from standard trigonometric functions

Declination (°)	Drift rate (arc-sec/min)
0	900
10	886
20	846
30	779
40	689
50	579
60	450
70	308
80	156
85	78
89	16

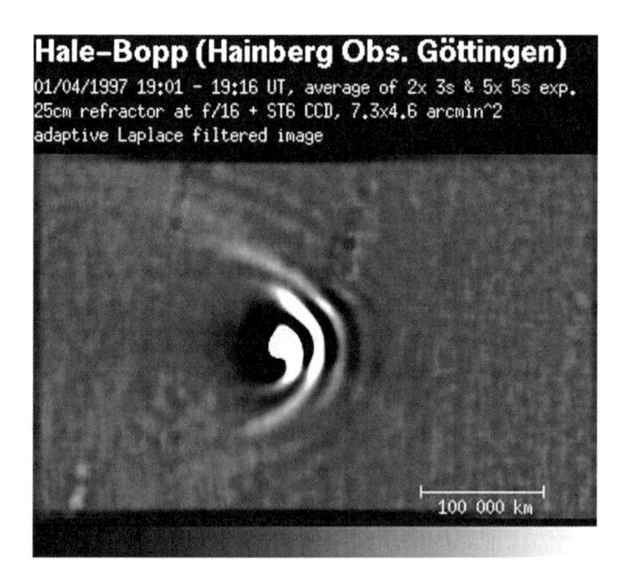

Fig. 4.26. The inner coma of Comet Hale-Bopp (C/1995 O1) made by Klaus Reinsch. The image spans 7.3 by 4.6 arc-min. Note the scale of 100,000 km at the *bottom* (credit: Klaus Reinsch).

Comet Observation Projects

There are several specific comet measurements or observations which one should be able to carry out with small telescopes. Table 4.8 lists various research areas that on which one may concentrate, together with how these areas may increase our knowledge of comets. Specific types of measurements and observations are also listed in each research area. I will describe each of these areas and, when appropriate, carry out example calculations.

Position Measurement

When a new comet is discovered, the top priority is to determine its orbit. Position measurements enable scientists to learn more about a comet's orbit and any

Table 4.8. Comet projects which one may undertake with telescopes having diameters of up to 12 in. or 0.30 m

Research area	What will we learn?	Specific measurements or observations
Position measurement	Comet orbit Changes in comet orbit due to non-gravitational forces	Measure the position of the central condensation Time occultations of stars by comets
Rotation of the nucleus	Affect of non-gravitational forces on rotation rate Orientation of spin axis of nucleus Relationship between rotation rate of the nucleus and tail and coma development	Measurements or drawings of arcs over time Measure the brightness of the central condensation – *look for periodic changes in brightness*
Fragmentation of the nucleus	How the nucleus broke apart	Draw or image the central condensation Measure coma brightness and color
Degree of condensation	Evolution of the coma as the comet-Sun distance changes	Estimate the degree of condensation in one or more wavelengths of light
Tail structure	Interaction between gas and dust tail and both the solar radiation and the solar wind Growth and development of tails	Draw or image gas and dust tails Image tails in specific wavelengths
Jet development	Growth and development of jets near the nucleus Number and size of jets	Draw or image jets Image jets through Swan band filter
Tail length and tail Geometry	Affect of sunlight on tail shapes and sizes Affect of solar wind on tail shapes and sizes	Measure angle between gas and dust tails Measure angular spread of tails Measure angular length of the tail
The gas-to-dust ratio	Production rates of gas and dust as the comet-Sun distance changes Composition and structure of the coma	Measure or estimate comet brightness in two different filters Image or photograph spectra (see Chap. 5)
Lightcurve	How a comet's brightness changes as the comet-Sun distance changes Causes of major or minor outbursts	Measure the brightness and/or color of the comet or central condensation

non-gravitational forces acting on the comet. One determines a comet's position by measuring the right ascension and declination of the central condensation. Ideally, these measurements are made often over a period of several weeks, but preliminary orbits may be based on data taken over a much shorter time period. The accuracy of a comet's orbit increases as the number of position measurements (including respective dates and times) increases.

It is best to report comet positions to an accuracy of 1.0 arc-min or better. Positions accurate to 1.0 arc-sec are ideal. The problem with attaining this level of accuracy is that star coordinates can change by several arc-seconds in both right ascension and declination each year. Star atlases generally list star coordinates for a specific year like 1950 or 2000. The *Astronomical Almanac* lists certain star positions each year; however, only a few hundred stars are contained in the listings. Alternatively, one may simply show comet positions relative to stars and let others compute the final positions.

One way of measuring a comet's position is to use a filar micrometer or an eyepiece having a reticule. For example, let's say that one uses a calibrated reticule where each division is 10.0 arc-sec. He or she measures the central condensation as being 22 divisions (or 220 arc-sec) from star A and 33.1 divisions (or 331 arc-sec) from star B. Then, with the use of a compass, he or she draws out arcs that are 220 arc-sec from star A and 331 arc-sec from star B. See Fig. 4.27. The comet's location is where the arcs intersect. Then one can compute the right ascension and declination of its central condensation. During measurements, it is critical to record the time and to move readily since some comets move quickly with respect to the star background. Don Parker and Michael Mooney used a filar micrometer to measure the position of the asteroid 747 Winchester. They measured both the distance of the asteroid from a nearby star and the position angle of the asteroid relative to the star. The results are shown in Fig. 4.28. They reported a standard error for these points of 0.115 arc-sec.

A second way of measuring a comet's position is from an image. One can measure the position relative to nearby stars on the image. One then looks up the star positions in a star atlas or computer database. The Tycho and Hubble Guidestar Catalogs contain the positions of certain stars down to about magnitude 15. There are also several software packages which have star atlases. Finally, there are several paper star atlases which contain star positions. *Sky Catalogue 2000.0, Volume*

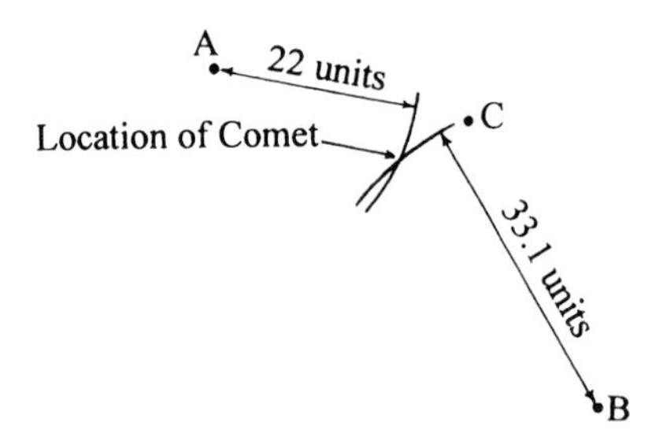

Fig. 4.27. If a comet is 22 units from Star A, 33.1 units from Star B and is close to Star C then the position is determined with a compass. Arcs with radii of 22 and 33.1 units are drawn and their intersection is the comet's location (credit: Richard W. Schmude, Jr.).

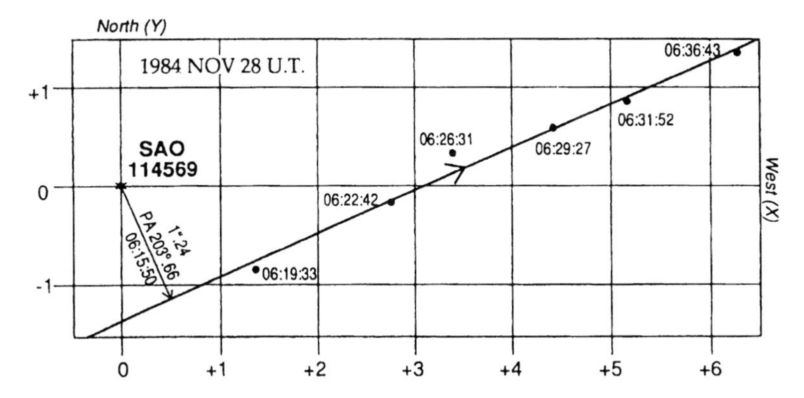

Fig. 4.28. Positions of the asteroid 747 Winchester (credit: courtesy of the Association of Lunar and Planetary Observers).

I: Stars to Magnitude 8.0© 1991 by Alan Hirshfeld, Roger W. Sinnott and François Ochsenbein is a good source. The comet positions can be measured either manually or with appropriate software. It is, of course, essential that one record the exact time of the image.

A third (and rare) method of measuring the position of a comet is to record a stellar occultation by the comet's central condensation. Stephen J. O'Meara observed such an event on Jan. 31, 1980. Essentially, he observed the central condensation (or possibly the nucleus) of Comet Bradfield (C/1979 Y1) occult a ninth magnitude star. According to O'Meara, the star flickered several times during this event. It is essential to record the exact time of such an occultation. With an accurate time, one can determine the comet's position at a specific time. One can get an accurate time either by calling WWV at 303-499-7111 and listening to the time signal or using a global positioning satellite (GPS) instrument.

Once a comet orbit is determined, should people continue to make position measurements? Yes for three reasons: (1) orbital calculations improve as the time interval of measurements increases and the number of position measurements increase, (2) the gravity of the planets can alter the course of a comet and hence, affect its position and (3) non-gravitational forces can change a comet's position. Various non-gravitational forces are described in Chap. 1.

Rotation of the Nucleus

One important comet characteristic is the rotation period of its nucleus. This can yield information on the non-gravitational forces acting on it, the orientation of the spin axis and the relationship between the rotation rate and tail/coma development. Drawings and images of the coma and any arcs can yield information on the rotation of the nucleus. In fact, Whipple was able to compute the rotation period of the nucleus of Comet Donati (C/1858 L1) from drawings made of it in 1858. An explanation of how he did this together with an example follows.

Figure 4.29 shows four drawings of a hypothetical comet with arcs similar to that of Comet Donati at four different times. Focus on the arcs. At 2:00 UT an arc is forming and at 5:00 UT a bright jet has starting forming which will soon be an arc. Now note that the closest arc to the nucleus at 0:00 UT has an arrow. Through careful measurements and the use of the times, one can confirm that *this same arc is in the other three drawings shown by the arrow.* This arc is spreading out. By knowing the comet distance, one can use the angular scale to compute a scale in kilometers or miles. The angular scale is shown in the first drawing in Fig. 4.29. In the figure, the bar is 1.0 arc-min long (or 1/60th of a degree long). Let's say that the comet-Earth distance is 6.9×10^7 km. Using this distance and 1.0 arc-min, one computes a size (or length) of 20,079 km from (3.13). The second bar in Fig. 4.29 shows the length of 20,000 km.

After determining the scale, the next step in solving for the rotation period is to measure the expansion velocity of one or more arcs. In many cases, the arcs are not symmetrical about the nucleus and, hence, the arc *diameter* is measured as is shown in Fig. 4.30. For consistency, measure the outer edge of the arc. One does

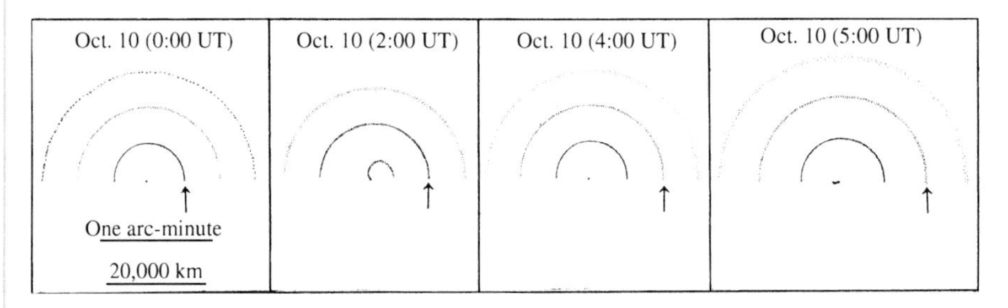

Fig. 4.29. Four drawings of a hypothetical comet. The *arrow points* to the same arc in all four drawings. One can compute the expansion velocity of this arc by measuring its diameter and using (4.8) in the text (credit: Richard W. Schmude, Jr.).

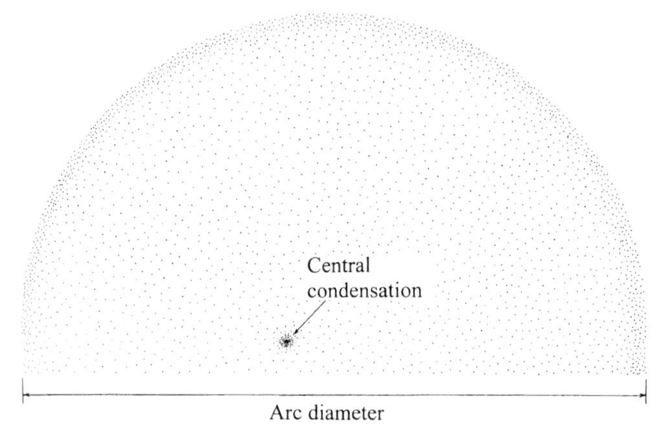

Fig. 4.30. Since the nuclear condensation is not always at the center of the arc, one measures the arc diameter as shown here (credit: Richard W. Schmude, Jr.).

this for the arc with the arrow in all four drawings in Fig. 4.29 and comes up with the values in Table 4.9. The expansion velocity is estimated from the first and second time 0:00 UT and 2:00 UT as:

$$\text{Expansion velocity} = \frac{0.5 \times (19{,}300\,\text{km} - 12{,}900\,\text{km})}{2\,\text{hours}} = \frac{1600\,\text{km}}{\text{hour}} \tag{4.7}$$

The factor 0.5 is needed since the diameter of the arc, instead of its radius, is used. The calculation is repeated for the second to third time 2:00 UT to 4:00 UT and the third to fourth time 4:00 UT to 5:00 UT. The results are listed in Table 4.9. The average speed is 1,760 km/h, and it will be used later to determine the rotation rate of the nucleus.

The next step is to compute the zero date or ZD for each arc. The ZD is the time when the arc had a diameter of 0 km. Figure 4.31 is a reproduction of Fig. 4.29 except that each arc is labeled A, B, C and D. Four different arcs are shown in this sequence. To compute the ZD of arc D one uses:

$$\text{ZD} = \text{decimal date of observation} - \frac{\text{arc diameter} \times 0.5}{\text{average expansion velocity}} \tag{4.8}$$

It is important to use the average expansion velocity, 1,760 km/h, rather than the expansion velocity of the individual arc. This is because there are uncertainties in

Table 4.9. Diameter and expansion velocities of the arc with the arrows in Fig. 4.29

Time (UT)	Diameter (km)	Expansion velocity (km/h)
0:00	12,900	–
2:00	19,300	1,600
4:00	25,800	1,625
5:00	29,900	2,050
Average		1,760 km/h

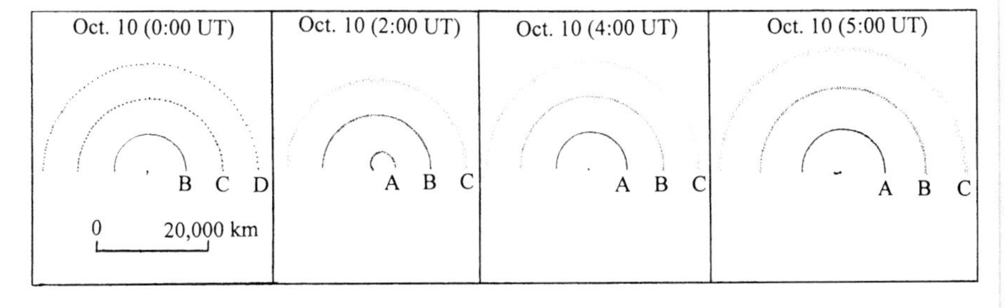

Fig. 4.31. Four drawings of the same hypothetical comet in Fig. 4.29 except that the arcs are labeled (credit: Richard W. Schmude, Jr.).

individual velocities. The average expansion velocity is the most reliable value since it is based on an average of two or more measurements. For arc D at 0:00 UT we have:

$$ZD = Oct.10.0 - \frac{38,800\,km \times 0.5}{1760\,km/hr}$$

$$ZD = Oct.10.0 - 11.02\,hours$$

$$ZD = Oct.10.0 - 0.46\,days$$

$$ZD = Oct.9.54$$

This result shows that arc D started to form on Oct. 9.54 or 11.02 h before the beginning of Oct. 10. The 38,800 km is measured using the scale. I used the same procedure for arcs B and C. For arc A, I had to use the drawing made at 5:00 UT. In this case, the calculation is the same as in the example except that Oct. 10.21 is used in place of Oct. 10.00 in (4.8). The results are summarized in Table 4.10. The first thing that is needed to compute the rotation of the nucleus is to assume that just one active area is responsible for arcs A, B, C and D. With this assumption, one looks for a periodic difference between the ZD values. The difference in ZD between arcs A and B is 0.18 days. The corresponding differences between arcs B and C and between arcs C and D are 0.16 and 0.15 days respectively. Since these arcs were successive features, the rotation period is computed as the average of the three differences as:

$$\text{Rotation period} = \frac{(0.18\,days + 0.16\,days + 0.15\,days)}{3} \tag{4.9}$$

$$\text{Rotation period} = 0.49\text{ days}/3 = 0.163\text{ days} \sim 0.16\text{ days.}$$

In past cases, Astronomers usually did not witness successive arcs but instead witnessed arcs with differently spaced ZD values. They used computer programs to look for a constant time interval between the ZD values. This procedure is beyond the scope of this Book and the reader should consult Whipple (1982) for more information about this topic. Nevertheless, drawings can still be used in determining the rotation rate of the nucleus provided one includes a scale with his or her drawings or images.

Table 4.10. Zero Date and time differences between successive zero date values

Arc	Zero Date (ZD)	Δ-Time difference between successive ZD values in days
A	Oct. 10.03	–
B	Oct. 9.85	0.18
C	Oct. 9.69	0.16
D	Oct. 9.54	0.15

In future studies, it is hoped that when a comet has arcs the observer will make observations every 30–60 min. In this way, he or she can measure in a couple of nights both the expansion speed and the rotation period.

A second way of measuring the rotation period of a nucleus is with photometry. Two Brazilian astronomers carried out photometry of the "inner coma" of Comet 17P/Holmes. They used a 0.3 m (12 in.) telescope along with an ST-7XME CCD camera (manufactured by Santa Barbara Instruments) and a Bessell V filter to image it. They made images between October 27 and Nov. 11, 2007, and extracted 523 brightness measurements of the "inner coma". They entered these measurements into the computer and used a Fourier analysis program to search for a period consistent with the data. They found that the "inner coma" went through a cyclic change in brightness with a period of 6.29 ± 0.01 h. If the nucleus had one dominant area of activity which caused the "inner coma" to brighten, the nucleus's rotation period would be 6.29 h. I believe that this is a very good piece of work which may serve as a guide to future comet studies.

If a single large jet is responsible for the brightness change of a central condensation, one could compute the rotation period without too much trouble. If more than one jet affects the brightness, the analysis becomes more complicated; however, it is still possible to compute a rotation period.

A third method of determining the rotation period may be accomplished by carrying out photometry on the bare nucleus. This will require a large telescope and a sensitive camera.

Fragmentation of the Nucleus

Over 20 comet nuclei have broken apart. Two well known examples are Comet Shoemaker-Levy 9 (D/1993 F2) and Comet 73P/Schwassmann-Wachmann 3. Figure 4.32 shows several fragments of Comet Shoemaker-Levy 9. Table 4.11 lists a few comets that have undergone fragmentation in the early twenty-first

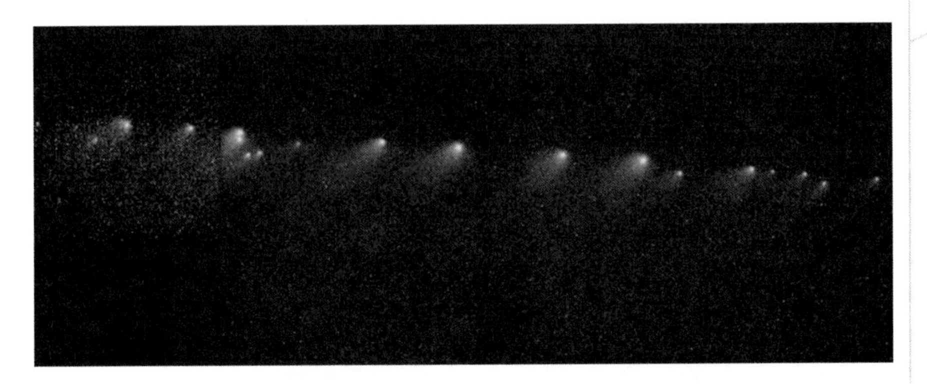

Fig. 4.32. Hubble Space Telescope image of Comet Shoemaker-Levy 9 (D/1993 F2) made in March of 1994 (credit: H. Weaver).

century. Three ways that comet nuclei can break apart are through tidal forces (Fig. 4.33), extreme centrifugal forces (Fig. 4.34) and through impact with a large object.

The break-up of a nucleus provides a rare opportunity for astronomers to learn more about the comet. By analyzing the movement of the fragments, one can gain information on how the nucleus broke apart. If the coma is bright enough, one may be able to record its spectrum. This, in turn, would yield information on the composition of the fresh surfaces of the nucleus. Spectroscopy is described in Chap. 5. Finally, brightness measurements can yield information on the size distribution of the fragments.

Table 4.11. A few comets that have underwent fragmentation in the early twenty-first century

Comet	Year that it fragmented	Source[a]
LINEAR (C/2001 A2)	2001	IAUC 7605, 7616
51P/Harrington	2001[b]	IAUC 7769 7773
LINEAR-Hill (C/2004 V5)	2001	IAUC 8440
LINEAR (C/2003 S4)	2004	IAUC 8434
LINEAR (C/2005 K2)	2005	IAUC 8545
LINEAR (C/2005 A1)	2005	IAUC 8559, 8562

[a]IAUC stands for the International Astronomical Union Circular
[b]This date is for the separation of fragment D; other fragments broke off in previous years

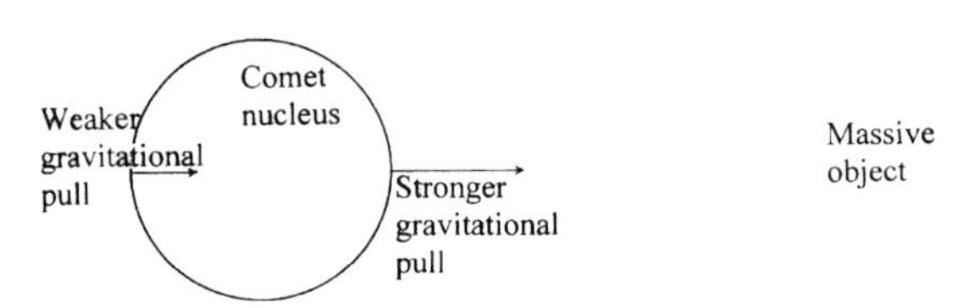

Fig. 4.33. When a comet nucleus approaches a massive object like Jupiter, one side feels a stronger gravitational pull than the other. The difference in the gravitational pull is the tidal force (credit: Richard W. Schmude, Jr.).

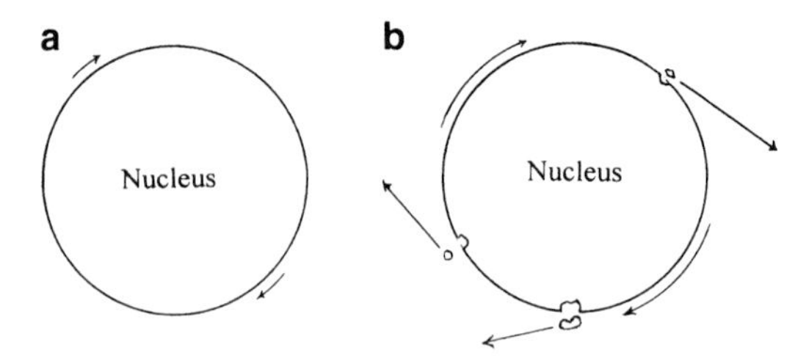

Fig. 4.34. (**a**) The centrifugal force is low on a slow rotating nucleus and hence, the nucleus stays in one piece. (**b**) The centrifugal force on a fast rotating nucleus is high and exceeds the gravitational force. In this case, material flies off of the nucleus (credit: Richard W. Schmude, Jr.).

One specific observation is to draw or image the central condensation with the aim of looking for fragments. This requires medium and high magnifications. If the central condensation becomes elongated, this could be a sign of a break-up. If the nucleus breaks apart, one should pay close attention and note the position of each fragment. A position measurement should include both the separation between the fragment(s) and the central condensation along with the position angle of the fragment(s). The positions of each fragment will allow astronomers to compute orbits. This, in turn, should lead to accurate predictions of fragment positions in the future. This was accomplished for Comet Shoemaker-Levy 9 (D/1993 F2) and Comet 73P/Schwassmann-Wachmann 3. Fragment positions may also yield information on what caused the nucleus to break up. As indicated earlier, one should include a scale for all drawings and images of comet fragments or describe the scale in the caption.

One can also monitor the brightness of the central condensation for a sudden increase in brightness. A brightness increase may be the result of the fragmentation of the nucleus. Two of the comets in Table 4.11, LINEAR (C/2001 A2) and LINEAR (C/2005 K2), underwent large brightness increases before fragmenting.

Degree of Condensation

Observers are also encouraged to estimate the degree of condensation (DC) of the coma. Such a study can yield information on the evaluation of the coma as the comet-Sun distance changes. Visual observers should consult Fig. 3.17 to estimate the DC value of a bright comet. (The situation is different for photographs and images.) Since the coma is usually the brightest part of the comet, it is often overexposed. This is because most people want to image faint portions of the tail. One can carry out a successful study of DC by imaging the coma and estimating the DC value from Fig. 3.17. The exposure time should be adjusted so that no part of the coma is overexposed. In addition, the exposure time should be kept nearly constant throughout the study period.

Another useful project would be to study the DC value in different wavelengths of light or through different filters. For example, one could image the coma through a Swan-Band Filter. Any changes in the DC value would be due to gas since this filter transmits light emitted mostly by gases. In addition, one could image the coma through a filter which transmits light reflected by dust. Such a study would show how dust affects the coma.

Tail Structure

Does the gas tail interact with the solar wind? Do gas tails change within 1 h? Do dust tails undergo changes as the comet-Sun distance changes? The answer to these questions is "yes." Detailed drawings or high resolution images of comet tails can reveal changes. The best techniques for drawing and imaging the tail are described below.

If you want to draw the faint gas tail, observe from a dark location and start with low magnification. A blue (or Swan-Band) filter will enhance this feature. I prefer to focus on the detail in the comet and shade after the observing session. If there is detail in the gas tail, pay close attention to it. If possible, make an observation every 30 min and look for changes.

One can also image the gas tail as the comet-Sun distance changes. A sequence of images made through red, green and blue filters may be added to yield a color image. The gas tail will have either a bluish or a greenish color. A diligent imaging campaign could also show rapid changes in this tail. People imaged large changes in Comet Bradfield (C/1979 Y1) in just a few minutes on Feb. 6, 1980. One can also study how the gas tail changes with solar activity. Finally, more specialized Astronomers may image the gas tail through a Swan-Band Filter and look for changes.

One can also draw or image a comet's dust tail over a period of time. Such a study may reveal changes as the comet approaches and recedes from the Sun. Longer exposures will pick up fainter parts of the tail that would not be visible in shorter exposures. For this reason, one should use a fairly consistent exposure time. Changes in the transparency may lead also to systematic error in a tail study. Therefore, drawings and images should only be made when the sky transparency is good or better. If the tail is larger than the telescope field-of-view, a piggy back arrangement may be appropriate. See Fig. 4.35. In this case, a camera and lens combination is attached to the top of a telescope. The telescope will move as the Earth rotates. If a tracking telescope is not available, one should be able to make a hand tracker or a "barn-door" tracking platform and use it to take pictures of bright comets and their tails. There are several designs of hand trackers on the Internet. One may look also at the instructions in the June 2007 issue of *Sky and Telescope* Magazine, p. 80, to make an electronic tracker.

Fig. 4.35. The camera is on top of a telescope that tracks. This is known as a piggy-back arrangement (credit: Truman Boyle).

One could also image a comet in a specific wavelength of light. For example, if one is interested in hydrogen, he or she could image a comet with either a hydrogen-alpha or a hydrogen-beta filter as the comet-Sun distance changes. These filters transmit light emitted only by hydrogen. A series of images taken over a few weeks could show how the amount of hydrogen changes with time. Professional astronomers at the University of Hawaii used a sodium light filter to image the sodium tail of Comet Hale-Bopp (C/1995 O1).

Jet Development

How fast do jets form? Can a nucleus have more than one jet? How long do jets last? What is the composition of jets? How big to jets get? How do jets evolve in the hours after formation? These and other questions can be answered by studying jets near a comet's central condensation. Veteran comet observer, John Bortle, reported that jets are better seen in deep twilight than in full darkness.

There are a few things to keep in mind if one plans to look for jets. They often develop when the nucleus is near perihelion. Filters may help in detecting these features. One should also use high magnifications to look for jets. If you find one, make a drawing of it or image it every 30 min. You should also estimate its length and position angle. The position angle is described in Chap. 2. Changes in the jets of Comet Hale-Bopp were visible in as little as 20 min. One should look also for multiple jets. Once again, Comet Hale-Bopp had at least two active jets. Knowledge of the number of jets would be important when attempting to determine the rotation rate of the nucleus.

One can use also a camera to image jets. They require an image scale of around 0.3 arc-sec/pixel. Images should be taken with several different exposure times. Several people used an exposure time of around 1 s to image the jets in Comet Hale-Bopp. In some cases, one may have to guide his or her images on the central condensation instead of stars because some comets move quickly. One should also use a cable release when taking highly magnified images because it helps to reduce vibration. This is a device that attaches to the camera. A sequence of images may reveal the growth of jets and the spreading out of arcs.

Tail Length and Tail Geometry

Both the gas and dust tails start at the nucleus and spread out. Both tails are often wider than the coma. Large tail widths are consistent with different trajectories of the tail particles. The gas tail points in nearly the opposite direction of the Sun as seen from the nucleus. This is usually not the case for the dust tail. Once again since each comet is different, the dust tail on each is unique.

Changes in the tails of a comet can lead to a better understanding of how sunlight and the solar wind affect their size and shape. Four specific measurements which may yield relevant information are (1) measurements of the angle between the gas and dust tails, (2) the angle between any antitail and the gas tail, (3) measurements of the angular spread of a tail and (4) measurements of tail lengths.

These measurements are described below. This is followed by a description of how to convert the angular length of a gas tail to a length in astronomical units.

Figure 4.36 shows the gas and dust tails of Comet LINEAR (C/2003 K4). One can measure the angle between the gas and dust tails with a protractor. One can use these measurements to determine the exact orientation of the dust tail in space and to look for small changes in tail orientation with respect to the nucleus and Sun. I measured the angle between the longest part of the bluish gas tail and the longest (or brightest) portion of the dust tail for Comet Hale-Bopp (C/1995 O1). I used over 300 images on the Internet in my study, and, from the measurements, I computed daily averages. See Fig. 4.37. The angle between the two tails increased in early April 1997, but dropped later that month. This may have been due to the angle between the Earth and the orbital plane of Hale-Bopp. This angle was relatively high in early April. This is close to the time of maximum angular separation of the tails. See Fig. 4.37.

One could measure also the angle between any antitail and the gas tail. In theory, this angle should be 180°; however, small deviations may occur. These deviations may give us information about the antitail. Table 2.9 lists a few measurements of the angle between the antitail and the gas tail which I made from photographs of Comet 1P/Halley. Table 4.12 lists the angle between the antitail and the gas tail for several different comets.

One may better understand also how sunlight and the solar wind affect comet tails by measuring the angular spread of the dust and/or gas tails. The angular spread is a measure of how spread out a tail has become. A tail with a constant width has an angular spread of 0°, whereas a fan-shaped tail has a large angular

Fig. 4.36. An image of Comet LINEAR (C/2003 K4) showing the long gas tail and a second short tail. Mike Holoway made this image on Aug. 17, 2004 (credit: Mike Holoway).

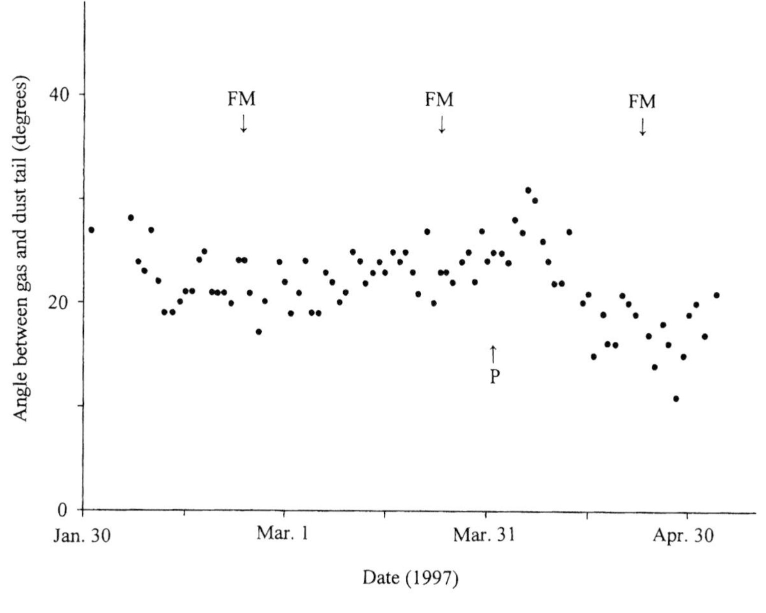

Fig. 4.37. Angle between the longest part of the bluish gas tail and the longest (or brightest) portion of the gas tail for Comet Hale-Bopp (C/1995 01). Each point is a daily average value. The full Moon dates are shown as FM and the perihelion date is shown as P (credit: Richard W. Schmude, Jr. and everyone who contributed images to the Comet Hale-Bopp website).

Table 4.12. Measurements from several sources of the angle between the antitail and the gas tail for several comets. I have also added the angle between the Earth and the comet's orbital plane (angle ECP) at the time of measurement

Comet	Dates	Angle between the antitail and the gas tail (°)	Angle ECP (°)	Source
1P/Halley	May 5 to June 2, 1986	179	0–5	Brandt et al. (1992)
19P/Borrelly	Nov. 4 to Dec. 15, 1994	178	0–23	[a]
19P/Borrelly	Jan. 16, 2002	161	14	[b]
Arend-Roland (C/1956 R1)	Apr. 25, 1957	179	1	[c]
Bradfield (C/1987 P1)	Dec. 22, 1987	180	1	Sky & Tel. 75: pp. 334–335
Hale-Bopp (C/1995 01)	Dec. 4 to Dec. 25, 1996	179	3–10	[d]
Lulin (C/2007 N3)	Jan. 7 to Jan. 30, 2009	175	1	[e]

[a] http://cometography.com/pcomets/019P.html
[b] http://www.castfvg.it/comete/19p/19p_12.htm
[c] http://cometography.com/lcomets/1956r1.html
[d] http://www2.jpl.nasa.gov/comet/images.html
[e] http://www.spaceweather.com/comets/lulin/08jan09/Paul-Mortfield1.jpg

spread. One measures the angular spread by holding the protractor parallel to one side of the tail and measuring the angle made by the other side. See Fig. 4.38. The angular spread of the gas tail depends on several factors, including how that tail interacts with the solar wind and sunlight. I measured the angular spread of the gas tails of several comets, and the results are presented in Table 4.13. In this table, the angular spread seems to be smaller near solar maximum (2003 and 2004) and larger near solar minimum (1986, 1997, 2006 and 2008). It would be interesting to confirm this trend in future years.

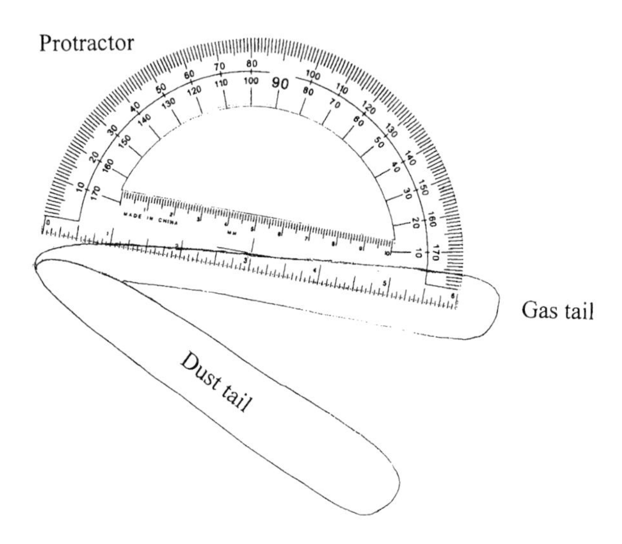

Fig. 4.38. The angular spread is the deviation from a constant width tail. One measures this quantity by lining up the protractor so that its base is parallel to one side of the tail and measures the angle of the other side. In this figure, the angular spread equals 4° (credit: Richard W. Schmude, Jr.).

Table 4.13. Angular spread measurements for a few comets

Comet	Date	Gas tail angular spread (°)	Source
West (C/1975 VI-A)	Mar. 9, 1976	5	Burnham (2000), p. 23
1P/Halley	Apr. 8–9, 1986	7	[a]
Hyakutake (C/1996 B2)	Mar. 24–26, 1996	2.5	[b]
Hyakutake (C/1996 B2)	Apr. 13, 1996	10	[b]
Hale-Bopp (C/1995 O1)	Mar. 31, 1997	7	[c]
NEAT (C/2002 V1)	Mar. 5, 2003	0	[d]
LINEAR (C/2003 K4)	Nov. 16, 2004	0	[d]
Pojmanski (C/2006 A1)	Feb. 28, 2006	2.5	[d]
Boattini (C/2007 W1)	July 4, 2008	4	[d]

[a]http://Commons.wikimedia.org/wiki/image:Comet_Halley.jpg
[b]http://www.makinojp.com/bekkoame/comet.htm
[c]http://www2.jpl.nasa.gov/comet/images.html
[d]http://www.yp-connect.net/~mmatti/

A fourth project which may yield information on how sunlight and the solar wind affect the appearance of comet tails is the measurement of the angular lengths of the gas and dust tails. Angular lengths are usually measured as an angle in units of degrees or arc-minutes. The angular length may be converted into a length in astronomical units; this is described in the next section. Lengths can lead to a better understanding of how tails form and interact with their environment. There are three ways that one can measure the angular length of a tail, namely, eye estimates, measurements from photographs or images, and computation from two points of right ascension and declination. The first two methods are described below. The third method is described in the Appendix.

One can estimate the angular length visually. I usually estimate the angular length of short tails in terms of the binocular or telescope field-of-view (FOV). If, for example, the tail covers half of a 0.8° telescope FOV, the length would equal $\frac{1}{2} \times 0.8° = 0.4°$. For bright comets, one can use a fully extended fist. I have found an average angular size of 9° for the fully extended fist of an adult regardless of height or gender. If a comet's tail is two fists long it would have a length of 18°. A more exact method of estimating a tail length would entail the use of an extended protractor. See Fig. 4.39. The extended protractor has two sticks which are

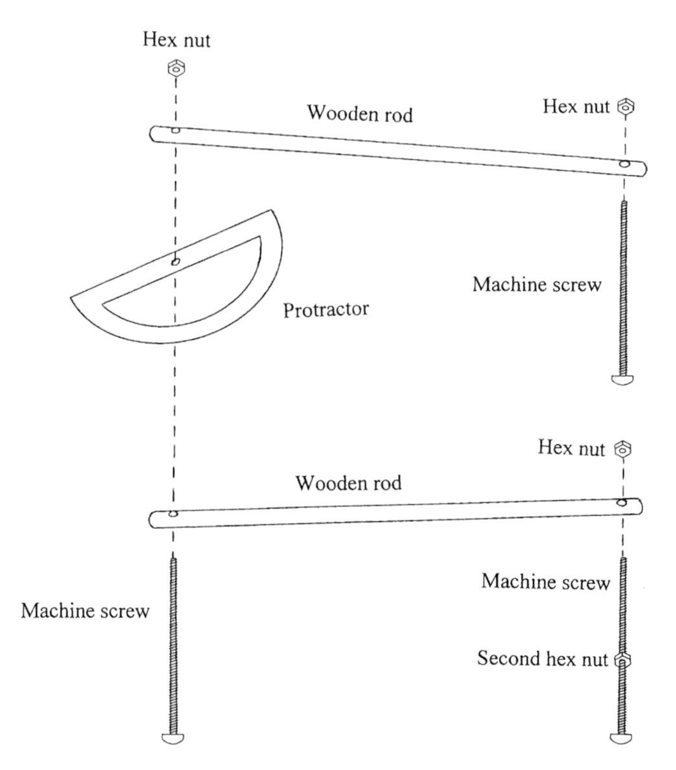

Fig. 4.39. One can measure a long tail to an accuracy of 0.5° with a calibrated extended protractor. In this device, two rods are attached to a protractor with a machine screw and a hex nut. The ends of the rods each have a machine screw. The tips of the screws are *white*; this helps with visibility at night. One simply aims one rod at the central condensation and the other rod at the end of the tail and then measures the angular length from the protractor (credit: Richard W. Schmude, Jr.).

attached to a protractor as shown in the figure. When calibrated, this device can yield angular lengths accurate to about 0.5°. Calibration and use of the extended protractor are described below.

The extended protractor must be calibrated. This is because the wooden sticks are not perfectly straight, holes may not be perpendicular to the wooden sticks, and one's eye will usually not be exactly at the vertex of the angle formed by the wooden sticks. Table 4.14 lists several pairs of stars and their distance. One may use these star distances to calibrate the extended protractor. For example, I made 20 angular length measurements of the distance between Alpha-Lyrae and Alpha-Aquilae. The average measured distance was 37.13°, and the actual distance was 34.2°. The calibration factor is $34.2° \div 37.13° = 0.921$. When I make a measurement with my extended protractor, I must multiply it by 0.921 to get the correct angular length. It is possible to measure the lengths of long comet tails to an accuracy of about 0.5° with this instrument provided that one makes at least ten measurements and computes an average value.

When making visual angular length estimates, one should be consistent. Use the same technique and strive to make measurements in dark skies. Record the limiting magnitude and, if the skies are hazy, do not make a measurement.

One may also measure the angular length of a tail from images and photographs. In order to do this, the scale must be available. Scale was discussed earlier in the Chapter. If one plans to go this route, it is imperative to use consistent exposure times and processing techniques. The sky transparency should be consistent. If film is used, use the same manufacturer and ASA number throughout the study.

If a comet has a very long tail, one can determine its angular length from the right ascension and declination of the two ends. An example calculation of this is presented in the Appendix.

Table 4.14. Distance between selected pairs of stars to be used to calibrate the extended protractor for observers in either the northern or southern hemisphere. I computed the distances using the technique in the appendix. The necessary right ascension and declination values are from *Sky Catalog 2000.0*, 2nd edition ©1991 by Alan Hirshfeld, Roger W. Sinnott and François Ochsenbein

Star pair	Hemisphere N = north, S = south	Distance (°)
Beta-Cassiopeiae and Epsilon-Cassiopeiae	N	13.3
Beta-Ceti and Zeta-Ceti	N or S	18.1
Alpha-Orionus and Beta-Orionus	N or S	18.6
Alpha-Canis Majoris and Zeta-Canis Majoris	N or S	14.5
Alpha-Leonis and Beta-Leonis	N or S	24.6
Alpha-Ursae Majoris and Eta-Ursae Majoris	N	25.7
Alpha-Virginis and Epsilon-Corvi	N or S	21.3
Alpha-Bootis and Gamma-Bootis	N or S	19.5
Gamma-Sagittarii and Sigma Sagittarii	S	11.6
Alpha-Lyrae and Alpha Aquilae	N	34.2

Computing the Tail Length in Astronomical Units

In the preceding section, I mentioned three ways of measuring the angular length of a tail. The angular length is measured in angular units, like degrees, and may be used to compute the tail length in astronomical units (au) or kilometers. This is easy for a gas tail since it is pointed away from the Sun. The length of the gas tail is computed from:

$$L = \frac{\Delta \times \sin(\ell)}{\sin(\alpha - \ell)} \tag{4.10}$$

In this equation, L is the gas tail length in astronomical units (au), Δ is the comet-Earth distance in au, ℓ is the angular length of the gas tail as measured from Earth in degrees and α is the solar phase angle of the comet in degrees. If the angular length is in arc-seconds or arc-minutes, it must be converted to degrees. There are 60 arc-min in 1.0°, and there are 3,600 arc-sec in 1.0°. The solar phase angle is the angle between the Earth and the Sun measured from the comet's nucleus. Equation (4.10) is valid only when the gas tail is pointed directly away from the Sun as seen from the nucleus. As a result, one should not use it for dust tails since they often do not point directly away from the Sun.

In the text below, I work out an example of how to compute the approximate length of the gas tail. Alssandro Dimai reported that the gas tail of Comet Hale-Bopp was 10° long on Feb. 28, 1997. At this time, the comet-Earth distance, Δ, was 1.504 au, and the solar phase angle, α, was 41°. The α and Δ values are from the Horizons on-Line Ephemeris System. The tail length for Feb. 28, 1997 was:

$$L(\text{in au}) = \frac{1.504\,\text{au} \times \sin(10^\circ)}{\sin(31^\circ)}$$

or

$$L(\text{in au}) = \frac{1.504\,\text{au} \times \sin(10^\circ)}{\sin(31^\circ)}$$

or

$$L = \frac{1.504\,\text{au} \times 0.17365}{0.51504} = \frac{0.26117}{0.51504}$$

$$L = 0.5071 \text{ au or } 0.51 \text{ au.}$$

Gas tail lengths for Hale-Bopp are listed in Table 4.15. In all cases, the same group (Piergiorgio Cusinato, Alessandro Dimai, Diego Gaspari, Davide Ghirardo, Giuseppe Menardi, Renzo Volcan and Alessandro Zardini) made the angular lengths. Their lengths are posted on several dates in the image archive on http://www2.jpl.nasa.gov/comet/. I have also plotted these lengths in Fig. 4.40. In the last section of this chapter, I describe some factors which may affect tail length measurements.

The dust tail is usually not directly opposite from the Sun and, hence, its orientation is usually poorly understood. To make matters more complicated, this tail is often curved. These factors make it difficult to calculate accurate lengths.

Table 4.15. Length of the gas tail of Comet Hale Bopp (C/1995 O1) based on the angular length estimates made by Piergiorgio Cusinato, Alessandro Dimai, Diego Gaspari, Davide Ghirardo, Giuseppe Menardi, Renzo Volcan and Alessandro Zardini. I computed all gas tail lengths from (4.10)

Date (1997)	Gas tail (ℓ) (°)	Comet-Earth distance, Δ (au)	Solar phase angle, α, (°)	Gas tail length (au)
Jan. 18	1.5	2.26	20	0.18
Jan. 31	3	2.02	26	0.27
Feb. 6	5	1.906	29	0.41
Feb. 8	5	1.866	30.	0.38
Feb. 10	5	1.827	31	0.36
Feb. 14	6	1.749	33	0.40
Feb. 15	8	1.730	34	0.55
Feb. 16	8	1.711	34	0.54
Feb. 18	10	1.674	36	0.66
Feb. 28	10	1.504	41	0.51
Mar. 2	12	1.474	42	0.61
Mar. 6	12	1.421	44	0.56
Mar. 8	14	1.398	45	0.65
Mar. 10	14	1.377	46	0.63
Mar. 11	10	1.368	47	0.39
Mar. 12	15	1.360	47	0.66
Mar. 13	15	1.352	47	0.66
Mar. 19	16	1.320	49	0.67
Mar. 20	16	1.318	49	0.67
Mar. 26	18	1.320	49	0.79
Mar. 27	18	1.323	49	0.79
Mar. 30	18	1.338	48	0.83
Apr. 3	15	1.367	47	0.67
Apr. 6	15	1.396	46	0.70
Apr. 7	14	1.406	45	0.66
Apr. 9	12	1.430	44	0.56

The Gas-to-Dust Ratio

One important characteristic of a comet is the rate which it produces gas and dust. This information can lead to a better understanding of the composition and structure of the coma. One way to determine gas and dust production rates is to estimate a comet's gas-to-dust ratio.

The spectrum of Comet 17P/Holmes is shown in Fig. 4.41. In this figure, the intensity is plotted on the vertical axis, and the wavelength is plotted on the horizontal axis. The double peak near 5,150 Å (or 515 nm) is due to light given off by gaseous C_2. This light has a blue-green color. If a comet has lots of gas, it probably will give off lots of light with wavelengths of 511 and 514 nm. If a comet does not give off much light at these wavelengths, it probably has little gas. An explanation of how one can measure a comets' gas-to-dust ratio is presented below.

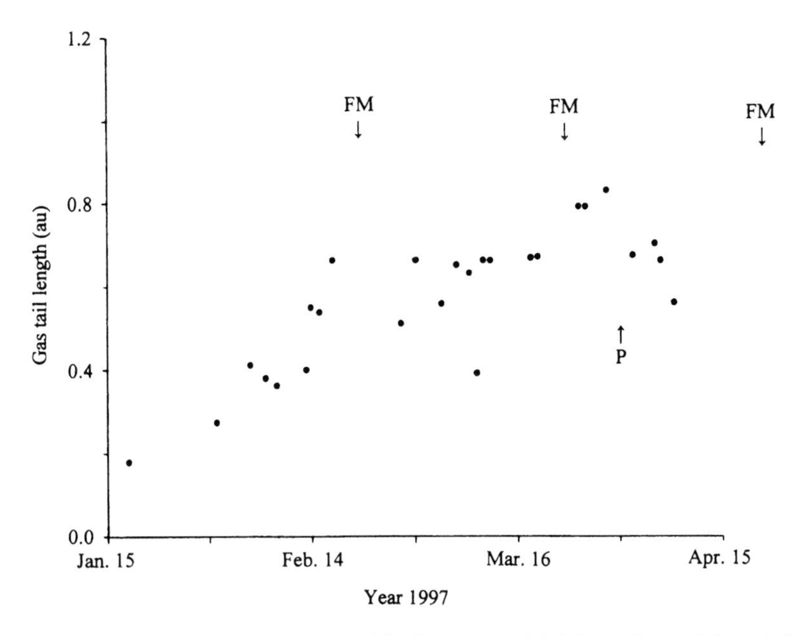

Fig. 4.40. Gas tail lengths of Comet Hale-Bopp (C/1995 O1) based on estimates made by P. Cusinato, A. Dimai, D. Gaspari, D. Ghirardo, G. Menardi, R. Volcan and A. Zardiri and (4.10). The full Moon dates are shown as FM and the perihelion date is shown as P (credit: Richard W. Schmude, Jr.).

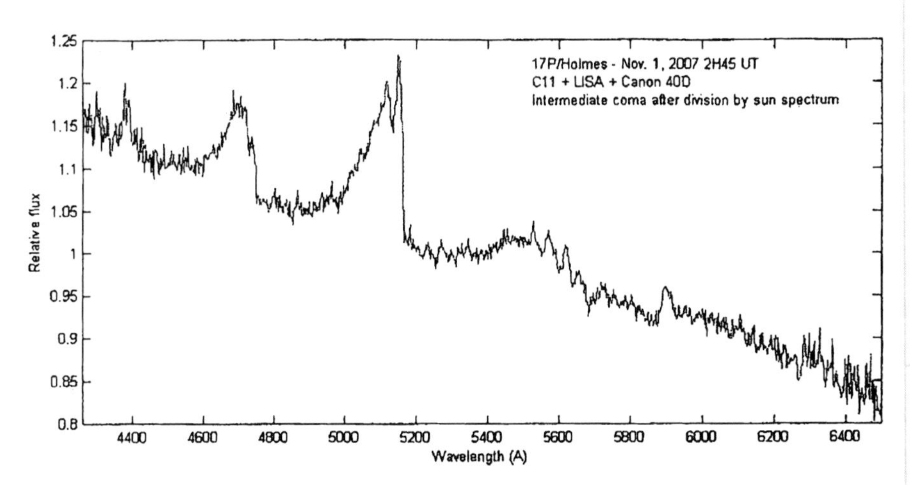

Fig. 4.41. A spectrum of the coma of Comet 17P/Holmes. The double peak near 5,150 Å is due to C_2 emissions (credit: Christian Buil).

One way to get quantitative information on the gas-to-dust ratio is to image a comet with two different filters. The first image is taken with a yellow or green filter along with an infrared blocking filter. The second image is taken with a Swan Band (or comet) Filter along with the same infrared blocking filter. Then one corrects for the difference in the FWHT (full-width-at-half-transmission) for the two filters. If the comet is much fainter in the second image, most of its light would be the result of dust. If the second image, however, is brighter than the first one, most of the comet's light would come from gas. An interesting project would be to make both images over a period of a few weeks and determine how the gas-to-dust ratio changes. One may obtain useful information on comets fainter than tenth magnitude with modern CCD cameras and a medium sized telescope.

Another way of measuring the gas-to-dust ratio is to use a photoelectric photometer along with color filters, like those transformed to the Johnson B and V system. Table 4.16 lists my brightness measurements of Comet 17P/Holmes. The B–V color index is computed by subtracting the V filter magnitude from the B filter magnitude. The V-R color index is computed in a similar manner. The B–V and V–R color indexes of the Sun are 0.65 and 0.54 respectively. These values are close to those of Comet 17P/Holmes in late October of 2007. This is evidence that dust was largely responsible for this comet's dramatic brightening at that time. Hence, its gas-to-dust ratio was low. If the Sun would have been redder than the comet, this would have been evidence of lots of gas.

Lightcurve

A lightcurve is a graph of brightness versus time. In the case of comets, a lightcurve can yield information about how the comet's brightness changes as it approaches the Sun. We can also learn about the causes of major and minor outbursts form lightcurves. Lightcurves are presented for several comets in Chap. 2. Before one can construct this type of graph, he or she must have brightness measurements over an extended period of time. There are three methods of measuring a comet's brightness, namely, with the eye, with a CCD camera and with a photoelectric photometer. These methods are described below.

I described visual brightness estimates in the previous Chapter. One can use a telescope to estimate the brightness of comets too faint to be seen with binoculars. I want to stress that one should use the *smallest* instrument possible to

Table 4.16. Brightness measurements of Comet 17P/Holmes made by the Author

Date (2007)	Filter	Measured magnitude[a]	B–V	V–R
Oct. 29.048	V	2.50	–	–
Oct. 29.065	B	3.25	0.75	–
Oct. 29.081	R	1.94	–	0.55[a]
Oct. 30.092	V	2.52	–	–
Oct. 30.113	B	3.32	0.79[a]	–
Oct. 30.135	R	1.99	–	0.53

[a] These values were computed from magnitudes computed to three places beyond the decimal point, but are rounded off to two places. Likewise, the magnitudes are rounded off to two places beyond the decimal

estimate a comet's brightness. If a comet is too faint to be seen in binoculars, a telescope must be used. In this case, the same general procedures described in Chap. 3 for binoculars should be used. When using a telescope, one should use a low magnification to estimate the brightness. The magnification should also be reported along with the telescope characteristics. Correction factors ((3.4) and (3.5)) must be computed if a telescope is used in making visual brightness estimates. Visual brightness estimates are only accurate to about 0.5 magnitudes. If one desires more accurate measurements, a CCD camera or a photoelectric photometer should be used.

Most brightness measurements today are done with CCD cameras. The CCD camera contains an array of 10,000 or more pixels arranged in a square or rectangular shaped chip. Each pixel can be thought of as a small detector. When carrying out brightness measurements, one must include the entire coma and some space around it. There are three calibration corrections which must be made to a CCD image before one can use it for photometry. These are flat-field, dark field and color corrections. These corrections are described in standard references on imaging. Table 4.17 lists references of product reviews for several types of cameras and electronic eyepieces. Many of these cameras are ideal for photoelectric

Table 4.17. Product reviews of cameras and electronic eyepieces in *Sky & Telescope* (S&T) and *Astronomy* (Ast.)

Vendor	Camera (s)	Reference
Astrovid, Meade, and Supercircuits	StellaCam-EX, Electronic Eyepiece, PC 164C and PC 165C	(S&T) Feb. 03, p. 57
Santa Barbara	ST-2000XM and ST-9XE	(Ast.) Mar. 03, p. 84
Starlight Xpress	MX916	(Ast.) Mar. 03, p. 84
Santa Barbara Instruments	SBIG STL-11000M CCD camera	(S&T) Jul. 04, p. 96
Canon USA Inc.	10D digital camera	(Ast.) Sept. 04, p. 84
Adirondack Video Astronomy	StellaCam II	(S&T) Oct. 04, p. 86
Meade Instruments Corp.	Lunar planetary imager (LPI)	(Ast.) June 05, p. 80
Meade	DSI color CCD	(S&T) Oct. 05, p. 76
Canon	EOS 20Da digital SLR	(S&T) Nov. 05, p. 84
Lumenera	SkyNyx 2-0	(S&T) Jun. 06, p. 76
Canon	EOS 20Da digital SLR	(Ast.) July 06, p. 90
Meade	DSI II Pro monochrome CCD	(S&T) Sept. 06, p. 76
Apogee	Apogee Alta U9000 CCD	(S&T) Jun. 07, p. 64
Finger Lakes Instrumentation Inc.	FLI MaxCam ME2 CCD	(Ast.) July 07, p. 70
Adirondack	Astrovid StellaCam^3	(S&T) Sept. 07, p. 64
The Imaging Source	DMK 21AF04.AS Monochrome	(S&T) Oct. 07, p. 36
Adirondack	Atik Instruments ATK 16IC	(S&T) Nov. 07, p. 36
Orion	StarShoot Deep Space Color Imager II	(S&T) Apr. 08, p. 32
Nikon	D300 Digital SLR	(S&T) Apr. 08, p. 36
Meade	Meade Deep Sky Imager III Camera	(Ast.) Sept. 08, p. 64
Orion	StarShoot Autoguider	(S&T) Nov. 08, p. 43
Orion	StarShoot Pro CCD	(S&T) Feb. 09, p. 34
Finger Lakes Instrumentation	FLI MicroLine ML8300	(S&T) Apr. 09, p. 34
Quantum Scientific Imaging	QSI 540 wsg CCD Camera	(S&T) Jun. 09, p. 36

photometry while others are better suited for imaging. Check with the vendor before making a purchase.

Transformation coefficients must be measured if one uses filters with a full-width at half transmission (FWHT) value exceeding ~50 nm for photometry. If one desires an accuracy of 0.001 magnitudes, he or she should measure also transformation coefficients even if they are using filters with a FWHT value below 50 nm. A description of how to measure transformation coefficients is given in *Uranus, Neptune and Pluto and How to Observe Them* (Schmude 2008) or in *Photoelectric Photometry of Variable Stars* (Hall and Genet 1988).

Once the proper images are made, one must make flat field and dark field corrections. No other processing should be carried out, like contrast enhancement. Additional processing may introduce systematic error into the photometric data. Transformation corrections are applied *after* the raw photometric measurement is made. When purchasing a camera, make sure that the software will allow the making of photometric measurements. The aperture size should be variable so that different size objects can be measured.

One can also use a photoelectric photometer to construct comet lightcurves. A photoelectric photometer contains a detector instead of a camera chip. One can think of this instrument as having one giant pixel. As a result, one can not use it to take a picture. One commercially available photoelectric photometer is the SSP-3 model manufactured by Optec Inc. See Fig. 4.42.

There are three calibration corrections that one must make with a photoelectric photometer, namely, extinction corrections, color transformation corrections and sky and noise corrections. These corrections along with a detailed magnitude calculation are described in *Uranus, Neptune, and Pluto and How to Observe Them* ©2008 by Richard Schmude Jr.

Fig. 4.42. An SSP-3 photometer inserted into the 1¼ inch eyepiece hole of a telescope (credit: Richard W. Schmude, Jr.).

Factors which Affect Measured Tail Lengths

There are at least five factors which affect the perceived angular length of a comet's tail. The first is the sky transparency. As the sky becomes more opaque, less and less light from the tail makes its way to the eye or detector. In short, the tail will appear fainter and shorter in low sky transparency. This is why the limiting magnitude should be reported for all tail length measurements. A second factor which affects the perceived angular length is scattered light. The Moon and artificial light are examples of scattered light. Scattered light contributes to sky brightness. If lots of scattered light is present, a comet's tail will appear shorter than in a dark sky. Two additional factors which affect the perceived angular length of a tail are telescope aperture and magnification. I believe that if one wants to make a series of angular length measurements over time, the same instrument and magnification should be used. The position of the tail in relation to the Sun and observer is a fifth factor which affects the perceived angular length. An explanation of this factor follows.

Comet tails are almost always brightest near the coma, and grow fainter with increasing distance from it. At some point, the tail is too faint to see. This point is shown as the dashed curve in Fig. 4.43a. The area to the right of the dashed curve is not visible and, hence, the tail's length is reported as ℓ degrees. The tail in Fig. 4.43a is perpendicular to the line of sight. In Fig. 4.43b, the situation is different. The tail is now 60° from being perpendicular to the line of sight. I have drawn the location of the dashed curve for the geometry in Fig. 4.43a. The area beyond the dashed curve is oriented in such a way that the observer is seeing a much thicker portion of the tail. As a result, the observer sees a good portion of the tail *beyond* the dashed curve and reports a length of ℓ' degrees. After geometrical factors are considered, the reported length in Fig. 4.43b will be *longer* than in Fig. 4.43a.

The tail position relative to the Sun line can affect also the brightness and the perceived length of a tail. Backscattered and forward-scattered light are shown in

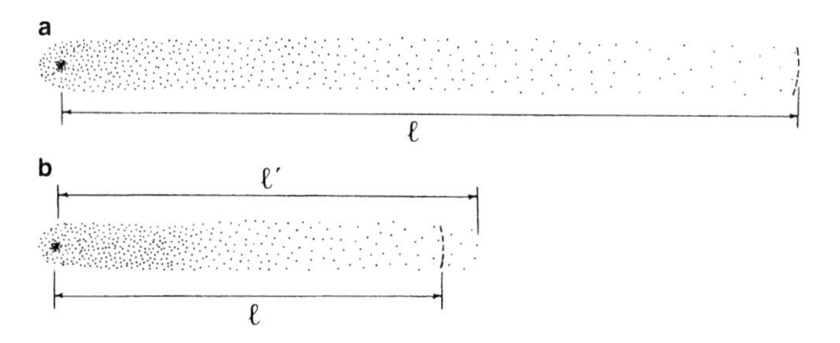

Fig. 4.43. (**a**) The tail is perpendicular to the observer's line of sight. The length is estimated to be l. Faint parts beyond the *dashed curve* are invisible at this orientation. (**b**) The tail is about 60° from being perpendicular to the observer's line of sight. As a result of this, the observer looks through more of the tail and estimates its length as l′ which is longer than l after foreshortening is taken into account (credit: Richard W. Schmude, Jr.).

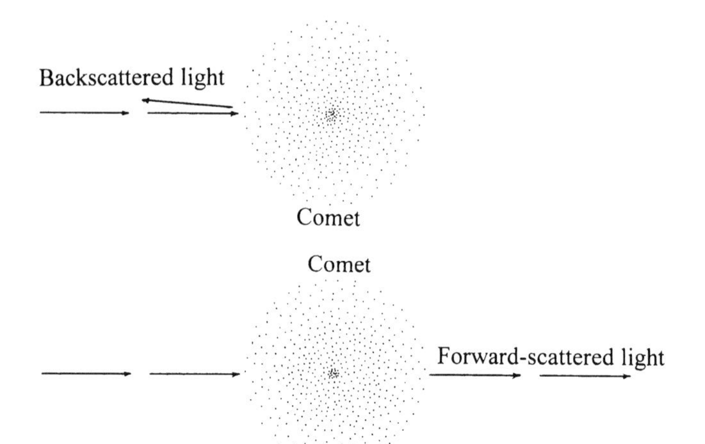

Fig. 4.44. Backscattered light is shown in the top frame and forward-scattered light is shown in the bottom frame. Backscattered light starts out in one direction but the target reflects it back into the opposite direction. Forward-scattered light, on the other hand, strikes the target and continues in the same direction (credit: Richard W. Schmude, Jr.).

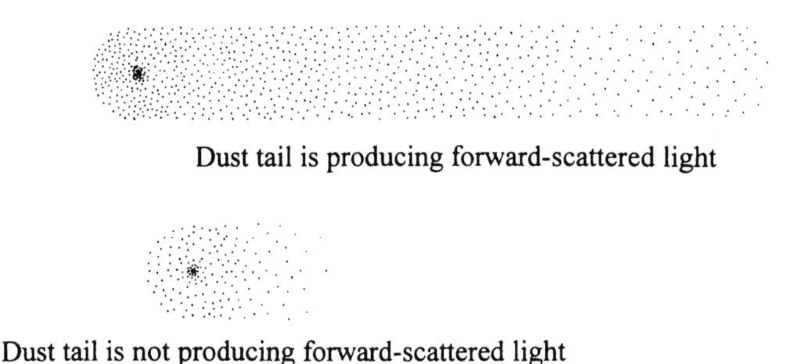

Fig. 4.45. A dusty tail will be long and bright in forward-scattered light but will grow smaller and dimmer when it is not in the correct position to send forward-scattered light to the observer (credit: Richard W. Schmude, Jr.).

Fig. 4.44. Backscattered light occurs when light strikes an object and is reflected backwards. In the case of forward-scattered light, light strikes material but continues in a forward direction. If the tail is nearly between the Sun and Earth, we would see mostly forward-scattered light. If, on the other hand, the Earth is nearly between the Sun and the tail, we would see mostly backscattered light from the tail. Particles with sizes near the wavelength of light are much brighter in forward-scattered light than in backscattered light. Objects much larger than the wavelength of light are brighter in backscattered light. Dust particles having the same size as the wavelength of visible light will be bright in forward-scattered light but dim in backscattered light. A tail made up of this type of dust will be long and bright in forward-scattered light. See Fig. 4.45. When the comet shifts position, such that this same tail is no longer producing forward-scattered light, it will appear short and dim.

Chapter 5

Observing with Large Telescopes

Introduction

There are several comet observation projects which one can carry out with a large telescope (aperture larger than 0.30 m or 12 in.). For example, one can use a large telescope to record spectra of a comet, or to make polarization measurements of the coma. One can determine also when the coma first forms and dissipates, along with other information which may yield clues to a comet's age. He or she may be able to measure the brightness of the bare nucleus or carry out several of the projects described in the previous chapter. In this chapter, I will describe several observing projects which may be carried out with large telescopes, namely, spectroscopy, radio studies, photometry of comets and lightcurves, photometry of the bare nucleus, polarization measurements, searching for new comets, recovery of periodic comets and measuring the opacity of comets.

Spectroscopy

Substances can absorb or emit specific wavelengths of electromagnetic radiation (or radiation). Spectroscopy is the sub-discipline of science which involves the recording and interpretation of spectra. In many cases, a spectrum can reveal information on the composition, temperature and expansion speed of gases within a comet.

By way of background, I will discuss the different types of spectra and how atoms and molecules interact with radiation, beginning with how comets emit and reflect light. This is followed by a discussion of the three types of spectra (absorption, continuous and emission spectra), a discussion of electronic, vibrational and rotation transitions and a discussion of spectral resolution and the spectra of comets.

R. Schmude, *Comets and How to Observe Them*, Astronomers' Observing Guides,
DOI 10.1007/978-1-4419-5790-0_5, © Springer Science+Business Media, LLC 2010

Emission and Reflection of Light by Comets

Light can interact with matter in five ways, namely, absorption, reflection, scattering, dispersion and emission. In absorption, molecules absorb light with specific wavelengths, with, each type of molecule absorbing its own unique set of wavelengths. If many different molecules are present, or if the molecules are so close together that they are not independent of one another, they will absorb broad ranges of wavelengths. Reflection, on the other hand, is a process whereby light strikes matter and bounces off of it. Reflected light can be in one direction or in many directions depending on the type of reflection. Scattering is a process whereby light strikes matter and, as a result, moves in random directions. In a few situations, matter can disperse light into different wavelengths. Microscopic water droplets which are suspended in the air can disperse sunlight into a rainbow. Finally matter can also emit light. This emission can be as blackbody radiation, fluorescence or phosphorescence. Of the five ways which light interacts with matter, the two most important for comets are emission and reflection. Note, however, that reflected light is important in viewing the comet and dust tail which reflect all colors of sunlight. Since sunlight has a yellow hue, the dust in a comet often will have this hue.

Blackbody radiation is emitted because of the thermal motion of atoms making up an object. The hotter an object, the more radiation it emits. The wavelength at which a body's blackbody emission is highest depends on the body's temperature. Essentially Wein's Law describes this relationship as:

$$\lambda_{max} = (0.0029\,\text{meter} \times \text{Kelvin}) \div T \tag{5.1}$$

In this equation, λ_{max} is the most intense wavelength given off by the object and T is the temperature in degrees Kelvin. Therefore, if a comet nucleus has a temperature of 200°K (−99°F or −73°C), the most intense wavelength given off is 1.45×10^{-5} m or 14.5 μm. This lies in the infrared portion of the electromagnetic spectrum. In order for an object to have a peak blackbody emission in visible light (at say a wavelength of 500 nm), its temperature must be near 5,800°K (~10,000°F or ~5,500°C).

Astronomers have measured the brightness of several comets at wavelengths near 10 μm. They have used this information to determine blackbody temperatures. Table 5.1 lists a few comets and their measured blackbody temperatures.

The processes of fluorescence and phosphorescence are the other ways that emitted light may occur. Fluorescence occurs when matter absorbs high-energy

Table 5.1. Blackbody temperatures of a few recent comets

Comet	Blackbody temperature (°K)	Year measured	Source
153P/Ikeya-Zhang (C/2002 C1)	270	2002	IAUC 7921
Kudo-Fujikawa (C/2002 X5)	340	2003	IAUC 8062
NEAT (C/2001 Q4)	315	2004	IAUC 8360
LINEAR (C/2003 K4)	235	2004	IAUC 8361
LINEAR (C/2006 VZ_{13})	275	2007	IAUC 8855

light and almost instantaneously (within a few nanoseconds) emits light at lower energy. Phosphorescence is similar except that there is a much greater delay in the emission of lower-energy light. Many comets undergo fluorescence. An example of fluorescence is the emission of visible light by the dicarbon radical (C_2) resulting in the Swan Bands.

Finally, a description of the three different types of spectra is in order.

Absorption, Continuous and Emission Spectra

The three types of spectra are absorption, continuous and emission. An absorption spectrum has narrow dips of intensity. Figure 5.1 shows an absorption spectrum of hydrogen. Note the narrow intensity dips. The dips are due to hydrogen absorbing specific wavelengths of light. Hydrogen gas, for example, absorbs light with wavelengths of 434, 486 and 656 nm (Hydrogen can either absorb or emit light at wavelengths of 434, 486 and 656 nm depending on its temperature). On the other hand, an emission spectrum is the opposite of an absorption spectrum. Essentially, it shows intensity maxima (or peaks). Figure 5.2 shows an emission spectrum of hydrogen. The emission peaks are at 434, 486 and 656 nm. Finally, the continuous spectrum shows all of the wavelengths of light without any dips or peaks. See Fig. 5.3. With three types of spectra, one may ask: What type of spectrum will a comet display? This depends on the nature of the observation, the region being studied and what process is taking place in the comet.

Kirchhoff's three laws describe the three types of spectra which an object will give off if it is very hot. These laws are:

1. A hot solid, liquid or compressed gas produces a continuous spectrum.
2. A hot transparent gas which is not compressed *and* is in front of a cooler background produces an emission spectrum.

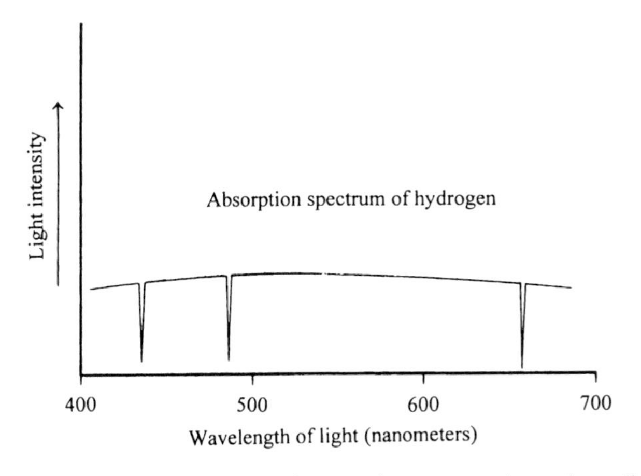

Fig. 5.1. A hypothetical absorption spectrum of hydrogen. The *three narrow dips* in intensity are absorption features. These features are due to hydrogen absorption. (Credit: Richard W. Schmude, Jr.).

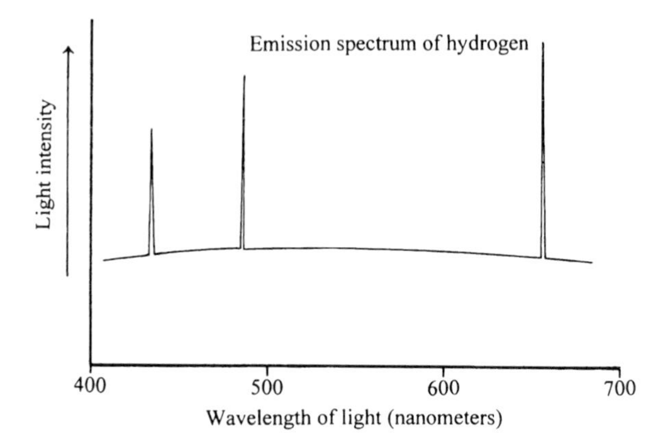

Fig. 5.2. A hypothetical emission spectrum of hydrogen. The *three intensity maxima* (or peaks) are emission features. These features are due to hydrogen emitting extra light at wavelengths of 434, 486 and 656 nm. Hydrogen can either absorb or emit light at these wavelengths depending on its temperature. (Credit: Richard W. Schmude, Jr.).

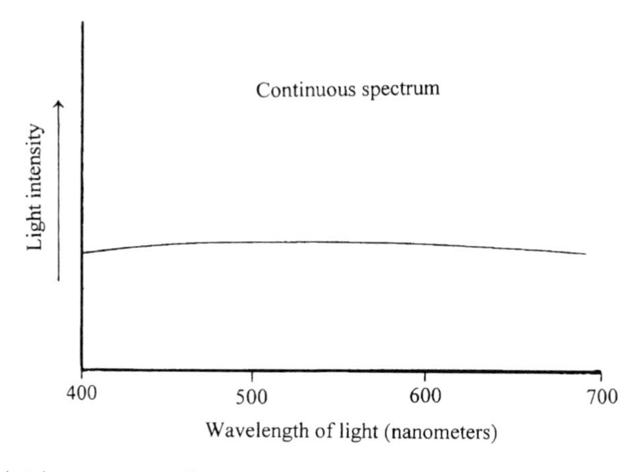

Fig. 5.3. A hypothetical continuous spectrum. There are no dips or peaks in the spectrum. (Credit: Richard W. Schmude, Jr.).

3. A hot transparent gas which is not compressed *and* is in front of a hotter source of continuous radiation produces an absorption spectrum.

A gas-rich comet at a distance of 1.0 au from the Sun will produce an emission spectrum at visible wavelengths due to fluorescence. Part of the spectrum in visible wavelengths will resemble a continuous spectrum due to the reflection of sunlight by dust. If this comet moves in front of a bright star, one may be able to detect also an absorption spectrum.

What causes an object to give off a spectrum? Much of its spectrum is due to the reflection and scattering of sunlight. The more interesting portions of the spectrum, however, are due to relatively independent atoms or molecules moving from one energy state to another. When they occur, they emit or absorb specific wavelengths of light. If enough of these transitions occur, emission or absorption peaks will show up in the spectrum.

Molecules have three types of energy levels, namely, electronic, vibrational and rotational energy levels. Each of these levels is discussed below.

Electronic, Vibrational and Rotational Transitions

The electrons in an atom or molecule lie in specific energy levels. When an electron moves from a higher energy level to a lower one, a specific wavelength of light is given off. This is illustrated with an example.

Carbon has four valence electrons and oxygen has six and, hence, the carbon monoxide molecule (CO) has ten electrons in molecular orbitals. These orbitals are the same as electronic energy levels in atoms. The other four electrons in CO are not valence electrons and do not lie in orbitals. They are not considered further. The ten electrons lie in specific orbitals. The electrons must obey the Pauli Exclusion Principle which states: "*No more than two electrons may occupy any given orbital and, if two do occupy one orbital, then their spins must be paired.*" (*Physical Chemistry*, seventh edition ©2002 by Peter Atkins and Julio de Paula). When CO is at room temperature the electrons almost always lie in the lowest energy state possible, which is called the ground state. Figure 5.4 shows the arrangement of the ten electrons in the ground state of CO. Since the energy increases upwards, the electrons lie in the lowest energy combination possible. Keep in mind that each bar in Fig. 5.4 is an orbital, each arrow is an electron and that one cannot place more than two electrons in an orbital.

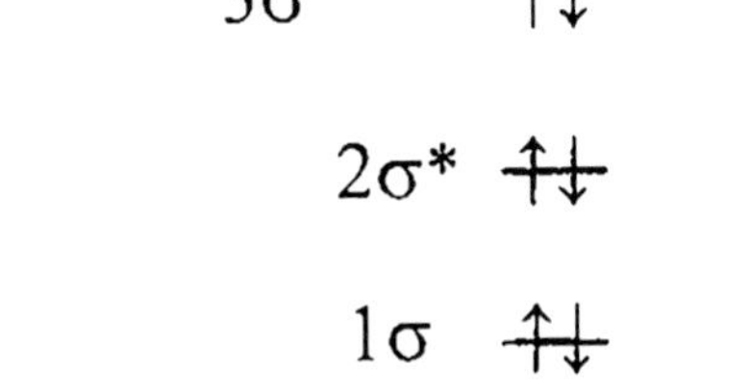

$4\sigma^*$

$2\pi_x^*$ $2\pi_y^*$

$1\pi_x$ $1\pi_y$

3σ

$2\sigma^*$

1σ

Fig. 5.4. Each *arrow* represents a valence electron and each *line* represents an electronic energy level or orbital. The 1σ level has the lowest energy and the $4\sigma^*$ level has the highest energy. Since carbon has four valence electrons and oxygen six, the carbon monoxide molecule (CO) has ten valence electrons in molecular energy levels (orbitals). In this drawing, the electrons lie in the lowest energy configuration possible and, hence, we say that this is the ground state configuration for CO. (Credit: Richard W. Schmude, Jr.).

One photon of visible or ultraviolet light can bump an electron to a higher orbital. In Fig. 5.5, when ground state CO (electron arrangement on the left) absorbs light, an electron moves to the next highest energy level (electron arrangement on the right). The arrangement on the right is no longer the ground state since one of the ten electrons is not in the lowest possible energy configuration. When an electron falls from a higher energy state to a lower one, such as the ground state, the molecule emits light at a specific wavelength. See Fig. 5.6. Visible and ultraviolet wavelengths are often associated with electronic transitions.

In addition to electronic energy levels, molecules can also have vibrational energy levels. The molecule CO has two atoms which are bonded to each other. At times these atoms will move closer and farther apart, much like a spring. Essentially, one can think of a spring joining the C and O atoms, and that the spring vibrates. This movement is called vibration. See Fig. 5.7. CO has different vibration

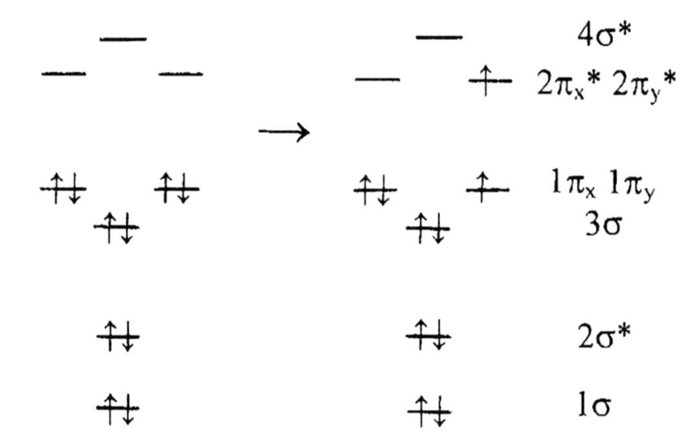

Fig. 5.5. When a CO molecule absorbs light of sufficient energy, an electron is bumped to a higher energy level as shown below. When this occurs, the arrangement of electrons in the CO molecule changes. This process is called absorption. (Credit: Richard W. Schmude, Jr.).

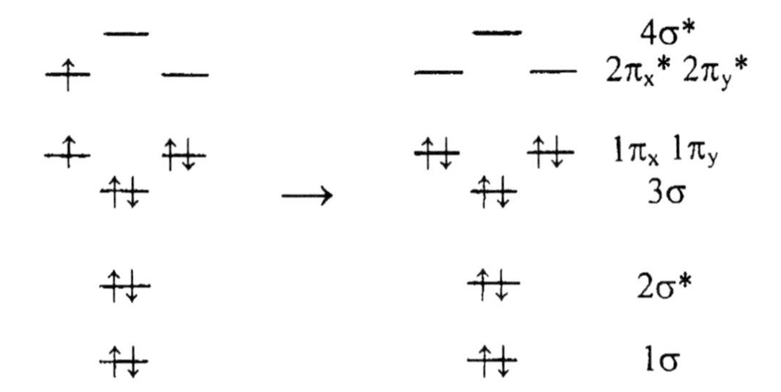

Fig. 5.6. When an electron moves from a higher to a lower energy level, light is given off. This process is called emission. (Credit: Richard W. Schmude, Jr.).

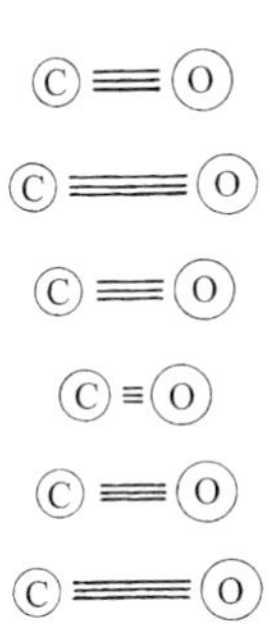

Fig. 5.7. When the CO molecule vibrates the bond stretches and contracts as shown. The more the bond stretches and contracts, the higher is the vibrational energy level. There are specific vibrational energy levels. (Credit: Richard W. Schmude, Jr.).

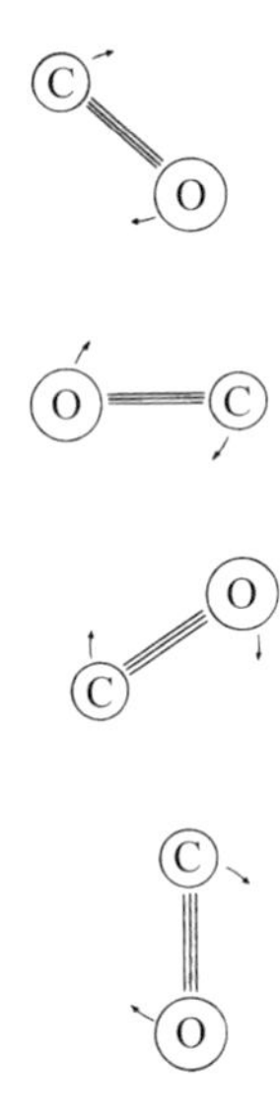

Fig. 5.8. During rotation, a molecule literally rotates as is shown here. There are specific rotational energy levels for this molecule. The faster the CO molecule rotates, the higher is the rotational energy level. (Credit: Richard W. Schmude, Jr.).

energy levels, with each electronic state having different vibrational levels. Most pure vibrational transitions involve light with wavelengths of between 3.0 and 30 μm which is within the infrared region of the electromagnetic spectrum.

In addition to electronic and vibrational energy levels, molecules also have rotational energy levels. See Fig. 5.8. Different rotational levels correspond to different rotational speeds. Pure rotational transitions for small molecules often lie in the frequency range of 100–1,000 GHz. This lies in the radio (or microwave) region of the electromagnetic spectrum.

It should be realized that a CO molecule has the same energy levels as another CO molecule, but it has different energy levels than an H_2O molecule. In fact, each type of molecule has its own unique set of energy levels and, hence, has its own unique spectrum. This concept is central to the understanding of how Astronomers detect different substances in a comet.

Therefore, and as shown above, a molecule like CO has three types of energy levels. In many cases, an energy change may involve two or all three levels (electronic, vibrational and rotational). Therefore, one can have "vibronic" transitions which involve a change in both electronic and vibrational energy levels. For example, a molecule may move from the electronic level 2 and vibrational level 3 to electronic level 1 and vibrational level 2. If a spectrum has suitable resolution, one could even pick out rotational transitions in vibrational spectra. This is one of the reasons why the interpretation of spectra can be very complex. Spectra also become more complex as the resolution improves but, at the same time, these complex spectra can reveal a wealth of information about comets. A description of spectral resolution is in order.

Spectral Resolution

The resolution of a spectroscope is defined as its ability to distinguish adjacent absorption or emission peaks. This is called spectral resolution. Resolution is affected by the amount of light reaching the detector, the nature of the prism or grating, the optical system and the slit width. One way to define specifically the resolution is by application of the Rayleigh Criterion. This states that there must be a valley between two equally intense emission lines of at least 19% of the height of the peaks. See Fig. 5.9. The resolution, R is defined as:

$$R = \lambda \div \Delta\lambda \tag{5.2}$$

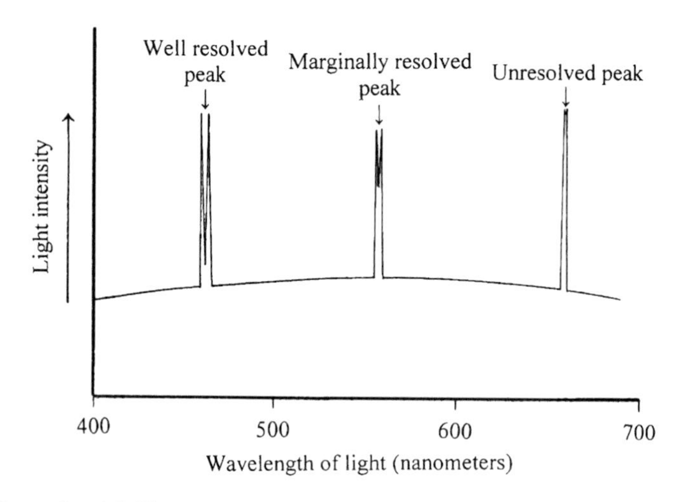

Fig. 5.9. The two peaks on the far *left* at a wavelength of 460 nm are well resolved because the valley between them is almost equal to the height of the peaks. The two peaks near the *center* at a wavelength of 560 nm are marginally resolved The feature on the *right* at a wavelength of 660 nm does not show two resolved peaks because the valley between them is less than 19% of the height of the peaks. (Credit: Richard W. Schmude, Jr.).

In this equation, λ is the wavelength of light and Δλ is the smallest distance between two peaks that can be resolved. If a spectroscope is operating at 600 nm and is able to just distinguish two peaks that are 2.0 nm apart, its spectral resolution would be 600 ÷ 2.0 = 300 nm. The higher the resolution of a spectrum, the more likely it is to have lots of peaks and, hence, the more information it will possess.

Table 5.2 summarizes the different types of spectra that professional astronomers collected of Comet 9P/Tempel 1, together with the telescope diameter and respective spectral resolutions.

Spectra of Comets

A visible-light emission spectrum is the easiest type of comet spectrum to image because our atmosphere absorbs less visible light than other types of light. Table 5.3 lists a few molecules that Chrstian Buil detected in the emission spectrum of Comet LINEAR (C/2001 A2) near its central condensation. Table 5.4 lists the equipment and procedure that he used in obtaining his spectrum. The exact transition wavelengths may change just a little due to several factors. One may use the wavelengths of the peaks in the spectrum as a guide for identifying species in a comet.

One may image also an emission spectrum in the mid-infrared region. The Earth's atmosphere, however, has a low transmission for mid-infrared wavelengths. Professional Astronomers have been able to collect spectra in this area with large telescopes at elevations above 2 km (6,600 ft).

Table 5.2. Recent spectra of Comet 9P/Tempel 1

Region	Approximate resolution	Telescope diameter (m)	Source
Ultraviolet and visible	47,000	10	Icarus, Vol. 191, p. 360
Ultraviolet and visible	1,000	0.9	Icarus, Vol. 191, p. 526
Ultraviolet, visible and infrared	1,000	2.2	Icarus, Vol. 191, p. 389
Mid-infrared	300	8	Icarus, Vol. 191, p. 432
Near-infrared	18,500	3.8	Icarus, Vol. 191, p. 371
Radio	300,000	305, 100	Icarus, Vol. 191, p. 469
Radio	600,000	10.4, 30, 35 × 300	Icarus, Vol. 191, p. 494

Table 5.3. Compounds detected in Comet LINEAR (C/2001 A2) by Christian Buil

Species	Characteristic wavelength (nm)
CN	388
C_3	405.6
C_2	438.0, 473.8, 516.5, and 563.5
$C_2 + NH_2$	610.0
$O + NH_2$	630.0

Table 5.4. Procedure and equipment that Christian Buil used in imaging the emission spectrum of Comet LINEAR (C/2001 A2) near its central condensation

Telescope	Takahashi FSQ-106-F/D = 5
Mount	Vixen GP-DX
Spectroscope	Littrow Spectrograph
CCD camera	Audine KAF-0401 E
Dispersion	0.8 nm/pixel
Resolution (at a wavelength of 550 nm)	340
Acquisition software	Pisco
Processing software	Iris
Spectral analysis software	VisualSpec

A sensitive radio telescope can pick up rotational transitions. Professional Astronomers have used this technique to study individual substances in several comets. Radio telescopes and a little of what we can learn from them are discussed below.

Radio Studies

Almost all objects in the universe give off radio waves. The radio region of the electromagnetic spectrum covers wavelengths larger than 10^{-4} m. The shorter wavelengths in this region often are called microwaves. Veteran Radio Astronomer G.W. Swenson said that wavelengths between 0.5 and 6 m were most useful for amateur radio telescopes.

The radio astronomer should be aware of both spectral and spatial resolution. Spectral resolution deals with how well one can distinguish between two nearby peaks in a spectrum. Spatial resolution deals with how well a telescope is able to distinguish two objects in the sky. This is similar to the Dawes Limit described in the previous chapter for optical telescopes. Radio telescopes often have excellent spectral resolution but poor spatial resolution. A general equation for spatial resolution is

$$\text{Spatial Resolution} = (250{,}000\,\text{arc - seconds} \times \lambda) \div \text{D} \qquad (5.3)$$

In this equation, λ is the wavelength and D is the telescope diameter. For example, if one uses a 4.0 m radio telescope and is studying radio emissions at a wavelength of 0.5 m, the spatial resolution would be computed as

$$\text{Spatial Resolution} = (250{,}000\,\text{arc - seconds} \times 0.5\,\text{m}) \div 4.0\,\text{m}, \text{ with the result:}$$

$$\text{Spatial Resolution} = 31{,}250\,\text{arc - seconds} = 8.7\,\text{degrees}$$

This value is much worse than the resolution of our eyes. One may improve spatial resolution by building a bigger telescope, working at a smaller wavelength or linking up several radio telescopes together. The Very Large Array (VLA) telescope in New Mexico is an array of 27 telescopes that can be moved to cover a maximum separation of 40 km. The separation essentially becomes the telescope diameter. Therefore, if one is interested in radio waves with a wavelength 0.5 m, he/she can attain a spatial resolution of just over three arc-seconds with the VLA.

What can radio data tell us? With a radio telescope, one can measure the production rate of a specific substance like H_2O in a comet. In addition, he or she can measure the speed at which this or other material is moving away from the nucleus and can search for different compounds in a comet. For example, one group detected OH, H_2O, HCN, CH_3OH (methanol) and H_2S in comet 9P/Tempel 1 in 2005. One can measure also the size and rotation rate of the nucleus.

Perhaps the most useful project for the amateur radio astronomer would be to study meteor showers which are associated with comets. Meteors in several of the major showers, including the Perseids, Orionids and Leonids are from comet debris. Radio studies may be carried out under a full Moon and under overcast skies. More information about radio telescopes and how to use them to study meteor showers can be found at the UK amateur radio astronomy website and the Society of Amateur Radio Astronomers website. These websites are listed in Table 5.5.

Photometry and Lightcurves

Most comets undergo two stages – nuclear and coma. Some comets undergo also a third stage which I will call the outburst stage. One can detect transitions between these stages by carrying out brightness measurements. These comet stages, together with the complete lightcurve, which shows the history of a comet's brightness, are described below. This is followed by a description of what we know about Comet Hale-Bopp's lightcurve.

Table 5.5. Important websites related to comet studies

Purpose or organization	Website
Palomar Sky Surveys I and II	http://stdatu.stsci.edu/cgi-bin/dss_form
Palomar Sky Survey	http://aps.umn.edu/
UK amateur radio astronomy	http://www.ukaranet.org.uk
Society of Amateur Radio Astronomers	http://www.qsl.net/SARA
SOHO	http://sohowww.nascom.nasa.gov/data/realtime-images.html
NEO confirmation page	http://www.cfa.harvard.edu/iau/NEO/ToConfirm.html
Minor Planet Comet Ephemeris Service	http://www.cfa.harvard.edu/iau/MPEph/MPEph.html, *Scroll down a little*
BAA comets section	http://www.ast.cam.ac.uk/~jds/
Gary Kronk (ALPO comet section coordinator)	http://cometography.com/index.html
Royal Astronomical Society of Canada	http://www.rasc.ca
Cometary Archive for Amateur Astronomers (CARA)	http://www.cara-project.org

Nuclear Stage

During the nuclear stage, the comet is too cold to have a coma. As a result, the comet's only source of light is that emitted from its nucleus. During this stage, the brightness follows (1.4). Essentially, the nucleus brightens by a factor of 4 when its distance to the Sun is cut in half. Similarly, the brightness increases also by a factor of 4 when its distance to the Earth is cut in half. Both the solar phase angle and the opposition surge may affect its brightness by over a factor of 2. Finally, the brightness of the nucleus may change as it rotates. This change will follow a repeatable pattern if the nucleus rotates around a single axis. If, on the other hand, the nucleus has chaotic rotation, its brightness pattern, due to rotation, will be unpredictable.

Before 1990, almost no comets were detected during their nuclear stage. This is due largely to the combination of the feeble light given off by comet nuclei and the limited sensitivity of cameras before 1990. Starting in the 1990s, more sensitive cameras became available. As a result, astronomers discovered some comets during their nuclear stage. In many cases, comets were discovered as bare nuclei and catalogued as asteroids. As these objects approached the Sun, they developed comas and were re-classified as comets. Astronomers have followed a few comets as they went from the coma to the nuclear stages.

As the nucleus approaches the Sun, volatile substances begin leaving it. At this point, a thin coma begins to develop. At first, the coma is so thin that the nucleus reflects most of the light coming from the comet. Equation (1.4) is still the best equation for modeling the comet's brightness at this stage. As the nucleus gets closer to the Sun, more and more material enters the coma. At some point, the escaping gas forces dust particles into the coma. At this point, the coma may be two to six times brighter than the nucleus. Many new comet discoveries made today are at this stage. I would say that at this point, the comet has entered the coma stage. At this stage, (1.4)–(1.6) are inadequate models for the comet's brightness because the nucleus and the coma reflect a large fraction of the total amount of light reflected. As the nucleus gets closer to the Sun the coma thickens. At this point, the coma reflects almost all of the light reflected, and (1.5) and (1.6) may be adequate models for use in determining the comet's brightness.

Coma Stage

During the coma stage, the comet brightens rapidly as it approaches the Sun. In many cases, the comet may brighten by a factor of 16 when its distance to the Sun is cut in half. The nucleus, on the other hand, brightens only by a factor of 4 when its distance to the Sun is halved. Equations (1.5) and (1.6) in Chap. 1 are the two most common ways of modeling the brightness of large comas. Any brightness changes due to changing solar phase angles and the opposition surge would be small. During the coma stage, comets may undergo random outbursts. During 1985–1986, for example, Comet 1P/Halley underwent several small brightness changes which were on the order of 20–40% or 0.2–0.4 stellar magnitudes.

Outburst Stage

During the outburst stage, a comet's brightness does not follow (1.4), (1.5) or (1.6). During this stage, a comet's brightness may increase or decrease suddenly. Outbursts may occur as the comet approaches the Sun or moves away from it. Astronomers may detect outbursts only after the comet's brightness in the coma and/or nuclear stages are well established. Comets LINEAR (C/2001 A2), LINEAR (C/2005 K2) and 17P/Holmes have undergone outbursts since 2000. Comet 1P/Halley underwent a large outburst in February of 1991. During this time, it was in its nuclear stage. Hence, outbursts may occur when a comet is in either its nuclear or coma stage.

Complete, Nearly Complete and Partial Lightcurves

A complete lightcurve is a graph of a comet's brightness during its entire nuclear and coma stage. This is a difficult graph to construct because most comets are very faint during their nuclear stage. A good alternative is the nearly complete lightcurve. This shows the brightness of the entire coma stage and at least part of the nuclear stage. One astronomer has constructed nearly complete lightcurves for Comets 1P/Halley, 81P/Wild 2, 19P/Borrelly, 21P/Giacobini-Zinner, 9P/Tempel 1, 67P/Churyumov-Gerasimenko, 26P/Grigg-Skjellerup and 28P/Neujmin 1. These lightcurves are published in Icarus, Vol. 178, pp. 493–516 and Icarus, Vol. 191, pp. 567–572.

A comet of recent interest is Hale-Bopp (C/1995 O1). I analyzed data for its coma phase. I corrected all telescopic and binocular data to a standard aperture of 6.8 cm according to (3.4) and (3.5) and corrected all unaided-eye brightness estimates to the standard aperture according to (3.6). With these data, I constructed lightcurves for Hale-Bopp that are all described in detail below.

Figures 5.10 and 5.11 are graphs of $M_c - 5\log[\Delta]$ versus log[r] for Comet Hale-Bopp. Figure 5.10 covers the time before perihelion, and Fig. 5.11 covers the time after perihelion. Comet Hale-Bopp reached perihelion on April 1, 1997. In both graphs, the points are measurements made when the comet was in its coma stage. The sloping line near the bottom of Figs. 5.10 and 5.11 represent the $M_n - 5.0\log[\Delta]$ values for the bare nucleus and, hence, correspond to the nuclear stage. Notice that the slope is steeper in the coma stage than in the nuclear stage. The difference in slope is due to the fact that, as Hale-Bopp approaches the Sun during the coma stage, it brightens because (1) it reflects more sunlight and (2) more material enters the coma which reflects even more light. This second reason for brightening does not occur in the nuclear stage and, hence, this is the reason for the steeper slopes during the coma stage.

Figure 5.12 shows a partial lightcurve of Comet Hale-Bopp (C/1995 O1). It is partial because it does not show any brightness measurements of the bare nucleus. Essentially I combined Figs. 5.10 and 5.11 to create Fig. 5.12. This lightcurve tells us at least four characteristics of the comet. These are the time when the coma begins to develop, the rate that the coma brightens as the comet

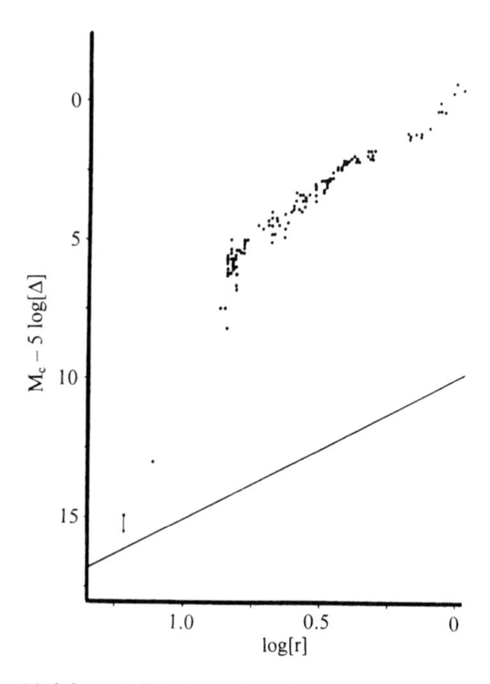

Fig. 5.10. A graph of $M_c - 5\log[\Delta]$ versus log[r] for Comet Hale-Bopp (C/1995 O1). All data are based on measurements made on or before the perihelion date of April 1, 1997. (Credit: Richard W. Schmude, Jr.).

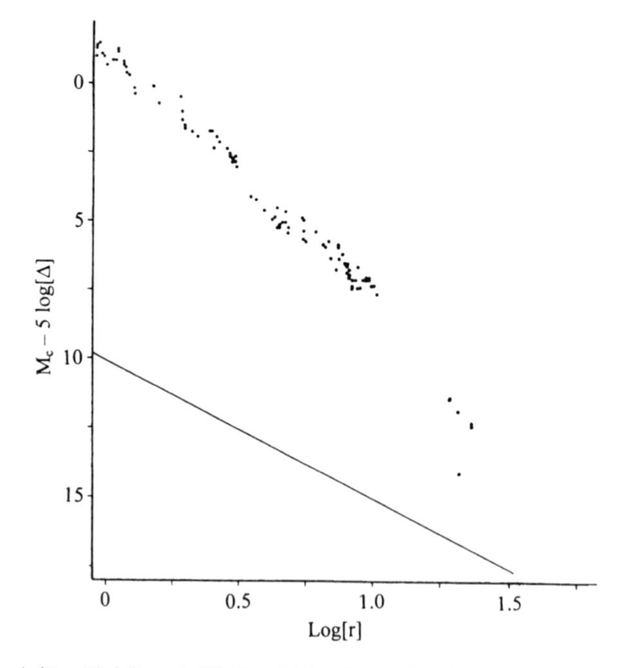

Fig. 5.11. A graph of $M_c - 5\log[\Delta]$ versus log[r] for Comet Hale-Bopp (C/1995 O1). All data are based on measurements made on or after the perihelion date of April 1, 1997. (Credit: Richard W. Schmude, Jr.).

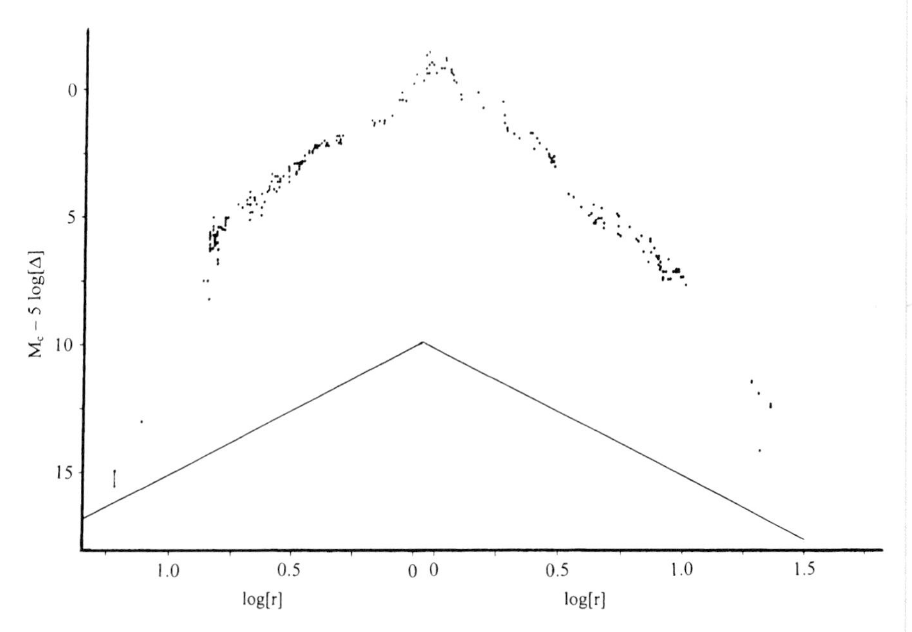

Fig. 5.12. Figures 5.10 and 5.11 are combined to yield a lightcurve showing the coma stage of Comet Hale-Bopp (C/1995 O1). Since Hale-Bopp reached a perihelion distance of $r = 0.91$ au, the center of the horizontal axis is $\log[0.91] = -0.04$. This value is in between the two *tick marks* on the horizontal axis labeled as 0 0. All data in Fig. 5.10–Fig. 5.14 are from the IAU circulars and from the Author's personal records (Credit: Richard W. Schmude, Jr.).

approaches the Sun, the enhancement factor (described in Chap. 1) and the presence of outbursts. Each of these characteristics is discussed below.

Figure 5.13 is the same as Fig. 5.12 except that I drew lines through the coma-stage data both before and after perihelion passage. Before perihelion, the data followed at least two slopes. The first line is drawn through all data between August 1995 when $\log[r] = 0.84$ up to April 1, 1997, when $\log[r] = -0.04$. It is obvious that this line does not run through the single pre-discovery measurement made in 1993 when $\log[r] = 1.12$. Therefore, the data in 1993 must follow a second slope. Accordingly, I assumed that the slope changed in August 1995. The second line runs close to the 1993 data point and the data collected in August 1995. Note that the 1993 point is an upper limit and, hence, I drew a small downward-pointing line from it. This point is near the lower left corner in the figure. This second line intersects the curve defined by just the nucleus when $\log[r] = 1.25$ or when r (the comet-Sun distance) ~18 au shown as R_{on} in Fig. 5.13. Therefore, Hale-Bopp probably started to develop a coma when it was about 18 au from the Sun. The estimated uncertainty of R_{on} is ±3 au. Therefore, $R_{on} = 18 \pm 3$ au for this comet. It reached this point in early 1991. The intersection between the lines representing the coma stage after perihelion and the nuclear stage is called "R_{off}". As of 2005–2006, Hale-Bopp was still about ten times brighter than just its nucleus. The data after perihelion follow a linear trend at least through January 2000 when $\log[r] = 1.01$. The behavior after this date is unknown. I assumed that the slope changed in January 2000 and drew a line connecting the 2000 and 2005–2006 data. This final line intersects that of

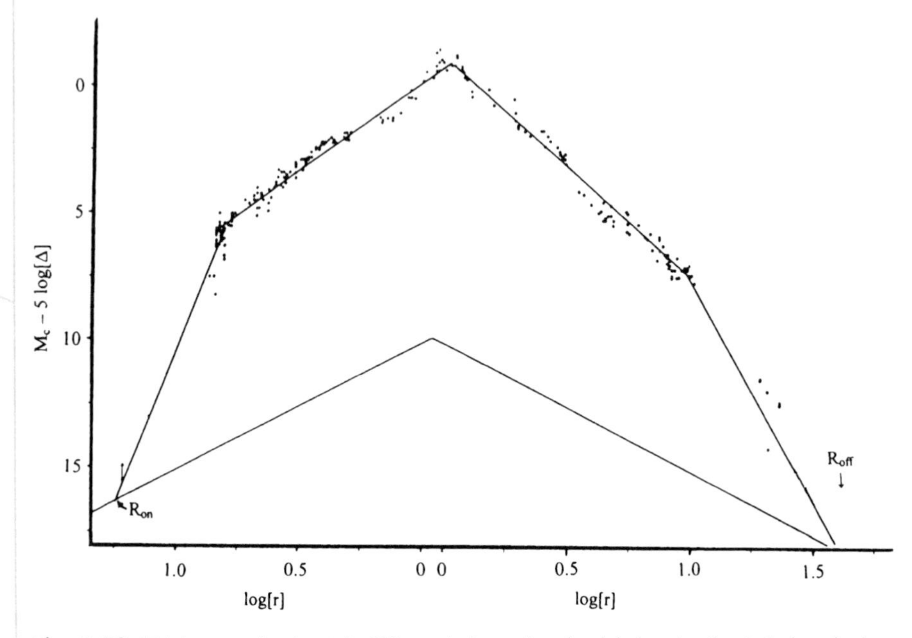

Fig. 5.13. This is the same graph as shown in Fig. 5.12 except that lines are drawn through the data points. The point R_{on} shows when Comet Hale-Bopp went from the nuclear to the coma stage. The point R_{off} shows when Comet Hale-Bopp will go from the coma to the nuclear stage. (Credit: Richard W. Schmude, Jr.).

the nuclear stage at R_{off} in Fig. 5.13. A rough estimate of R_{off} is 32 au, and Hale-Bopp should reach this point in late 2011.

The coma stage of Hale-Bopp probably started in early 1991, or about 6 years before perihelion passage. The coma stage will last probably until about 2011, which is about 14 years after perihelion. This is not symmetrical. Why? The best explanation may be that it took awhile for the frigid nucleus to warm up as it approached the Sun and, hence, R_{on} was delayed. Furthermore, heat form the Sun sunk into deeper layers of the nucleus as the comet reached perihelion. The nucleus retained its heat even after it began to pull away from the Sun. As a result, gases continued to escape from it long after it had passed the Sun. Another possible reason for the asymmetry may be the large size of the nucleus and, hence, its higher escape speed. Since the escape speed is higher than for a smaller nucleus, more of the dust may have remained near the nucleus that would have escaped otherwise.

The lightcurve in Fig. 5.13 shows also how fast Comet Hale-Bopp brightened as it approached the Sun and how fast it dimmed as it receded from it. Data before perihelion with $0.88 < \log[r] < -0.04$ follow the equation:

$$M_c - 5\log[\Delta] = 7.75\log[r] - 0.67 \tag{5.4}$$

This equation is of the same form as (3.9) in Chap. 3. By comparing (3.9) and (5.4), it is evident that the pre-exponential factor, 2.5n, equals 7.75 and $H_0 = -0.67$. The pre-exponential factor expresses how quickly the comet brightened, and is equal to the slope of the dots in Fig. 5.10. The −0.67 is the value of $M_c - 5\log[\Delta]$ when $\log[r] = 0$. One can re-arrange (5.4) as:

$$M_c = 5\log[\Delta] + 7.75\log[r] - 0.67 \quad (5.5)$$

This equation predicts the brightness of Comet Hale-Bopp during its coma phase for $0.88 > \log[r] > -0.04$, before perihelion. The data after perihelion for $-0.04 < \log[r] < 1.01$ follows (5.6)

$$M_c - 5\log[\Delta] + 8.60\log[r] - 1.10 \quad (5.6)$$

The corresponding values of 2.5n and H_0 equal 8.60 and −1.10. One can rearrange (5.6) as:

$$M_c = 5\log[\Delta] + 8.60\log[r] - 1.10 \quad (5.7)$$

Equations (5.5) and (5.7) may be used as a baseline for searching for comet outbursts or periodic brightness variations. Comet Hale-Bopp did not have any outbursts exceeding a factor of 6, or 2 stellar magnitudes, when log[r] was less than 0.88 before perihelion or when $\log[r] < 1.01$ after perihelion. It may have had an outburst near January of 2006. After all, it was about a factor of 6 brighter than what it was a year earlier.

The enhancement factor for Comet Hale-Bopp is 11.3 ± 0.3 stellar magnitudes, or a factor of 33,000. Comets with lots of volatile substances will brighten more than those that lack these compounds. Comets which get very close to the Sun will brighten also more than those that do not do so. Table 5.6 lists several comets, their perihelion distances, enhancement factors and ages. A comet that has reached perihelion a few times is considered to be "young", and one that has reached perihelion many, even hundreds of times, is considered to be "old".

Table 5.6. Difference in brightness between the coma and the nucleus of various comets at perihelion

Comet	Perihelion distance (au) [a]	Enhancement factor at perihelion (stellar magnitudes)	Age [b]
1P/Halley	0.5871	10.5 ± 0.2[c]	Young
81P/Wild 2	1.590	7.7 ± 0.4[c]	Young
19P/Borrelly	1.355	7.7 ± 0.2[c]	Young
21P/Giacobini-Zinner	1.04	8.0 ± 0.3[d]	Young
9P/Tempel 1	1.506	5.9 ± 0.3[c]	Young
67P/Churyumov-Gerasimenko	1.25	4 ± 1[d]	Young
26P/Grigg-Skjellerup	1.12	5 ± 0.5[d]	Old
28P/Neujmin 1	1.55	2 ± 1[d]	Old
Hale Bopp (C/1995 01)	0.91	11.3 ± 0.3[e]	Young

[a] Most recent passage as of mid-2008
[b] From Ferrín (2005)
[c] Computed by the author; see Chap. 2
[d] Estimated by the author from data in Ferrín (2005)
[e] Estimated by the author from his data

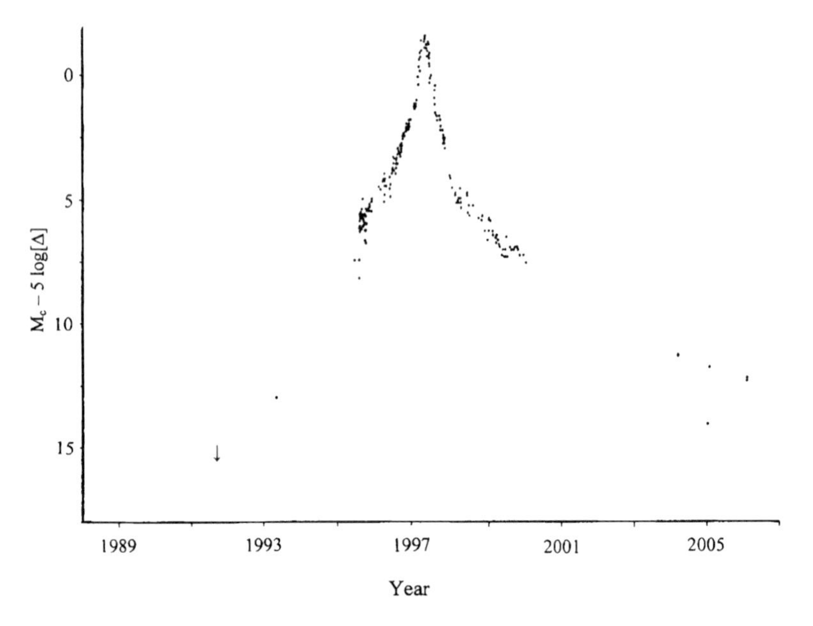

Fig. 5.14. A graph of $M_c - 5\log[\Delta]$ versus the year for Comet Hale-Bopp (C/1995 O1). (Credit: Richard W. Schmude, Jr.).

The comets with the largest enhancement factors are Comet 1P/Halley and Comet Hale-Bopp. The huge brightness increases stem mainly from the fact that they are relatively young comets. Comet 26P/Grigg-Skjellerup, on the other hand, is old and brightens by only 5 ± 0.5 stellar magnitudes even though its perihelion distance is similar to that of Hale-Bopp. In addition, the brightness increases are due partly to the close perihelion distance of Comets 1P/Halley and Hale-Bopp.

One can also look at the lightcurve in terms of time. Figure 5.14 shows a graph of the $M_c - \log[r]$ versus year for Comet Hale-Bopp. The advantage of this graph is that it shows the brightness history in real time. There was a change in the slope in February 1997 (before perihelion) and a second change in slope toward the end of 1997 (after perihelion). These changes are due largely to the comet's changing distance.

Photometry of the Bare Nucleus

What can we learn by studying the bare nucleus of a comet? We can learn its color and other characteristics. One may measure color by measuring the brightness of the nucleus through different color filters, like those transformed to the Johnson B and V system. Brightness measurements of the bare nucleus may yield also information on the nature of its rotation and, in many cases, its rotation rate. Specifically, one may plot the normalized magnitude of the nucleus versus time. If there is a repeatable brightness change, the nucleus would be rotating around a single axis. If there are no repeatable brightness changes, the nucleus would be undergoing chaotic rotation. One could measure albedo from brightness data, provided that the size of the nucleus is known. One could look also for outbursts. Comet 1P/Halley had a large outburst when it was over 14 au from the Sun. In this project, one may use unfiltered images to increase the sensitivity.

Professional astronomers have measured the brightness, color and solar phase angle coefficients of several nuclei. A typical normalized magnitude of a nucleus, V(1,0) = 17th magnitude. Because of this, bare nuclei are seldom brighter than a stellar magnitude of 20. The brightest nucleus is probably Comet 95P/Chiron which can reach a stellar magnitude of ~17. A telescope-camera system which can reach a stellar magnitude of ~19.5 will be able to yield reliable measurements of the nucleus of this comet.

The best time to study the bare nucleus is when it is brightest. Many astronomers have carried out studies by subtracting light scattered by the coma. With modern software packages, those with modest equipment should be able to do this. Alternatively, one can also carry out brightness measurements of the nucleus just before it reaches R_{on}, or just after it reaches R_{off}.

Polarization Measurements

One may measure the amount of polarized light given off by different parts of a comet. Polarization measurements may yield information on a comet's composition, the mean particle size of the coma (or tail), the gas-to-dust ratio and the fractions of reflected, emitted and scattered light. Polarization measurements made of the central condensation may yield information on the nucleus. Finally, polarization measurements of the bare nucleus may yield information on the nature of the surface material. A discussion of polarized light, along with an example of how to measure it is given in *Uranus, Neptune and Pluto and How to Observe Them* ©2008 Richard Schmude, Jr. I will discuss three projects which one might carry out involving the measurement of polarized light.

The easiest project is to measure the amount of polarized light given off by different parts of a comet. For example, two groups of astronomers at the Central Astronomical Observatory and at the Special Astrophysical Observatory reported that the amount of polarized light for Comet Hale-Bopp (C/1995 O1) was much higher in the tail than in the coma. They suggested that this was consistent with strong molecular bond polarization. These groups also reported that the V filter polarization values are 10–20% in the gas tail, but less than 1% in the coma.

One can also measure the polarization value of the coma as the solar phase angle changes. In doing so, he or she may combine several of these measurements made at different solar phase angles to yield a phase curve. This is a graph that shows the polarization value versus the solar phase angle. The phase curve yields information on the amounts of gas and dust in the coma.

Another project which one might carry out would be to measure the phase curve of a comet's bare nucleus. These measurements may yield information on the mean particle size of the material covering the nucleus. One may be able also to use polarization measurements to detect a faint coma.

Searching for New Comets

With today's technology and data bases, there are several ways of carrying out comet searches. Before beginning a search, one should be aware of some trends over the last two centuries. Four trends of comet discoveries over the last two centuries are described below. These trends are the number of comets discovered per decade, the average brightness of newly discovery comets in stellar magnitudes, the percentage of comets discovered visually and the percentage of comets discovered in the northern hemisphere.

Figure 5.15 shows the number of comets discovered each decade between 1780 and 1999. During the nineteenth century, about two to three comets were discovered each year. This number doubled during the first half of the twentieth century, and it doubled again during the second half of the twentieth century. Between 2000 and 2008, over 1,500 comets were discovered; this is over 150 comets discovered each year. This result is not shown in Fig. 5.15 because it would be off of the scale. Why is this the case? The answer is simple – due to advances in technology and astronomical interest, comets are more likely to be discovered just after they have developed a coma than in earlier times. In many cases, they are discovered at a stellar magnitude of 18 or fainter. In short, comets are being discovered earlier in their coma stages and, in some cases, are being discovered in their nuclear stage. As a result, there are fewer opportunities of discovering comets brighter than 12th magnitude today than there were in the 1970s.

Figure 5.16 shows the average brightness of newly discovered comets in each decade between 1800 and 2008. During the nineteenth century, the typical newly discovered comet was as bright as a sixth to eighth magnitude star. During the early part of the twentieth century when high quality photographs became

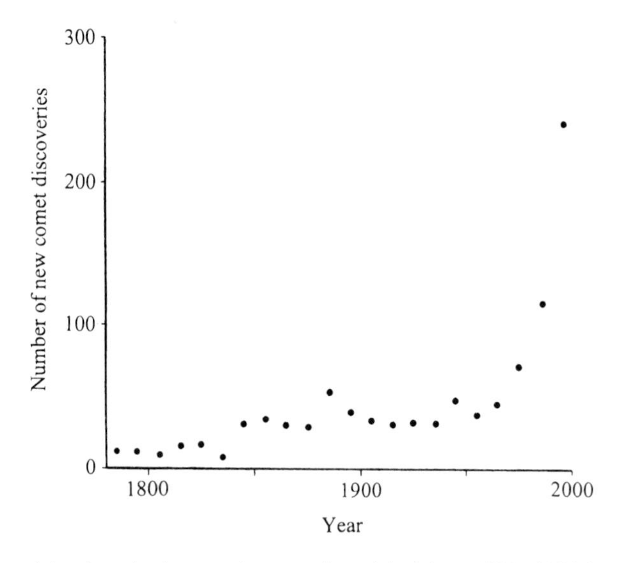

Fig. 5.15. This graph shows the number of new comet discoveries made in each decade between 1780 and 1999. Between 2000 and 2008, Astronomers had discovered over 1,500 comets. This result is not shown on the graph. (Credit: Richard W. Schmude, Jr.).

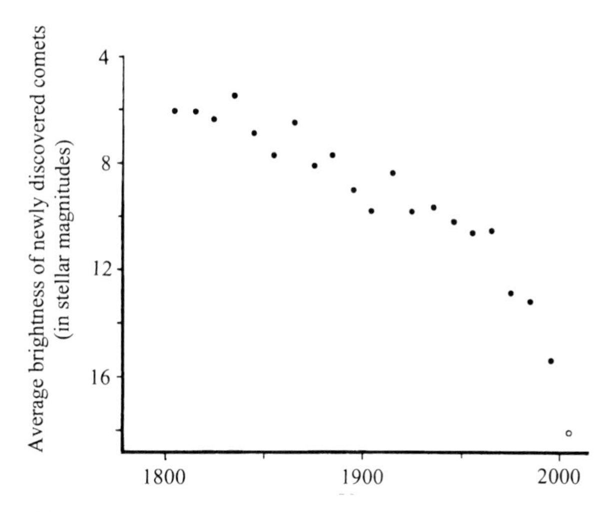

Fig. 5.16. This graph shows the average brightness of newly discovered comets in stellar magnitudes in each decade since 1800. The *open circle* corresponds to the average brightness of comets discovered between 2000 and 2008. Since this is not quite a decade, an open circle is used. Data if Figs. 5.15–5.18 are based on data in cometagrophy, vols. II–IV © 2003, 2004, 2009 by Kroak; the IAU circulars and the catalogue of cometary Orbits 17th edition © 2008 by Marsden and Gareth. This figure does not include Comets SOHO, SOLWIND, SMM and STEREO. (Credit: Richard W. Schmude, Jr.).

available, the typical newly discovered comet was about as bright as a ninth or tenth magnitude star. By the 1980s, the typical newly discovered comet was as bright as a star of between 13th and 14th magnitude – about as bright as Pluto at opposition. Excluding comets discovered on SOHO images, newly discovered comets between 2000 and 2008 had an average brightness equivalent to about an 18th magnitude star.

Figure 5.17 shows the percentage of comets discovered "visually" since 1870. Visual discoveries include those made with just a telescope, binoculars and/or the unaided eye. This percentage remained at 100% until 1892. This was the year when E.E. Barnard discovered a comet from a photograph. For the next century, photography was the best way to discover faint comets. Since the 1990s, astronomers have used either digital images or images made by SOHO to make most of the comet discoveries. Between 2000 and 2008, the percentage of comets discovered visually plummeted to less than 3%.

Figure 5.18 shows the percentage of comets discovered in the Northern Hemisphere since 1800. During the nineteenth century, the percentages were high, however, by the 1920s, the percentage had dropped to 55%. This means that almost half of the comet discoveries were made in the Southern Hemisphere in that decade. Since 1990, the percentage of comet discoveries in the Northern Hemisphere has risen. This is due mainly to the large number of comet discoveries which have taken place as a result of all-sky surveys mostly in the Northern Hemisphere.

The trend in Fig. 5.17 shows that the odds of finding a comet visually have decreased steadily since 1980. In spite of this, the amateur astronomer has much more sensitive equipment today and comet discovery should not be disregarded. I will describe five ways of making a discovery.

The traditional way of searching for new comets is to use either large binoculars or a telescope at a magnification of around 50×. David Levy, who has discovered over 20 comets, recommends looking for new comets within 90° of the Sun. Figure 5.19 shows the positions of ten comets discovered visually since 2000. Note that most of them are within 90° of our Sun. One should make horizontal sweeps of either the western sky after sunset or of the eastern sky before sunrise. Even before all-sky surveys, it often took about 1,000 h of searching to discover a comet. This is equivalent to searching 1 h each day for 33 months. Don Machholz uses a patch to cover one eye during his comet searches. See Fig. 5.20.

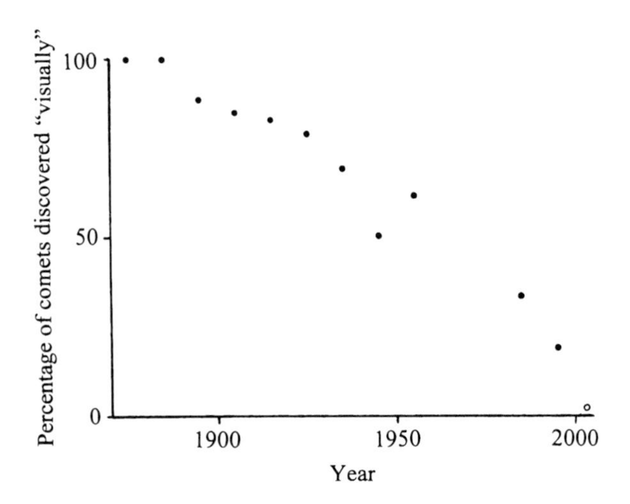

Fig. 5.17. The percentage of comets that were discovered "visually" in each decade since 1870 is plotted against the years indicted. Note that since the 1890s, this percentage has dropped due mainly to the advent of enhanced photography and digital imaging. This figure does not include comets named SOHO, SOLWIND, SMM and STEREO. (Credit: Richard W. Schmude, Jr.).

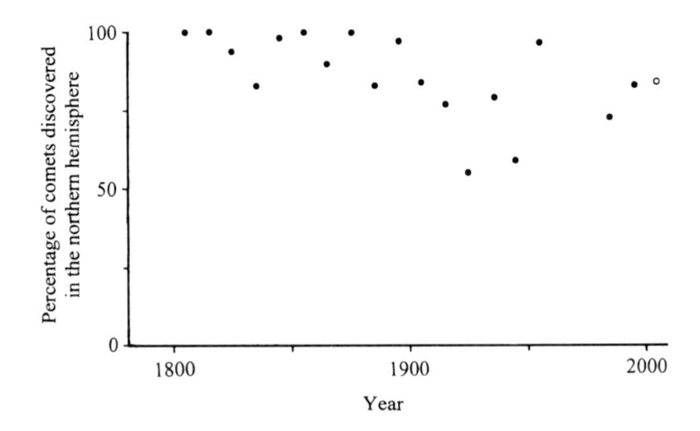

Fig. 5.18. This graph shows the percentage of comets discovered in the northern hemisphere for each decade from 1800 to 1999. The *open circle* represents the percentage between 2000 and the end of 2008. Since this is not quite a decade, an open circle is used. The graph does not include comets named SOHO, SOLWIND, SMM and STEREO. (Credit: Richard W. Schmude, Jr.).

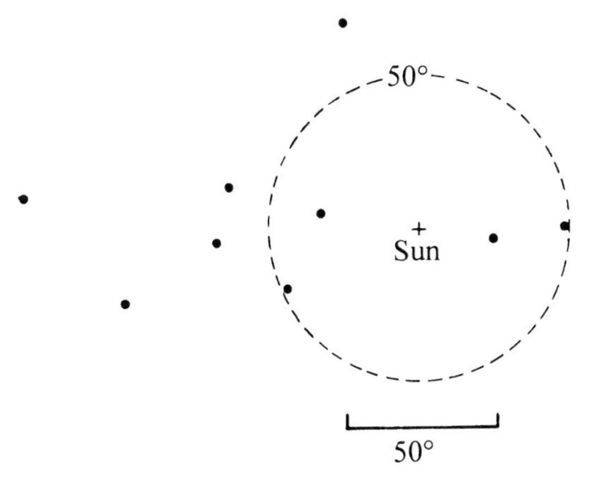

Fig. 5.19. Positions of ten comets discovered visually since 2000 in relation to the Sun. Each *dark circle* represents the location of a comet when it was discovered. The *dashed circle* is centered on the Sun with an angular radius of 50° (Credit: Richard W. Schmude, Jr.).

Fig. 5.20. This picture shows Don Machholz with his eye patch and telescope. This is the telescope that Don used to discover Comet Machholz (C/2004 Q2). (Credit: Don Machholz).

One can improve his or her chances of a discovery by using a digital camera and software which identifies moving objects. Gary Hug and Graham Bell took a 60 min exposure with an SBIG ST-9E CCD camera and a 12-in. (0.30 m) telescope and discovered a faint object. On subsequent images, they found that this object moved and was fuzzy. They had discovered a new comet! It was a very faint comet equivalent to a star of magnitude of 18.8. This new comet was named initially Comet P/1999 X1 Hug-Bell. It soon became apparent that this comet would return in a few years. It was recovered in 2006 and was renamed Comet 178P/Hug-Bell.

One can search also for comets in the daytime. L. Swift looked westward to a mountain that blocked part of the Sun on September 21, 1896. In spite of the fact that the Sun had not set, he saw a comet nearby. Comets are brightest when near the Sun. Thus one may carry out a search with a small telescope provided that the Sun is behind an opaque object. I used this technique to study Venus when it was about 15° from the Sun. If a suspicious object is noted, then it is important to determine its position along with the time. One way to measure a comet's approximate position in the daytime is to estimate its position with respect to the Sun. Essentially one estimates the angular distance between the object and the Sun and notes the position angle. Then one can estimate the right ascension and declination from the Sun's coordinates. A more exact method of estimating position is to use a telescope on an equatorial mount with setting circles. One follows the procedure in Chap. 4 for adjusting the right ascension and then points the telescope at the suspicious object. Finally, one reads off the right ascension and declination from the telescope's setting circles.

One may also search for comets within about 30° of the Sun during twilight or in darkness. Fainter comets may be detected as the sky darkens.

Another way of searching for comets is during a solar eclipse. Gary Kronk reports that Egyptian astronomers discovered a naked-eye comet near the totally eclipsed Sun on May 17, 1882. During a total eclipse, the sky darkens, and this allows many objects to become visible that would not be visible in the Sun's glare. One may search for fainter comets by taking wide-field images of the areas near the Sun. These images will reveal not only the beauty of the Sun's corona, but may reveal also a new comet. The next two total solar eclipses which will be visible from the United States will be on August 21, 2017, and April 8, 2024. The closest total solar eclipse for people living in the United Kingdom will be on the Faeroe Islands, which are north of Scotland, on March 20, 2015.

One can also make a comet discovery by examining images taken by the Solar and Heliospheric Observatory (SOHO). Images are posted on SOHO's website which may be analyzed by anyone with an internet connection. The SOHO website is listed in Table 5.5. Several people have discovered comets near the Sun from SOHO images. British astronomer Michael Oates describes his technique for discovering comets from SOHO images in the October 2000 issue of *Sky and Telescope* Magazine on page 89.

One can also discover a comet on an old photograph or image. Astronomers found five different comets on photographs taken as part of the Palomar Sky Survey I more than 20 years after the photographs were taken. These comets were all about as bright as stars of magnitude 18.5–19.5. As of 2008, there are several sky surveys that Astronomers have completed or are carrying out currently. These are listed in Table 5.7. One may be able to search for comets on archived images from one or more of these surveys.

One may image known asteroids or suspicious objects. In some cases, these objects develop comas when they approach the Sun. The majority of them are fainter than 17th magnitude and, hence, a large telescope would be needed for study purposes. Several suspected objects are posted on the NEO confirmation page. This website is listed in Table 5.5. For example, in IAU circular number 7546, it was reported that the Asteroid 2000 WM_1 was originally listed as a new asteroid. It, however, was placed in the NEO confirmation page because of its peculiar orbit. A few weeks later, T.B. Spahr imaged this object and noticed a coma around it.

Table 5.7. A summary of all-sky surveys carried out at visible and near-infrared wavelengths

Survey name	Location(s)	Time period of survey
Palomar all sky survey I (POSS-I)	117°W, 33°N	1950s
Palomar all sky survey II (POSS-II)	117°W, 33°N	1990s
Sloan Digital Sky Survey I and II	106°W, 33°N	2000s
Sloan Digital Sky Survey I and II	106°W, 33°N	2000s and 2010s
Spacewatch	112°W, 32°N	1990s and 2000s
Catalina Sky Survey (CSS)	111°W, 32°N and 149°E, 31°S	2000s
Lowell Observatory Near-Earth Object-Search (LONEOS)	112°W, 35°N	1990s and 2000s
Southern Sky Survey	70°W, 25°S	1990s
Near-Earth Asteroid Tracking Program (NEAT)	156°W, 21°N and 112°W, 32°N	2000s
Lincoln Near Earth Asteroid Research (LINEAR)	107°W, 34°N	2000s
Panoramic Survey Telescope and Rapid Response System (Pan-STARRS)[a]	156°W, 21°N	2010s

[a]Preliminary images taken in early 2008

Consequently, it was reclassified at a comet and was given the name LINEAR (C/2000 WM_1). It is named after the LINEAR team which discovered it while it was in its nuclear stage.

"Recovery" of Periodic Comets

An important activity is "to recover" periodic comets. This is vital to our understanding of the non-gravitational forces acting on comets, including their orbits. The best way to "recover" a comet is to look at its predicted location(s). Predicted positions are based on previous observations. Software packages, like Voyager, may be of great help.

Measuring the Opacity of Comets

One may monitor also the brightness of a star as a comet moves in front of it. Several early astronomers watched stars as comets moved in front of them. For example, Gary Kronk reported that J.R. Hind watched Comet C/1846 O1 de Vico-Hind move in front of a 12th magnitude star in *Cometography* Vol. II ©2003 by Gary Kronk. Hind reported that during the eclipse, the star's appearance was hardly affected.

More recently, in IAU Circular 3751, it was reported that two groups of Astronomers at the Observatoire de Meudon and at Institut d'Astrophysique at Liege reported that Comet Bowell (C/1980 E1) moved in front of a 15th magnitude star, and that the star's brightness "fell by 0.5". Light with a wavelength of 600–900 nm was used for the experiment. Regions 10,000–20,000 km from the Comet's nucleus moved in front of that star. This showed that Comet Bowell had at least some opaque regions within 20,000 km of its nucleus.

Appendix

Computing Tail Length from Two Points of Right Ascension and Declination

One degree of declination is the same regardless of where it is measured in the sky, but this is not the case for right ascension. One hour of right ascension equals an angle of 15°, but this is not necessarily equal to 15° of declination because, as one approaches the celestial pole, the right ascension circles get progressively smaller. Declination, however, is measured always along a great circle. At the north celestial pole, 1° of declination equals 1°, but 360° of right ascension equals zero. As a result, if any part of a tail is more than 20° from the celestial equator, and one desires to determine its length from its right ascension and declination, he/she must use (A.1)

$$\ell = \text{inv } \cos[\sin(\delta_1)\sin(\delta_2) + \cos(\delta_1)\cos(\delta_2)\cos(a_1 - a_2)] \tag{A.1}$$

In this equation, ℓ is the tail length in degrees; a_1 and δ_1 are the respective right ascension and declination of the central condensation (one end of the tail); and a_2 and δ_2 are the respective right ascension and declination of the other end of the tail. If the entire tail lies close to the celestial equator, there would be no need to use (A.1).

By way of example, let's say that the right ascension (RA) and declinations (Dec.) are: central condensation (RA = 02 h 05 min 21 s or 2.089 h or 31.3°, Dec. = 44.0°N); tip of tail (RA = 0 h 43 min 12 s or 0.72 h or 10.8°, Dec. = 40.0°N). In both cases, the RA is converted to degrees where each hour of RA equals 15°, each minute = 0.25° and each second = 15/3,600 = 0.004167°. In this example, the declination is already in degrees. Since the comet is north of 20°N, one uses (A.1) to determine ℓ.

$$\ell = \text{inv } \cos\,[\sin(44.0^\circ)\sin(40.0^\circ) + \cos(44.0^\circ)\cos(40.0^\circ)\cos(31.3^\circ - 10.8^\circ)]$$

$$\ell = \text{inv } \cos[(0.6947)(0.6428) + (0.7193)(0.7660)\cos(20.5^\circ)]$$

$$\ell = \text{inv } \cos[0.4465 + (0.7193)(0.7660)(0.9367)]$$

$$\ell = \text{inv } \cos[0.4465 + 0.5161]$$

$$\ell = \text{inv } \cos[0.9626] = 15.7^\circ$$

If I had determined the length of the tail from the Pythagorean Theorem and assumed that 1° of right ascension equaled 1° of declination, I would have wound up with a length of 20.9°. However, the true length, 15.7°, is the one that must be reported.

Bibliography

A'Hearn MF, Belton MJS, Delamere WA et al (2005) 'Deep Impact: Excavating Comet Tempel 1,' Science 310: 258–264.

A'Hearn MF, Combi MR (2007) 'Introduction: Deep Impact at Comet Tempel 1,' Icarus 191: 1–3.

Aguirre E (1997) 'Comet Hale-Bopp at its Peak,' Sky and Telescope 93 (4): 28–33.

Aguirre EL (2000) 'Amateurs Find Superfaint Comet,' Sky and Telescope 99 (4): 84.

Allen MM, Delitsky W, Huntress Y et al (1987) 'Evidence for methane and ammonia in the coma of Comet P/Halley,' Astronomy and Astrophysics 187: 502–512.

Anonymous (1990) 'Giant dust in Comet Tempel 2,' Sky and Telescope 79: 128–129.

Anonymous (1999) 'Cometary constancy,' Sky and Telescope 98 (3): 19.

Ashford AR (2005) 'Orion's IntelliScope XT10,' Astronomy 33 (1): 82–85.

Ashford AR (2006a) 'Celestron's Revamped 8-inch SCT,' Sky and Telescope 111 (3): 74–77.

Ashford AR (2006b) 'Filling an Aperture Void: Celestron's New C6-SGT,' Sky and Telescope 112 (2): 86–89.

Atkins P, de Paula J (2002) Physical Chemistry, 7th edn, W. H. Freeman and Company, New York.

Balsiger H, Altwegg K, Bühler F et al (1986) 'Ion composition and dynamics at comet Halley,' Nature 321: 330–334.

Bar-Nun A, Heifetz E, Prialnik D (1989) 'Thermal evolution of Comet P/Tempel 1 - Representing the group of targets for the *CRAF* and *CNSR* missions,' Icarus 79: 116–124.

Bar-Nun A, Pat-El I, Laufer D (2007) 'Comparison between the findings of Deep Impact and our experimental results on large samples of gas-laden amorphous ice,' Icarus 191: 562–566.

Barber RJ, Miller S, Stallard T et al (2007) 'The United Kingdom Infrared Telescope Deep Impact observations: Light curve, ejecta expansion rates and water spectral features,' Icarus 191: 371–380.

Bauer JM, Weissman PR, Choi YJ et al (2007) 'Palomar and Table Mountain observations of 9P/Tempel 1 during the Deep Impact encounter: First results,' Icarus 191: 537–546.

Beatty JK (1986) 'An inside look at Halley's comet,' Sky and Telescope 71: 438–443.

Beatty JK (2001) 'Meet Comet Borrelly,' Sky and Telescope 102 (6): 18–19.

Beatty JK, Bryant G (2007) 'McNaught's passing fancy,' Sky and Telescope 113: 32–36.

Bennett J, Donahue M, Schneider N et al (2009) The Cosmic Perspective, 5th edn, Pearson Education, Inc., San Francisco.

Bensch F, Melnick GJ, Neufeld DA et al (2007) 'Submillimeter Wave Astronomy Satellite observations of Comet 9P/Tempel 1 and Deep Impact,' Icarus 191: 267–275.

Benton JL Jr (2005) Saturn and How to Observe It, Springer, London.
Berry R (1986) 'Giotto encounters comet Halley,' Astronomy 14 (6): 6–22.
Biver N, Bockelée-Morvan D, Boissier J et al (2007) 'Radio observations of Comet 9P/Tempel 1 before and after Deep Impact,' Icarus 191: 494–512.
Bockelée-Morvan D, Biver N, Colom P et al (2004) 'The outgassing and composition of Comet 19P/Borrelly from radio observations,' Icarus 167: 113–128.
Bonev T, Jockers K, Karpov N (2008) 'A dynamical model with a new inversion technique applied to observations of Comet C/2000 WM_1 (LINEAR),' Icarus 197: 183–202.
Bortle JE (1981) 'How to observe comets,' Sky and Telescope 61: 210–214.
Bortle JE (1982a) 'Comet Digest,' Sky and Telescope 63: 315.
Bortle JE (1982b) 'Comet Digest,' Sky and Telescope 64: 102.
Bortle JE (1983a) 'Comet Digest,' Sky and Telescope 66: 175–178.
Bortle JE (1983b) 'Comet Digest,' Sky and Telescope 66: 473.
Bortle JE (1984a) 'Comet Digest,' Sky and Telescope 67: 483.
Bortle JE (1984b) 'Comet Digest,' Sky and Telescope 68: 482.
Bortle JE (1984c) 'Comet Digest,' Sky and Telescope 68: 583.
Bortle JE (1985a) 'Comet Digest,' Sky and Telescope 70: 394.
Bortle JE (1985b) 'Comet Digest,' Sky and Telescope 70: 509.
Bortle JE (1985c) 'Comet Digest,' Sky and Telescope 70: 629.
Bortle JE (1986) 'Comet Digest,' Sky and Telescope 71: 221.
Bortle JE (1987a) 'Comet Digest,' Sky and Telescope 73: 114.
Bortle JE (1987b) 'Comet Digest,' Sky and Telescope 73: 456–457.
Bortle JE (1988a) 'Comet Digest,' Sky and Telescope 75: 226.
Bortle JE (1988b) 'Comet Digest,' Sky and Telescope 75: 334–335.
Bortle JE (1990) 'An observer's guide to Great Comets,' Sky and Telescope 79: 491–492.
Bortle JE (1995) 'Borrelly's strange apparition,' Sky and Telescope 90 (2): 108–109.
Bortle JE (1997a) 'Estimating a comet's brightness,' Sky and Telescope 93 (4): 31–32.
Bortle JE (1997b) 'How important is forward-scatter geometry?' Sky and Telescope 94 (3): 12, 14.
Bortle JE (2008) 'The Astounding Comet Holmes,' Sky and Telescope 115 (2): 24–28.
Brandt JC, Caputo FM, Hoeksema JT et al (1999) 'Disconnection Events (DEs) in Halley's comet 1985–1986: The correlation with crossings of the Heliospheric Current Sheet (HCS),' Icarus 137: 69–83.
Brandt JC, Niedner MB Jr, Rahe J (1992) The International Halley Watch Atlas of Large-Scale Phenomena, Johnson Printing Co., Boulder, CO.
Brandt JC, Panther RW, Green D (1981) 'Twisting a comet's tail,' Sky and Telescope 61: 107.
Britt DT, Boice DC, Buratti BJ et al (2004) 'The morphology and surface processes of Comet 19/P Borrelly,' Icarus 167: 45–53.
Brownlee D, Tsou P, Aléon J et al (2006) 'Comet 81P/Wild 2 under a microscope,' Science 314: 1711–1716.
Bryant G (2005) 'Targeting Comet Tempel 1,' Sky and Telescope 109 (6): 67–69.
Bryant G (2006) 'A very close comet flyby,' Sky and Telescope 111 (5): 60–65.
Bryant G, MacRobert A (2005) 'Comet Machholz in the Evening Sky,' Sky and Telescope 109 (1): 84–87.
Buratti BJ, Hicks MD, Soderblom LA et al (2004) 'Deep Space 1 photometry of the nucleus of Comet 19P/Borrelly,' Icarus 167: 16–29.
Burnell J (2008) 'A wide-field imager's dream scope,' Astronomy 36 (1): 74–77.
Burnett DS (2006) 'NASA returns rocks from a comet,' Science 314: 1709–1710.
Burnham R (2000) Great Comets, Cambridge University Press, Cambridge.
Busko I, Lindler D, A'Hearn MF et al (2007) 'Searching for the Deep Impact crater on Comet 9P/Tempel 1 using image processing techniques,' Icarus 191: 210–222.

Cain L (1984) 'A 17½-inch Binocular Reflector,' Sky and Telescope 68: 460–463.
Campins H, Rieke MJ, Rieke GH (1989) 'An infrared color gradient in the inner coma of comet Halley,' Icarus 78: 54–62.
Chaple G (2003) 'SkyQuest: Easy exploring,' Astronomy 31 (5): 90–93.
Chernova GP, Kiselev NN, Jockers K (1993) 'Polarimetric characteristics of dust particles as observed in 13 comets: Comparisons with asteroids,' Icarus 103: 144–158.
Clairemidi J, Moreels G, Krasnopolsky VA (1990) 'Gaseous CN, C_2, and C_3 jets in the inner coma of Comet P/Halley observed from the Vega 2 spacecraft,' Icarus 86: 115–128.
Cochran AL, Jackson WM, Meech KJ et al (2007) 'Observations of Comet 9P/Tempel 1 with the Keck 1 HIRES instrument during Deep Impact,' Icarus 191: 360–370.
Combes M, Moroz VI, Crifo JF et al (1986) 'Infrared sounding of comet Halley from Vega 1,' Nature 321: 266–268.
Combes M, Moroz VI, Crovisier J et al (1988) 'The 2.5–12 μm spectrum of comet Halley from the IKS–VEGA experiment,' Icarus 76: 404–436.
Combi MR (1989) 'The outflow speed of the coma of Halley's comet,' Icarus 81: 41–50.
Combi MR, McCrosky RE (1991) 'High-resolution spectra of the 6300-Å region of Comet P/Halley,' Icarus 91: 270–279.
Cox AN, editor (2000) Allen's Astrophysical Quantities, Fourth edition, Hamilton Printing Co., Rensselaer.
Cremonese G, Fulle M (1989) 'Photometrical analysis of the neck-line structure of comet Halley,' Icarus 80: 267–279.
Crovisier J, Encrenaz T (2000) Comet Science: The Study of Remnants form the Birth of the Solar System (Translated by S Lyle), Cambridge University Press, Cambridge.
Curdt W, Keller HU (1990) 'Large dust particles along the Giotto trajectory,' Icarus 86: 305–313.
Davidsson BJR, Gutiérrez PJ (2004) 'Estimating the nucleus density of Comet 19P/Borrelly,' Icarus 168: 392–408.
Davidsson BJR, Gutiérrez PJ (2006) 'Non-gravitational force modeling of Comet 81P/Wild 2 I. A nucleus bulk density estimate,' Icarus 180: 224–242.
Delsemme HA (1987) 'Galactic tides affect the Oort cloud: an observational confirmation,' Astronomy and Astrophysics 187: 913–918.
di Cicco D (2002a) 'Sky window: A novel mount for binocular astronomy,' Sky and Telescope 103 (1): 55–57.
di Cicco D (2002b) 'The NexStar 11 GPS: Beauty and brains,' Sky and Telescope 103 (2): 49–54.
di Cicco D (2002c) 'Meade's LXD55 Schmidt-Newtonians,' Sky and Telescope 104 (6): 48–54.
di Cicco D (2003a) 'Orion's SkyView Pro equatorial mount,' Sky and Telescope 106 (2): 56–57.
di Cicco D (2003b) 'TEC's 5½-inch Apochromat,' Sky and Telescope 106 (6): 54–58.
di Cicco D (2004a) 'King of the chips: SBIG's STL-11000M,' Sky and Telescope 108 (1): 96–102.
di Cicco D (2004b) 'Pint-size powerhouse: Tele Vue's TV-60,' Sky and Telescope 108 (6): 102–106.
di Cicco D (2006a) 'Meade's RCX400: Raising the bar,' Sky and Telescope 111 (2): 78–83.
di Cicco D (2006b) 'Triple play: The ZenithStar 66 Refractors,' Sky and Telescope 111 (5): 76–80.
di Cicco D (2006c) 'Simply elegant: Meade's LightBridge Dobsonians,' Sky and Telescope 112 (4): 80–84.
di Cicco D (2006d) 'The Bigha StarSeeker,' Sky and Telescope 112 (6): 90–92.
di Cicco D (2007a) 'Apogee's Alta U9000 CCD camera,' Sky and Telescope 113 (6): 64–67.
di Cicco D (2007b) 'Tele Vue's Flagship Imaging System,' Sky and Telescope 114 (1): 66–70.

di Cicco D (2007c) 'Staying on track,' Sky and Telescope 114 (6): 37–40.
di Cicco D (2008a) 'Astro-Tech Voyager Mount,' Sky and Telescope 116 (1): 36.
di Cicco D (2008b) 'Astro-Tech AT80EDT Refractor,' Sky and Telescope 116 (3): 39.
di Cicco D (2009) 'Signature 20×110 Binoculars,' Sky and Telescope 117 (5): 37.
DiSanti MA, Fink U, Schultz AB (1990) 'Spatial distribution of H_2O^+ in Comet P/Halley,' Icarus 86: 152–171.
DiSanti MA, Villanueva GL, Bonev BP et al (2007) 'Temporal evolution of parent volatiles and dust in Comet 9P/Tempel 1 resulting from the Deep Impact experiment,' Icarus 191: 481–493.
Dobbins T, Sheehan W (2000) 'Beyond the Dawes Limit: Observing Saturn's Ring Divisions,' Sky and Telescope 100 (5): 117–121.
Dobbins TA (2005) 'AVA's Atmospheric Dispersion Corrector,' Sky and Telescope 109 (6): 88–91.
Dobbins TA, Parker DC, Capen CF (1988) Observing and Photographing the Solar System, Willmann-Bell Inc., Richmond.
Dolciani MP, Wooton W, Beckenbach EF et al (1968) Modern School Mathematics Algebra 2 and Trigonometry, Houghton Mifflin Co., Boston.
Dollfus A (1961) 'Polarization Studies of Planets,' in Planets and Satellites (Kuiper GP and Middlehurst BM, editors) The University of Chicago, Chicago, pp. 343–399.
Donn B, Rahe J, Brandt JC (1986) Atlas of Comet Halley 1910 II, NASA SP-488, NASA, Washington, DC.
Dyer A (2002a) 'Premium refractors: Having it all,' Sky and Telescope 103 (5): 44–51.
Dyer A (2002b) 'Premium refractors: Having it all,' Sky and Telescope 103 (6): 48–55.
Dyer A (2003a) 'Brains and Brawn: Meade's LX200GPS,' Sky and Telescope 105 (3): 50–57.
Dyer A (2003b) 'A Stellarvue Duo,' Sky and Telescope 106 (3): 50–55.
Dyer A (2004) 'Another Dream Apo Refractor,' Sky and Telescope 107 (6): 94–98.
Dyer A (2005a) 'Binocular viewing on a budget,' Sky and Telescope 109 (3): 88–93.
Dyer A (2005b) 'Vixen's Sphinx "Go To" Mount,' Sky and Telescope 110 (1): 84–89.
Dyer A (2005c) 'Celestron's Advanced Series "Go To" Mount,' Sky and Telescope 110 (2): 82–85.
Dyer A (2005d) 'Canon's Astrocamera: The EOS 20Da,' Sky and Telescope 110 (5): 84–88.
Dyer A (2007) 'Celestron's Grab-'n'-Go 6-inch,' Sky and Telescope 114 (6): 34–35.
Dyer A (2008a) 'Joining the Borg,' Sky and Telescope 115 (3): 40–43.
Dyer A (2008b) 'Guiding on a budget,' Sky and Telescope 116 (5): 43–46.
Dyer A (2009) 'Short and sweet,' Sky and Telescope 117 (3): 36–39.
Dyer A, Walker S (2007) 'Two New Apos from the Same Tree,' Sky and Telescope 113 (5): 74–78.
Eaton N, Scarrott SM, Warren-Smith RF (1988) 'Polarization images of the inner regions of comet Halley,' Icarus 76: 270–278.
Edberg S (2003a) 'An upgraded classic,' Astronomy 31 (8): 96–100.
Edberg SJ (2003b) 'Choosing an eyepiece,' Astronomy 31 (9): 110–115.
Edberg S (2004a) 'Star power,' Astronomy 32 (7): 88–91.
Edberg S (2004b) 'The Maksutov revolution,' Astronomy 32 (10): 82–85.
Edenhofer P, Bird MK, Brenkle JP et al (1986) 'First results from the Giotto Radio-Science experiment,' Nature 321: 355–357.
Eicher DJ (1986a) 'Halley fades in early April,' Astronomy 14 (7): 42–47.
Eicher DJ (1986b) 'Halley brightens one last time,' Astronomy 14 (8): 38–42.
Ellis TA, Neff JS (1991) 'Numerical simulation of the emission and motion of neutral and charged dust from P/Halley,' Icarus 91: 280–296.
Ernst CM, Schultz PH (2007) 'Evolution of the Deep Impact flash: Implications for the nucleus surface based on laboratory experiments,' Icarus 191: 123–133.

Farnham TL, Schleicher DG (2005) 'Physical and compositional studies of Comet 81P/Wild 2 at multiple apparitions,' Icarus 173: 533–558.
Farnham TL, Wellnitz DD, Hampton DL et al (2007) 'Dust coma morphology in the Deep Impact images of Comet 9P/Tempel 1,' Icarus 191: 146–160.
Feaga LM, A'Hearn MF, Sunshine JM et al (2007) 'Asymmetries in the distribution of H_2O and CO_2 in the inner coma of Comet 9P/Tempel 1 as observed by Deep Impact,' Icarus 191: 134–145.
Feldman PD, Lupu RE, McCandliss SR et al (2006) 'Carbon Monoxide in Comet 9P/Tempel 1 before and after the *Deep Impact* encounter,' The Astrophysical Journal 647: L61–L64.
Feldman PD, McCandliss SR, Route M et al (2007a) 'Hubble Space Telescope observations of Comet 9P/Tempel 1 during the Deep Impact encounter,' Icarus 191: 276–285.
Feldman PD, Stern SA, Steffl AJ et al (2007b) 'Ultraviolet spectroscopy of Comet 9P/Tempel 1 with Alice/Rosetta during the Deep Impact encounter,' Icarus 191: 258–262.
Fera B (2008) 'Speedy Austrian Astrograph,' Sky and Telescope 115 (6): 37–38.
Fernández YR, Lisse CM, Kelley MS et al (2007a) 'Near-infrared light curve of Comet 9P/Tempel 1 during Deep Impact,' Icarus 191: 424–431.
Fernández YR, Meech KJ, Lisse CM et al (2007b) 'The nucleus of *Deep Impact* target Comet 9P/Tempel 1,' Icarus 191: 11–21.
Ferrín I (2005) 'Secular light curve of Comet 28P/Neujmin 1 and of spacecraft target Comets 1P/Halley, 9P/Tempel 1, 19P/Borrelly, 21P/Giacobinni-Zinner, 26P/Grigg-Skjellerup, 67P/Churyumov-Gerasimenko and 81P/Wild 2,' Icarus 178: 493–516.
Ferrín I (2007) 'Secular light curve of Comet 9P/Tempel 1,' Icarus 191: 567–572.
Festou MC, Feldman PD, A'Hearn MF et al (1986) 'IUE observations of comet Halley during the Vega and Giotto encounters,' Nature 321: 361–363.
Fienberg RT (2009) 'The CCDelightful QSI 540wsg,' Sky and Telescope 117 (6): 36–38.
Fink U (2009) 'A taxonomic survey of comet composition 1985–2004 using CCD spectroscopy,' Icarus 201: 311–334.
Fix JD (2008) Astronomy: Journey to the Cosmic Frontier, 5th edn, McGraw Hill Higher Education, Boston.
Flynn GJ, Bleuet P, Borg, J et al (2006) 'Elemental compositions of Comet 81P/Wild 2 samples collected by Stardust,' Science 314: 1731–1735.
Furusho R, Ikeda Y, Kinoshita D et al (2007) 'Imaging polarimetry of Comet 9P/Tempel 1 before and after the Deep Impact,' Icarus 191: 454–458.
Gehrz RD, Johnson CH, Magnuson SD et al (1995) 'Infrared observations of an outburst of small dust grains from the Nucleus of Comet P/Halley 1986 III at perihelion,' Icarus 113: 129–133.
Goidet-Devel B, Clairemidi J, Rousselot P et al (1997) 'Dust spatial distribution and radial profile in Halley's inner coma,' Icarus 126: 78–106.
Grard R, Pedersen A, Trotignon JG et al (1986) 'Observations of waves and plasma in the environment of comet Halley,' Nature 321: 290–291.
Green DWE (1995) 'Brightness-variation patterns of recent long-period comets vs. C/1995 O1,' International Comet Quarterly, 17: 168–178.
Green DWE, Morris CS (1987) 'The visual brightness behavior of P/Halley during 1981–1987,' Astronomy and Astrophysics 187: 560–568.
Green DWE, Nakano S, editors (2007/2008) Introduction International Comet Quarterly 29 (4a): H2–H14.
Gringauz KI, Gombosi TI, Remizov AP et al (1986) 'First *in situ* plasma and neutral gas measurements at comet Halley,' Nature 321: 282–285.
Groussin O, A'Hearn MF, Li JY et al (2007) 'Surface temperature of the nucleus of Comet 9P/Tempel 1,' Icarus 191: 63–72.

Gove PB, Editor-in-Chief (1971) Webster's Third New International Dictionary, G and C, Merriam Co., Springfield.
Gutiérrez PJ, Davidsson BJR (2007) 'Non-gravitational force modeling of Comet 81P/Wild 2 II. Rotational evolution,' Icarus 191: 651–664.
Haas S (2006) Double Stars for Small Telescopes, Sky Publishing Corp. Cambridge.
Hadamcik E, Levasseur-Regourd AC, Leroi V et al (2007) 'Imaging polarimetry of the dust coma of Comet Tempel 1 before and after Deep Impact at Haute-Provence Observatory,' Icarus 191: 459–468.
Hale A (1996) Everybody's comet: A Layman's guide to comet Hale-Bopp, High-Lonesome Books, Silver City.
Hall DS, Genet RM (1988) Photoelectric Photometry of Variable Stars, Willmann-Bell, Inc., Richmond, VA.
Hallas T (2007) 'Big Praise for Big Binos,' Sky and Telescope 114 (1): 12.
Hampel CA, Hawley GC, editors (1973) The Encyclopedia of Chemistry, 3rd edn, Van Nostrand Reinhold Company, New York, pp. 1030–1032.
Hanner MS, Hayward TL (2003) 'Infrared observations of Comet 81P/Wild 2 in 1997,' Icarus 161: 164–173.
Hanson M, Greiner RA (2004) 'Canon 10D digital camera,' Astronomy 32 (9): 84–87.
Harker DE, Woodward CE, Wooden DH et al (2007) 'Gemini-N mid-IR observations of the dust properties of the ejecta excavated from Comet 9P/Tempel 1 during Deep Impact,' Icarus 191: 432–453.
Harrington DM, Meech K, Kolokolova L et al (2007) 'Spectropolarimetry of the Deep Impact target Comet 9P/Tempel 1 with HiVIS,' Icarus 191: 381–388.
Harrington P (2002a) 'Russian-made telescopes,' Astronomy 30 (5): 62–65.
Harrington P (2002b) 'A happy medium,' Astronomy 30 (6): 66–69.
Harrington P (2002c) 'Refractor road test,' Astronomy 30 (10): 68–72.
Harrington P (2002d) 'Scoping out the new stargazers,' Astronomy 30 (11): 72–75.
Harrington P (2003a) 'Going global,' Astronomy 31 (1): 84–87.
Harrington P (2003b) 'Two eyes on the sky,' Astronomy 31 (4): 92–97.
Harrington P (2003c) 'High-power twin optics,' Astronomy 31 (5): 94–98, 102.
Harrington P (2003d) 'Off-axis vision,' Astronomy 31 (10): 82–85.
Harrington P (2004a) 'Orion's StarBlast,' Astronomy 32 (1): 84–87.
Harrington P (2004b) 'JMI's RB-66 Binoscope,' Astronomy 32 (2): 90–93.
Harrington P (2004c) 'TAL's 150K and 200K,' Astronomy 32 (3): 90–93.
Harrington P (2004d) 'Orion's Atlas 8,' Astronomy 32 (5): 86–89.
Harrington P (2004e) 'Celestron's advanced series telescopes,' Astronomy 32 (8): 88–91.
Harrington P (2005a) 'Backpack this scope,' Astronomy 33 (2): 92–94.
Harrington P (2005b) 'Secret weapons,' Astronomy 33 (8): 82–85.
Harrington P (2006a) 'Have lens, will travel,' Astronomy 34 (3): 86–88.
Harrington P (2006b) 'Head of the glass,' Astronomy 34 (10): 80–83.
Harrington P (2007a) 'Celestron's new Schmidt-Cassegrain,' Astronomy 35 (3): 76–77.
Harrington P (2007b) 'Orion's new 4-inch powerhouse,' Astronomy 35 (4): 70–73.
Harrington P (2007c) 'Astrolight reflectors offer quality optics,' Astronomy 35 (11): 70–73.
Harrington P (2007d) 'The Skypod mount performs superbly,' Astronomy 35 (12): 98, 100–101.
Harrington P (2008) 'Vixen's giant binoculars among largest sold,' Astronomy 36 (11): 72–73.
Healy D (2002) 'Have scope will travel,' Astronomy 30 (9): 66–68.
Healy D (2003) 'Testing a CCD trio,' Astronomy 31 (3): 84–87.
Healy D, Gary B (2007) 'MaxCam gets images started,' Astronomy 35 (7): 70–72.
Hergenrother CW, Mueller BEA, Campins H et al (2007) '*R*- and *J*-band photometry of Comets 2P/Encke and 9P/Tempel 1,' Icarus 191: 45–50.

Hicks MD, Bambery RJ, Lawrence KJ et al (2007) 'Near-nucleus photometry of comets using archived NEAT data,' Icarus 188: 457–467.
Hirao K, Itoh T (1986) 'The Planet-A Halley encounters,' Nature 321: 294–297.
Hirshfeld A, Sinnott RW (1985) Sky Catalogue 2000.0, vol. 2, Sky Publishing Corp., Cambridge.
Hirshfeld A, Sinnott RW, Ochsenbein F (1991) Sky Catalogue 2000.0, vol. 1, 2nd edn, Sky Publishing Corp., Cambridge.
Hoban S, A'Hearn MF, Birch PV et al (1989) 'Spatial structure in the color of the dust coma of Comet P/Halley,' Icarus 79: 145–158.
Hodapp KW, Aldering G, Meech KJ et al (2007) 'Visible and near-infrared spectrophotometry of the Deep Impact ejecta of Comet 9P/Tempel 1,' Icarus 191: 389–402.
Hodgman CD, Editor-in-Chief (1955) C.R.C. Standard Mathematical Tables, 10th edn, Chemical Rubber Publishing Company, Cleveland.
Horne J (2003a) 'Four low-cost astronomical video cameras,' Sky and Telescope 105 (2): 57–62.
Horne J (2003b) 'Losmandy's Gemini System,' Sky and Telescope 106 (4): 50–55.
Horne J (2004a) 'Celestron's CGE 1400 Telescope,' Sky and Telescope 107 (3): 54–61.
Horne J (2004b) 'Stella Cam II: Taking video into the deep sky,' Sky and Telescope 108 (4): 86–89.
Horne J (2005) 'Deep-sky imaging for everyone,' Sky and Telescope 110 (4): 76–81.
Horne J (2006) 'Good gets better: Meade's DSI II CCD cameras,' Sky and Telescope 112 (3): 76–80.
Horne J (2007) 'Next-Generation Video: Adirondack's StellaCam^3,' Sky and Telescope 114 (3): 64–67.
Horne J (2008) 'Raising the bar for entry-level imaging,' Sky and Telescope 115 (4): 32–36.
Horne J (2009) 'Powerful performer: Orion's StarShoot Pro,' Sky and Telescope 117 (2): 34–37.
Hörz F, Bastien R, Borg J et al (2006) 'Impact features on Stardust: Implications for Comet 81P/Wild 2 dust,' Science 314: 1716–1719.
Howell ES, Lovell AJ, Butler B et al (2007) 'Radio OH observations of 9P/Tempel 1 before and after Deep Impact,' Icarus 191: 469–480.
Howell SB (2006) Handbook of CCD Astronomy, 2nd edn, Cambridge University Press, Cambridge.
Howington-Kraus E, Kirk RL, Duxbury TC et al (2005) 'Topography of the 81P/Wild 2 nucleus from Stardust stereoimages,' Asia-Oceania Geosciences Society 2nd Annual Meeting, Singapore, Poster presentation 58-PS-A0956.
Hughes DW (2002) 'The magnitude distribution and evolution of short-period comets,' Monthly Notices of the Royal Astronomical Society 336: 363–372.
Hughes DW, Green DWE (2007) 'Halley's first name: Edmond or Edmund,' International Comet Quarterly 29 (1): 7–14.
International Astronomical Union (IAU) Circulars; several hundred between number 2700 and 9000.
International Comet Quarterly, various issues listing visual data of Comets 1P/Halley, 9P/Tempel 1, 19P/Borrelly and 81P/Wild 2.
Jorda L, Lamy P, Faury G et al (2007) 'Properties of the dust cloud caused by the Deep Impact experiment,' Icarus 191: 412–423.
Julian WH, Samarasinha NH, Belton MJS (2000) 'Thermal structure of cometary active regions: Comet 1P/Halley,' Icarus 144: 160–171.
Kalemjian E (2003) 'A Siberian Achromatic Refractor,' Sky and Telescope 105 (4): 56–60.
Kaufmann WJ III (1985) Universe, W H Freeman and Co., New York.
Kawakita H, Jehin E, Manfroid J et al (2007) 'Nuclear spin temperature of ammonia in Comet 9P/Tempel 1 before and after the Deep Impact event,' Icarus 191: 513–516.

Keller HU, Arpigny C, Barbieri C et al (1986) 'First Halley multicolour camera imaging results from Giotto,' Nature 321: 320–326.

Keller HU, Delamere WA, Huebner WF et al (1987) 'Comet P/Halley's nucleus and its activity,' Astronomy and Astrophysics 187: 807–823.

Keller HU, Küppers M, Fornasier S et al (2007) 'Observations of Comet 9P/Tempel 1 around the Deep Impact event by the OSIRIS cameras onboard Rosetta,' Icarus 191: 241–257.

Keller LP, Bajt S, Baratta GA et al (2006) 'Infrared spectroscopy of Comet 81P/Wild 2 samples returned by Stardust,' Science 314: 1728–1731.

Kelly P (2007) Observer's Handbook 2008, The Royal Astronomical Society of Canada, Toronto.

Keppler E, Afonin VV, Curtis CC et al (1986) 'Neutral gas measurements of comet Halley from Vega 1,' Nature 321: 273–274.

Kilburn KJ (2000) 'Hunting for SOHO Comets using the internet,' Sky and Telescope 100 (4): 89–92.

Kirk RL, Howington-Kraus E, Soderblom LA (2004) 'Comparison of USGS and DLR topographic models of Comet Borrelly and photometric applications,' Icarus 167: 54–69.

Kissel J, Brownlee DE, Büchler K et al (1986a) 'Composition of comet Halley dust particles from Giotto observations,' Nature 321: 336–337.

Kissel J, Sagdeev RZ, Bertaux JL et al (1986b) 'Composition of comet Halley dust particles from Vega observations,' Nature 321: 280–282.

Klavetter JJ, A'Hearn MF (1994) 'An extended source for CN jets in Comet P/Halley,' Icarus 107: 322–334.

Klimov S, Savin S, Aleksevich Y et al (1986) 'Extremely-low-frequency plasma waves in the environment of comet Halley,' Nature 321: 292–293.

Knight MM, Walsh KJ, A'Hearn MF (2007) 'Ground-based visible and near-IR observations of Comet 9P/Tempel 1 during the Deep Impact encounter,' Icarus 191: 403–411.

Korth A, Richter AK, Loidl A et al (1986) 'Mass spectra of heavy ions near comet Halley,' Nature 321: 335–336.

Krankowsky D, Lämmerzahl P, Herrwerth I et al (1986) '*In situ* gas and ion measurements at comet Halley,' Nature 321: 326–329.

Krasnopolsky VA, Gogoshev M, Moreels G et al (1986) 'Spectroscopic study of comet Halley by the Vega 2 three-channel spectrometer,' Nature 321: 269–271.

Kronk GW (1984) Comets: A Descriptive Catalog, Enslow Publishers, Inc., Hillside.

Kronk GW (1999) Cometography: A Catalog of Comets, Volume 1: Ancient-1799, Cambridge University Press, Cambridge.

Kronk GW (2003) Cometography: A Catalog of Comets, Volume 2: 1800–1899, Cambridge University Press, Cambridge.

Kronk GW (2007) Cometography: A Catalog of Comets, Volume 3: 1900–1932, Cambridge University Press, Cambridge.

Kronk GW (2009) Cometography: A Catalog of Comets, Volume 4: 1933–1959, Cambridge University Press, Cambridge.

Kuberek R (2006) 'A 12-inch powerhouse,' Astronomy 34 (2): 84–87.

Lamy P, Biesecker DA, Groussin O (2003) 'SOHO/LASCO observation of an outburst of Comet 2P/Encke at its 2000 perihelion passage,' Icarus 163: 142–149.

Lamy P, Toth I (2009) 'The colors of cometary nuclei – Comparison with other primitive bodies of the Solar System and implications for their origin,' Icarus 201: 674–713.

Lamy PL, Toth I, A'Hearn MF et al (2001) 'Hubble Space Telescope observations of the nucleus of Comet 9P/Tempel 1,' Icarus 154: 337–344.

Lamy PL, Toth I, A'Hearn MF et al (2007a) 'Hubble Space Telescope observations of the nucleus of Comet 9P/Tempel 1,' Icarus 191: 4–10.

Lamy PL, Toth I, A'Hearn MF et al (2007b) 'Rotational state of the nucleus of Comet 9P/Tempel 1: Results from Hubble Space Telescope observations in 2004,' Icarus 191: 310–321.
Lamy PL, Toth I, Weaver HA (1998) 'Hubble Space Telescope observations of the nucleus and inner coma of Comet 19P/1904 Y2 (Borrelly),' Astronomy and Astrophysics 337: 945–954.
Larson HP, Hu HY, Mumma MJ et al (1990) 'Outbursts of H_2O in Comet P/Halley,' Icarus 86: 129–151.
Laufer D, Pat-El I, Bar-Nun A (2005) 'Experimental simulation of the formation of non-circular active depressions on Comet Wild-2 and of ice grain ejection from cometary surfaces,' Icarus 178: 248–252.
Leet LD, Judson S, Schmitz EA (1965) Physical Geology, 3rd edn, Prentice-Hall Inc, Englewood Cliffs.
Li JY, A'Hearn MF, Belton MJS et al (2007) 'Deep Impact photometry of Comet 9P/Tempel 1,' Icarus 191: 161–175.
Lide DR, Editor-in-Chief (2008) Handbook of Chemistry and Physics, 89th edn, CRC Press, Boca Raton.
Lisse CM, Dennerl K, Christian DJ et al (2007a) 'Chandra observations of Comet 9P/Tempel 1 during the Deep Impact campaign,' Icarus 191: 295–309.
Lisse CM, Kraemer KE, Nuth JA III et al (2007b) 'Comparison of the composition of the Tempel 1 ejecta to the dust in Comet C/Hale-Bopp 1995 O1 and YSO HD 100546,' Icarus 191: 223–240.
Livitski R (1993) 'How I Built a 20-inch Binocular,' Sky and Telescope 85 (2): 89–91.
Loewenstein KL (1966) 'Glass Systems,' in Composite Materials (Holliday L, editor) Elsevier, Amsterdam, pp. 129–220.
Machholz D (1995) 'The 1989 apparition of periodic comet Brorsen-Metcalf (1989o = 1989 X),' Journal of the Association of Lunar and Planetary Observers 38: 75–78.
Machholz D (1996) 'The apparition of comet Okazaki-Levy-Rudenko (1989r=1989 XIX),' Journal of the Association of Lunar and Planetary Observers 39: 71–74.
Machholz D (1997) 'The apparition of comet Aarseth-Brewington (1989a1 = 1989 XXII),' Journal of the Association of Lunar and Planetary Observers 39: 131–134.
Machholz DE (1989) 'The apparition of comet Bradfield 1987s,' Journal of the Association of Lunar and Planetary Observers 33: 97–102.
Machholz DE (1991) 'The apparition of comet Wilson 1987 VII,' Journal of the Association of Lunar and Planetary Observers 35: 49–52.
MacRobert A (1985) 'Backyard Astronomy-11: Comet-Watching Tips,' Sky and Telescope 70: 20–21.
MacRobert A (2005) 'Comet Machholz on track for fine show,' Sky and Telescope 109 (2): 77.
MacRobert AM (1992) 'A pupil primer,' Sky and Telescope 83: 502–504.
MacRobert AM (2004) 'The pull of a "push to" telescope,' Sky and Telescope 108 (5): 86–91.
MacRobert AM (2005) 'So you want giant binoculars…' Sky and Telescope 110 (3): 96–98.
Magee-Sauer K, Scherb F, Roesler FL et al (1989) 'Fabry-Perot observations of NH_2 emission from comet Halley,' Icarus 82: 50–60.
Magee-Sauer K, Scherb F, Roesler FL et al (1990) 'Comet Halley $O(^1D)$ and H_2O production rates,' Icarus 84: 154–165.
Mäkinen JTT, Combi MR, Bertaux JL et al (2007) 'SWAN observations of 9P/Tempel 1 around the Deep Impact event,' Icarus 187: 109–112.
Malivoir C, Encrenaz T, Vanderriest C et al (1990) 'Mapping of secondary products in comet Halley from bidimensional spectroscopy,' Icarus 87: 412–420.

Manfroid J, Hutsemékers D, Jehin E et al (2007) 'The impact and rotational light curves of Comet 9P/Tempel 1,' Icarus 191: 348–359.
Marcotte M (2004) 'Konus's new Mak-Cass,' Astronomy 32 (4): 84–86.
Marcotte M (2005a) 'Meade's new 14-inch SCT: an instant classic,' Astronomy 33 (3): 78–81.
Marcotte M (2005b) 'Easy imaging for everyone,' Astronomy 33 (6): 80–83.
Marcotte M (2006) 'Eyes wide open,' Astronomy 34 (1): 94–95.
Marcotte MM (2007) 'Deep-sky-object hunter,' Astronomy 35 (5): 72–74.
Marcus J (1986) 'Halley in the daytime,' Sky and Telescope 71: 125.
Mason KO, Chester M, Cucchiara A et al (2007) 'Swift ultraviolet photometry of the Deep Impact encounter with Comet 9P/Tempel 1,' Icarus 191: 286–294.
Mayer EH (1984) 'Finder follies,' Sky and Telescope 67: 210.
Mazets EP, Aptekar RL, Golenetskii SV et al (1986) 'Comet Halley dust environment from SP-2 detector measurements,' Nature 321: 276–278.
McDonnell JAM, Alexander WM, Burton WM et al (1986) 'Dust density and mass distribution near comet Halley from Giotto observations,' Nature 321: 338–341.
McFadden LA, Weissman PR, Johnson TV (2007) Encyclopedia of the Solar System, 2nd edn, Elsevier, Amsterdam.
McKeegan KD, Aléon J, Bradley J et al (2006) 'Isotopic compositions of cometary matter returned by Stardust,' Science 314: 1724–1728.
McKinley DWR (1961) Meteor Science and Engineering, McGraw-Hill Book Co., New York.
Medkeff J (2005) 'A telescope mount for the 21st century,' Astronomy 33 (11): 94–97.
Meech KJ, Ageorges N, A'Hearn MF et al (2005) 'Deep Impact: Observations from a worldwide Earth-based campaign,' Science 310: 265–269.
Meisel DD, Morris CS (1982) 'Comet Head Photometry: Past, Present, and Future,' in Comets, (Wilkening LL and Matthews MS, editors) The University of Arizona Press, Tucson, pp. 413–432.
Merényi E, Földy L, Szeg K et al (1990) 'The landscape of comet Halley,' Icarus 86: 9–20.
Milani GA, Szabó GM, Sostero G et al (2007) 'Photometry of Comet 9P/Tempel 1 during the 2004/2005 approach and the Deep Impact module impact,' Icarus 191: 517–525.
Miles R (2007) 'Daytime Photometry of Comet McNaught,' Unpublished Report.
Moomaw B (2004) 'Stardust collects bits of Comet Wild 2,' Astronomy 32 (4): 24.
Moreels G, Gogoshev M, Krasnopolsky VA et al (1986) 'Near-ultraviolet and visible spectrophotometry of comet Halley from Vega 2,' Nature 321: 271–273.
Morris CS (1973) 'On aperture corrections for comet magnitude estimates,' Publications of the Astronomical Society of the Pacific 85: 470–473.
Mukai T, Miyake W, Terasawa T et al (1986) 'Plasma observation by Suisei of solar-wind interaction with comet Halley,' Nature 321: 299–303.
Münch RE, Sagdeev RZ, Jordan JF (1986) 'Pathfinder: accuracy improvement of comet Halley trajectory for Giotto navigation,' Nature 321: 318–320.
Nagler A (1991) 'Choosing your telescope's magnification,' Sky and Telescope 81: 553–559.
Nakano S, Green DWE, editors (2006) '2007 Comet Handbook,' International Comet Quarterly, Special Issue 28 (4a): H2–H12.
Nelson AE (2005) 'The big easy,' Astronomy 33 (5): 78–81.
Nelson RM, Rayman MD, Weaver HA (2004a) 'The Deep Space 1 encounter with Comet 19P/Borrelly,' Icarus 167: 1–3.
Nelson RM, Soderblom LA, Hapke BW (2004b) 'Are the circular dark features on Comet Borrelly's surface albedo variations or pits?' Icarus 167: 37–44.
Neubauer FM, Glassmeier KH, Pohl M et al (1986) 'First results from the Giotto magnetometer experiment at comet Halley,' Nature 321: 352–355.
Newton J (2006) 'Designed to shoot the sky,' Astronomy 34 (7): 90–93.

Oberc P (1999) 'Small-scale dust structures in Halley's coma: Evidence from the Vega-2 electric field records,' Icarus 140: 156–172.
Oberc P, Parzydlo W, Vaisberg OL (1990) 'Correlations between the Vega 2 Plasma Wave (APV-N) and Dust (SP-1) observations at comet Halley,' Icarus 86: 314–326.
Oberst J, Giese B, Howington-Kraus E et al (2004) 'The nucleus of Comet Borrelly: a study of morphology and surface brightness,' Icarus 167: 70–79.
Parker DC (1990) 'Position measurements of the minor planet 747 Winchester using a filar micrometer,' Journal of the Association of Lunar and Planetary Observers 34: 137–139.
Peale SJ (1989) 'On the density of Halley's comet,' Icarus 82: 36–49.
Peale SJ, Lissauer JJ (1989) 'Rotation of Halley's comet,' Icarus 79: 396–430.
Privett G (2002) 'The Sky-Watcher EQ6 Mount,' Sky and Telescope 104 (4): 45–48.
Ratcliffe M Ling A (2004) 'The deep sky,' Astronomy 32 (6): 59.
Ratcliffe M Ling A (2004) 'Comets and asteroids,' Astronomy 32 (6): 63.
Reeves R (2006a) 'Big results from a small package,' Astronomy 34 (9): 78–81.
Reeves R (2006b) 'Introduction to Webcam Astrophotography,' Willmann-Bell, Inc., Richmond.
Reinhard R (1986) 'The Giotto encounter with comet Halley,' Nature 321: 313–318.
Reitsema HJ, Delamere WA, Williams AR et al (1989) 'Dust distribution in the inner coma of Comet Halley: Comparison with models,' Icarus 81: 31–40.
Reynolds M (2006) 'Vixen's go-anywhere scope,' Astronomy 34 (6): 90–93.
Reynolds MD (2007a) 'Rebirth of a classic: the Porter Garden Telescope,' Astronomy 35 (6): 74–77.
Reynolds MD (2007b) 'Astronomy tests Celestron's CPC 1100 GPS,' Astronomy 35 (8): 72–74.
Reynolds MD (2008) 'Easy imaging with the DSI III,' Astronomy 36 (9): 64–65.
Richardson JE, Melosh HJ, Lisse CM et al (2007) 'A ballistics analysis of the Deep Impact ejecta plume: Determining Comet Tempel 1's gravity, mass and density,' Icarus 191: 176–209.
Richardson RS (1967) Getting Acquainted with Comets, McGraw-Hill Book Co., New York.
Riedler W, Schwingenschuh K, Yeroshenko YG et al (1986) 'Magnetic field observations in comet Halley's coma,' Nature 321: 288–289.
Rogers JH (1995) The Giant Planet Jupiter, Cambridge University Press, Cambridge.
Rogers JH (1996) 'The comet collision with Jupiter: II. The visible scars,' Journal of the British Astronomical Association 106: 125–150.
Roth J (2003) 'Breaking new ground in the Beginner's market,' Sky and Telescope 105 (6): 46–50.
Rumsey D (2005) Statistics Workbook for Dummies, Wiley Publishing, Inc., Hoboken.
Rupp W, Friedmann A, Farrell P (1989) Construction Materials for Interior Design, Whitney Library of Design, New York.
Sagdeev RZ, Blamont J, Galeev AA et al (1986a) 'Vega spacecraft encounters with comet Halley,' Nature 321: 259–262.
Sagdeev RZ, Szabó F, Avanesov GA et al (1986b) 'Television observations of comet Halley from Vega spacecraft,' Nature 321: 262–266.
Saladin KS (2007) 'Anatomy and Physiology: The Unity of Form and Function,' 4th edn, McGraw-Hill Higher Education, New York.
Samarasinha NH, Belton MJS (1994) 'The nature of the source of CO in Comet P/Halley,' Icarus 108: 103–111.
Sandford SA, Aléon J, Alexander CMO'D et al (2006) 'Organics captured from Comet 81P/Wild 2 by the Stardust spacecraft,' Science 314: 1720–1724.
Scherb F, Magee-Sauer K, Roesler FL et al (1990) 'Fabry-Perot observations of comet Halley H_2O^+,' Icarus 86: 172–188.

Schleicher DG (2007) 'Deep Impact's target Comet 9P/Tempel 1 at multiple apparitions: Seasonal and secular variations in gas and dust production,' Icarus 191: 322–338.
Schleicher DG, Millis RL, Birch PV (1998) 'Narrowband photometry of Comet P/Halley: Variation with heliocentric distance, season and solar phase angle,' Icarus 132: 397–417.
Schleicher DG, Woodney LM, Millis RL (2003) 'Comet 19P/Borrelly at multiple apparitions: seasonal variations in gas production and dust morphology,' Icarus 162: 415–442.
Schmude R Jr, Micciche S, Donegan A (1999) 'Observations of comet Hale-Bopp,' Journal of the Association of Lunar and Planetary Observers 41: 113–118.
Schmude RW Jr (2001) 'Full-disc wideband photoelectric photometry of the Moon,' Journal of the Royal Astronomical Society of Canada 95: 17–23.
Schmude RW Jr (2008) Uranus, Neptune and Pluto and how to observe them, Springer Science + Business Media, New York.
Schmude RW Jr and Dutton J (2001) 'Photometry and other characteristics of Venus,' Journal of the Association of Lunar and Planetary Observers 43 (4): 17–26.
Schultz D, Scherb F, Roesler FL (1993) 'H_2O^+ production rates of Comets Austin 1990 V and P/Halley 1986 III,' Icarus 104: 185–196.
Schultz PH, Eberhardy CA, Ernst CM et al (2007) 'The Deep Impact oblique impact cratering experiment,' Icarus 191: 84–122.
Schulz R, A'Hearn MF (1995) 'Shells in the C_2 coma of Comet P/Halley,' Icarus 115: 191–198.
Schulz R, A'Hearn MF, Samarasinha NH (1993) 'On the formation and evolution of gaseous structures in Comet P/Halley,' Icarus 103: 319–328.
Schwarz G, Craubner H, Delamere A (1987) 'Detailed analysis of a surface feature on Comet P/Halley,' Astronomy and Astrophysics 187: 847–851.
Sekanina Z (2008) 'On a forgotten 1836 explosion from Halley's comet, reminiscent of 17P/Holmes' outbursts,' International Comet Quarterly 30 (2): 63–74.
Sekanina Z, Brownlee DE, Economou TE et al (2004) 'Modeling the nucleus and jets of Comet 81P/Wild 2 based on the Stardust encounter data,' Science 304: 1769–1774.
Sekanina Z, Larson SM (1986) 'Dust jets in comet Halley observed by Giotto and from the ground,' Nature 321: 357–361.
Sen AK, Joshi UC, Deshpande MR et al (1990) 'Imaging polarimetry of Comet P/Halley,' Icarus 86: 248–256.
Seronik G (1998) 'Comet sleuthing on the internet,' Sky and Telescope 95 (6): 63–65.
Seronik G (2000) 'Image-stabilized binoculars aplenty,' Sky and Telescope 100 (1): 59–64.
Seronik G (2002a) 'Three New Maksutovs,' Sky and Telescope 103 (3): 44–48.
Seronik G (2002b) 'The Questar 50th anniversary edition telescope,' Sky and Telescope 104 (5): 49–54.
Seronik G (2004a) 'Hardin Optical's 10-inch Deep Space Hunter,' Sky and Telescope 107 (5): 96–100.
Seronik G (2004b) '10-inch Dobsonian reflectors,' Sky and Telescope 107 (5): 104–105.
Seronik G (2004c) 'Portable 4-inch Apochromatic Refractors,' Sky and Telescope 107 (6): 104–105.
Seronik G (2004d) 'Compact wide-field reflectors,' Sky and Telescope 108 (4): 96–97.
Seronik G (2004e) 'Compact wide-field refractors,' Sky and Telescope 108 (6): 112–113.
Seronik G (2005a) 'Clear viewing,' Sky and Telescope 109 (4): 88–93.
Seronik G (2005b) '90-mm Astronomical Maksutovs,' Sky and Telescope 109 (4): 98–99.
Seronik G (2005c) 'Low-cost starter scopes,' Sky and Telescope 110 (6): 86–90.
Seronik G (2007a) 'Build a tracking platform for your camera,' Sky and Telescope 113 (6): 80–83.
Seronik G (2007b) 'A modernized classic,' Sky and Telescope 114 (2): 64–67.

Seronik G (2008a) 'iOptron's capable cube,' Sky and Telescope 115 (2): 34–36.
Seronik G (2008b) 'FAR-sight binocular mount,' Sky and Telescope 115 (5): 34.
Seronik G (2008c) 'A bigger blast for the buck,' Sky and Telescope 116 (3): 36–39.
Seronik G (2008d) 'Following the Stars,' Sky and Telescope 116 (4): 38–40.
Seronik G (2009a) 'Obsession's 12 ½-inch Truss-tube Dob,' Sky and Telescope 117 (2): 38–39.
Seronik G (2009b) 'A 12-inch Star Cruiser,' Sky and Telescope 117 (5): 34–36.
Seronik G (2009c) 'A smart 12-inch Dob,' Sky and Telescope 118 (1): 34–36, 38.
Shanklin J (2002) Observing Guide to Comets, 2nd edn, The British Astronomical Association, London.
Shanklin J (2004) 'Visual observation of comets,' Journal of the British Astronomical Association 114: 158–160.
Shanklin J 'The Comet's Tail,' various issues of this newsletter.
Shibley J (2002) 'Get the Scoop on Italian Scopes,' Astronomy 30 (7): 66–69.
Shibley J (2003a) 'Test driving Meade's LX90,' Astronomy 31 (2): 82–85.
Shibley J (2003b) 'Focus on finders,' Astronomy 31 (3): 94–99.
Shibley J (2008) 'Obsession's new 18-inch scope,' Astronomy 36 (4): 68–69.
Shubinski R (2004) 'The Tele Vue-60,' Astronomy 32 (11): 90–93.
Shubinski R (2007) 'Sky-testing William Optics' new refractors,' Astronomy 35 (10): 70–72.
Simpson JA, Sagdeev RZ, Tuzzolino AJ et al (1986) 'Dust counter and mass analyser (DUCMA) measurements of comet Halley's coma from Vega spacecraft,' Nature 321: 278–280.
(1972) 'Comets lost and found,' Sky and Telescope 43: 155.
Schaoy F (2000) 'The near sky: Problems with Airglow,' Sky and Telescope 99 (6): 96.
Smith BA (2004) 'Stardust catches a comet,' Sky and Telescope 107 (4): 26.
Smyth WH, Marconi ML, Combi MR (1995) 'Analysis of hydrogen Lyman-α observations of the coma of Comet P/Halley near perihelion,' Icarus 113: 119–128.
Soderblom LA, Boice DC, Britt DT et al (2004a) 'Imaging Borrelly,' Icarus 167: 4–15.
Soderblom LA, Britt DT, Brown RH et al (2004b) 'Short-wavelength infrared (1.3–2.6 μm) observations of the nucleus of Comet 19P/Borrelly,' Icarus 167: 100–112.
Somogyi AJ, Gringauz KI, Szegő K et al (1986) 'First observations of energetic particles near comet Halley,' Nature 321: 285–288.
Stanton RH (1999) 'Visual magnitudes and the "average observer": The SS-Cygni field experiment,' Journal of the AAVSO 27 (2): 97–112.
Stein J, Editor-in-Chief (1980) The Random House Dictionary, Ballantine Books, New York.
Steinicke W, Jakiel R (2007) Galaxies and How to Observe Them, Springer, London.
Sterken C, Manfroid J, Arpigny C (1987) 'Photometry of P/Halley (1982i),' Astronomy and Astrophysics 187: 523–525.
Strazzulla G, Leto G, Gomis O et al (2003) 'Implantation of carbon and nitrogen ions in water ice,' Icarus 164: 163–169.
Sugita S, Ootsubo T, Kadono T et al (2005) 'Subaru telescope observations of Deep Impact,' Science 310: 274–278.
Sunshine JM, Groussin O, Schultz PH et al (2007) 'The distribution of water ice in the interior of Comet Tempel 1,' Icarus 191: 73–83.
Swenson GW (1978) 'An amateur radio telescope-1,' Sky and Telescope 55 (5): 385–390.
Tancredi G, Fernández JA, Rickman H et al (2006) 'Nuclear magnitudes and the size distribution of Jupiter family comets,' Icarus 182: 527–549.
Tate P (2009) Seeley's Principles of Anatomy and Physiology, McGraw-Hill Higher Education, Boston.
Tele Vue Eyepieces Brochure, Chester, NY.

Terrance G. (2002) 'An eye for luxary,' Astronomy 30 (12): 80–83.
Terrance G. (2003) 'The Paramount GT-1100 ME,' Astronomy 31 (4): 88–91.
Gerdon E. Taylor The Handbook of the British Astronomical Association for the Years 2004–2008, Burlington House, Piccadilly, London ISSN 0068-130-X.
Thomas PC, Veverka J, Belton MJS et al (2007) 'The shape, topography, and geology of Tempel 1 from Deep Impact observations,' Icarus 191: 51–62.
Ting E (2003) 'The essence of observing: Discovery's 12.5-inch Dobsonian,' Sky and Telescope 106 (5): 54–57.
Ting E (2004) 'An apochromat for the masses,' Sky and Telescope 107 (2): 60–63.
Tonkin S (2007) Binocular Astronomy, Springer, London.
Trusock T (2007) 'Meade's affordable large refractor,' Astronomy 35 (9): 86–87.
Trusock T (2008) 'Sky testing Orion's 102mm f/7 ED,' Astronomy 36 (10): 78–79.
Tsurutani BT, Clay DR, Zhang LD et al (2004) 'Plasma clouds associated with Comet P/Borrelly dust impacts,' Icarus 167: 89–99.
U.S. Govt. Printing Office Astronomical Almanac for the year 2009, U.S. Govt. Printing Office, Washington, DC, 2007.
Vaisberg OL, Smirnov VN, Gorn, LS et al (1986) 'Dust coma structure of comet Halley from SP-1 detector measurements,' Nature 321: 274–276.
Vsekhsvyatskii SK (1964) Physical Characteristics of Comets, Israel Program for Scientific Translations Ltd., Jerusalem.
Walker RG, Weaver WB, Shane WW et al (2007) 'Deep Impact: Optical spectroscopy and photometry obtained at MIRA,' Icarus 191: 526–536.
Walker S (2005a) 'Binocular viewers,' Sky and Telescope 109 (3): 98–99.
Walker S (2005b) 'German equatorial "Go To" Mounts,' Sky and Telescope 110 (1): 98–99.
Walker S (2006) 'Premier planetary imager,' Sky and Telescope 111 (6): 76–79.
Walker S (2007a) 'The STF Mirage 7 Mak,' Sky and Telescope 113 (2): 76–79.
Walker S (2007b) 'A new planetary camera,' Sky and Telescope 114 (4): 36–39.
Walker S (2007c) 'Atik Instruments ATK-16IC,' Sky and Telescope 114 (5): 36.
Walker S (2009) 'FLI's MicroLine ML8300: High resolution for small scopes,' Sky and Telescope 117 (4): 34–37.
Wallentinsen D (1982) 'Comet West 1976 VI: Observations of the Great Comet of 1976,' Journal of the Association of Lunar and Planetary Observers,' 29: 155–163.
Wallis MK, Swamy KSK (1987) 'Some diatomic molecules from comet P/Halley's UV spectra near spacecraft flybys,' Astronomy and Astrophysics 187: 329–332.
Weiler M, Rauer H, Knollenberg J et al (2007) 'The gas production of Comet 9P/Tempel 1 around the Deep Impact date,' Icarus 191: 339–347.
West RM, Pedersen H, Monderen P et al (1986) 'Post-perihelion imaging of comet Halley at ESO,' Nature 321: 363–365.
Whipple FL (1981) 'Rotation of comets,' Sky and Telescope 62: 20.
Whipple FL (1982) 'The Rotation of Comet Nuclei,' in Comets (Wilkening LL and Matthews MS, editors) The University of Arizona Press, Tucson, pp. 227–250.
Whipple FL, Green DWE (1985) The Mystery of Comets, Smithsonian Institution Press, Washington, DC.
Wikipedia, the free encyclopedia.
Winter V (2006) 'Two scopes in one,' Astronomy 34 (4): 90–93.
Woodward CE, Shure MA, Forrest WJ et al (1996) 'Ground-based near-infrared imaging of Comet P/Halley 1986 III,' Icarus 124: 651–662.
Yelle RV, Soderblom LA, Jokipii JR (2004) 'Formation of jets in Comet 19P/Borrelly by subsurface geysers,' Icarus 167: 30–36.
Yeomans D (1997) 'Orbit and Ephemeris Information for Comet Hale-Bopp (1995 O1),' obtained from the website http://www2.jpl.nasa.gov/comet/ephemjpl8.html.

Young DT, Crary FJ, Nordholt JE et al (2004) 'Solar wind interactions with Comet 19P/Borrelly,' Icarus 167: 80–88.
Young RV, Sessine S, editors (2000) World of Chemistry, Gale Group, Detroit, pp. 1021–1022.
Zolensky ME, Zega TJ, Yano H et al (2006) 'Mineralogy and petrology of Comet 81P/Wild 2 nucleus samples,' Science 314: 1735–1739.

Index